LONDON MATHEMATICAL SOCIETY LECTUI

Managing Editor: Professor Endre Süli, Mathematical Institute,
University of Oxford, Woodstock Road, Oxford OX2 6GG, United

The titles below are available from booksellers, or from Cambridg
www.cambridge.org/mathematics

"Positive definite matrices, kernels, sequences and functions, and operations on them that preserve their positivity, have been studied intensely for over a century. The techniques involved in their analysis and the variety of their applications both continue to grow. This book is an admirably comprehensive and lucid account of the topic. It includes some very recent developments in which the author has played a major role. This will be a valuable resource for researchers and an excellent text for a graduate course."

– Rajendra Bhatia, Ashoka University

"Khare demonstrates an extensive knowledge of this material and has taken considerable responsibility in presenting a blend of classical and contemporary topics in an appealing style opening the text to a broad readership with an interest in matrix positivity and their associated preservers. I expect this reference will be indispensable for both new and active researchers in this subject area."

– Dr. Shaun Fallat, University of Regina

"The opening notes of this symphony of ideas were written by Schur in 1911. Schoenberg, Loewner, Rudin, Herz, Hiai, FitzGerald, Jain, Guillot, Rajaratnam, Belton, Putinar, and others composed new themes and variations. Now, Khare has orchestrated a masterwork that includes his own harmonies in an elegant synthesis. This is a work of impressive scholarship."

– Roger Horn, University of Utah, Retired

"Classical chapters of matrix analysis and function theory are masterfully interlaced by the author with very recent advances. The resulting journey is distinguished by a careful balance between technical detail and historical notes. Relevant exercises and open problems challenge the reader. The joy of fresh discovery permeates every page of the book."

– Mihai Putinar, University of California, Santa Barbara and Newcastle University

"It has been known since the classical product theorem of Schur that the positivity properties of matrices can interact well with entrywise transformations. This monograph systematically explores this surprisingly rich interaction and its connections to such topics as metric geometry or the theory of Schur polynomials; it will be a useful resource for researchers in this area."

– Terence Tao, University of California, Los Angeles

Matrix Analysis and Entrywise Positivity Preservers

APOORVA KHARE
Indian Institute of Science

CAMBRIDGE
UNIVERSITY PRESS

CAMBRIDGE
UNIVERSITY PRESS

University Printing House, Cambridge CB2 8BS, United Kingdom

One Liberty Plaza, 20th Floor, New York, NY 10006, USA

477 Williamstown Road, Port Melbourne, VIC 3207, Australia

314–321, 3rd Floor, Plot 3, Splendor Forum, Jasola District Centre,
New Delhi – 110025, India

103 Penang Road, #05–06/07, Visioncrest Commercial, Singapore 238467

Cambridge University Press is part of the University of Cambridge.

It furthers the University's mission by disseminating knowledge in the pursuit of
education, learning, and research at the highest international levels of excellence.

www.cambridge.org
Information on this title: www.cambridge.org/9781108792042
DOI: 10.1017/9781108867122

First published 2022

Printed in the United Kingdom by TJ Books Limited, Padstow Cornwall

A catalogue record for this publication is available from the British Library.

Library of Congress Cataloging-in-Publication Data
Names: Khare, Apoorva, author.
Title: Matrix analysis and entrywise positivity preservers / Apoorva Khare.
Description: Cambridge ; New York, NY : Cambridge University Press, 2022. |
Includes bibliographical references and index. | Contents: The cone of
positive semidefinite matrices – The Schur product theorem and nonzero
lower bounds – Totally positive (TP) and Totally non-negative
(TN) matrices – TP matrices–generalized Vandermonde and Hankel moment
matrices – Entrywise powers preserving positivity in fixed dimension – Mid-convex
implies continuous, and 2×2 preservers – Entrywise preservers of positivity
on matrices with zero patterns – Entrywise powers preserving positivity,
monotonicity, superadditivity – Loewner convexity and single matrix encoders
of preservers – Exercises – History–Schoenberg, Rudin, Vasudeva, and metric

geometry – Loewner's determinant calculation in Horn's thesis – The stronger
Horn-Loewner theorem, via mollifiers – Stronger Vasudeva and Schoenberg
theorems, via Bernstein's theorem – Proof of stronger Schoenberg theorem
(part I)–positivity certificates – Proof of stronger Schoenberg theorem (part II)–
real analyticity – Proof of stronger Schoenberg theorem (part III)–complex
analysis – Preservers of Loewner positivity on kernels – Preservers of Loewner
monotonicity and convexity on kernels – Functions acting outside forbidden diagonal
blocks – The Boas-Widder theorem on functions with positive differences – Menger's
results and Euclidean distance geometry – Exercises – Entrywise polynomial preservers
and Horn-Loewner type conditions – Polynomial preservers for rank-one matrices, via
Schur polynomials – First-order approximation and leading term of Schur polynomials –
Exact quantitative bound–monotonicity of Schur ratios – Polynomial preservers
on matrices with real or complex entries – Cauchy and Littlewood's
definitions of Schur polynomials – Exercises.

Identifiers: LCCN 2021029851 (print) | LCCN 2021029852 (ebook) |
ISBN 9781108792042 (paperback) | ISBN 9781108867122 (epub)
Subjects: LCSH: Matrices. | Composition operators. | Positive operators. |
Kernel functions.
Classification: LCC QA188 .K45 2022 (print) | LCC QA188 (ebook) |
DDC 512.9/434–dc23
LC record available at https://lccn.loc.gov/2021029851
LC ebook record available at https://lccn.loc.gov/2021029852

ISBN 978-1-108-79204-2 Paperback

To Amruta and Anandi

Contents

xi

Foreword

Matrix positivity in its broad sense was and remains a catalyst of algebraic statements with an analytic proof, and of analytic statements with an algebraic proof. Every working mathematician can provide relevant examples in both directions. One of the turning points in promoting matrix positivity as a substitute of analytic techniques is cast in the following quote from a letter of Hermite:

> *En poursuivant mes recherches sur le théorème de Mr. Sturm, j'ai reussi à traiter par les memes principes les équations à coefficients imaginaires; ce qui m'a conduit au théorème de Mr. Cauchy pour le cas du rectangle, du cercle, et d'une infinité d'autres courbes qui sont meme à branches infinies, comme l'hyperbole. La théorie des formes quadratiques vient ainsi donner pour ces théorèmes des demonstrations indépendantes de toute considération de continuité, comme celle que Vous avez déjà pu conclure Vous-meme de ce que j'ai dit au sujet du théorème de Mr. Sturm dans les Comptes rendus de l'Académie. (1853, 1er semestre p. 294).*
> Extrait d'une lettre de Mr. Ch. Hermite de Paris Mr. Borchardt de Berlin, sur le nombre des racines d'une équation algébrique comprises entre les limites données, *J. Reine Angew. Math.* 52(1856), 39–51

The long-lasting impact of Sturm's sequence method and Hermite's quadratic forms on the development of mathematics is unquestionable. Think of Tarski's elimination of quantifiers principle for real closed fields, in itself the foundation stone of modern real algebraic geometry, respectively the role of Hermitian forms in spectral analysis or modern differential geometry. As a sign of vitality of these concepts, and their underlying matrix positivity characteristic, every new generation of scientists has unveiled fresh facets and striking applications. For instance, the data of a bounded analytic interpolation problem is encoded in the positivity of a Nevanlinna–Pick matrix, the power moments of a mass distribution on the line are characterized by the positivity of a Hankel matrix, the characteristic function of a probability distribution is identified to a positive definite kernel, the stability of a dynamical system can be derived from a positive matrix certificate, and so on.

Coming closer to the topics in the present book, we single out two discoverers in the vast landscape of matrix positivity: Charles Loewner (1893–1968) and Isaac J. Schoenberg (1903–1990). To Loewner we owe the characterization of matrix monotone functions, that is, differentiable functions f which preserve via the natural functional calculus the monotonicity of self-adjoint matrices: if $A \leq B$, then $f(A) \leq f(B)$. Note that when allowing matrices A, B of arbitrary dimension, these matrix monotone functions are real analytic with an analytic continuation in the upper-half plane that contains a positive imaginary part. Matrix monotonicity continues to fascinate to the point that the recent Springer volume *Loewner's Theorem on Monotone Matrix Functions* by Barry Simon starts with the provocative sentence: "This book is a love poem to Loewner's theorem."

In a different vein, soon after defending his doctoral dissertation, Schoenberg started a lifelong exploration of total positivity and variation diminishing linear transformations. A motivation for his study was the classical Hermite–Poulain theorem which isolates linear difference operations preserving the class of polynomials with real roots. Schoenberg's elaborate construct, masterfully combining matrix (total) positivity and the harmonic analysis of Laplace transforms, is unanimously accepted as the foundation of the theory of splines. A counterpart to Loewner's theorem, this time for the entrywise functional calculus $f[A]$, is a byproduct of Schoenberg's work: if $f[A] \geq 0$ for all matrices $A \geq 0$, then f is a real analytic function with nonnegative Taylor coefficients at zero.

Note in both examples above how purely algebraic and positivity statements imply the analyticity of the transformation. Apoorva Khare's book resonates precisely with this string by adding pertinent advances to the theories of Loewner and Schoenberg. The main theme is the structure of "positivity preservers," that is, entrywise transformations of positive matrices, them being of a prescribed finite size, infinite, structured, or even depending on a continuum of variables (better known as kernels). The source of this quest comes from the years the author spent among statisticians, and their pragmatic need of operating on large correlation matrices. Soon, however, the expertise of Professor Khare on group representation theory intervened, by considerably enlarging the scope and perspective of the inquiry.

The result is a highly original synthesis of a century of matrix positivity studies interlaced with very recent findings due to the author and his collaborators. The volume collects for the first time, in a surprisingly natural coordination, topics such as: distance geometry, moment problems, complete monotonicity, total positivity, graphical models, network analysis, and Schur polynomials. Qualitative features of the determinant of a linear pencil of symmetric matrices form the gateway to an array of deep new results touching the representation theory of the symmetric group, but also very basic concepts of numerical matrix analysis. Two recurrent phenomena are worth mentioning: the striking rigidity of entrywise preservers of various positive cones of matrices or kernels and the existence of a critical exponent for fractional power transforms, that is, a boundary separating a discrete spectrum from a continuous one.

The text was forged during two rounds of semester-long courses delivered by the author. In this sense, the book is far from being a definitive monograph, rather marking the birth of a promising new interdisciplinary area of research. Most of the classical theorems are included with complete proofs. Informative historical notes and a careful selection of exercises, not excluding open questions, will delight and challenge the reader. The joy of fresh discovery permeates every page of the book.

Mihai Putinar
University of California at Santa Barbara
and University of Newcastle

Preface

This text arose out of the course notes for Math 341: Matrix Analysis and Positivity, a one-semester course on analysis and matrix positivity preservers – or, more broadly, composition operators preserving various kinds of positive kernels. The course was offered in Spring 2018 and Fall 2019 at the Indian Institute of Science (IISc). In the present text, we briefly describe some notions of positivity in matrix theory, followed by our main focus: a detailed study of the operations that preserve these notions (and, in the process, an understanding of some aspects of real functions). Several different notions of positivity in analysis, studied for classical and modern reasons, are touched upon in the text:

- Positive semidefinite and positive definite matrices.
- Entrywise positive matrices.
- A common strengthening of the first two notions, which involves totally positive (*TP*) and totally nonnegative (*TN*) matrices.
- Settings somewhat outside matrix theory. For instance, consider discrete data associated with positive measures on locally compact abelian groups G. E.g., for $G = \mathbb{R}$, one obtains moment-sequences, which are intimately related to positive semidefinite Hankel matrices. For $G = S^1$, the circle group, one obtains Fourier–Stieltjes sequences, which are connected to positive semidefinite Toeplitz matrices. (See works of Carathéodory, Hamburger, Hausdorff, Herglotz, and Stieltjes, among others.)
- More classically, functions and kernels with positivity structures have long been studied in analysis, including on locally compact groups and metric spaces (see Bochner, Schoenberg, von Neumann, Pólya, etc.). These include positive definite functions and Pólya frequency functions and sequences.

The text begins by discussing positive semidefinite and TN matrices and some of their basic properties, followed by some of the early results on preservers of positive semidefiniteness. The next two parts then study, in detail, classical and modern results on entrywise positivity preservers in fixed and all dimensions. Among other things, this journey involves going through many beautiful classical results by leading experts in analysis during the first half of the twentieth century.

The purpose of this text is to study the post-composition transforms that preserve positivity on various classes of kernels. When the kernel has finite domain – i.e., is a matrix, as is the case in most of this book – then this amounts to studying *entrywise* preservers of various notions of positivity. The question of why this entrywise calculus was studied – as compared to the usual holomorphic functional calculus – has a rich and classical history in the analysis literature, beginning with the work of Schoenberg, Rudin, Loewner, and Horn (these results are proved in Part II of the text), but also drawing upon earlier works of Menger, Schur, Bochner, and others. (In fact, the entrywise calculus was introduced, and the first such result proved, by Schur in 1911.) Interestingly, this entrywise calculus also arises in modern-day applications from high-dimensional covariance estimation; we elaborate on this in Section 7.1, and briefly also in Chapter 8. Furthermore, this evergreen area of mathematics continues to be studied in the literature, drawing techniques from – and also contributing to – symmetric function theory, statistics and graphical models, combinatorics, and linear algebra (in addition to analysis).

As a historical curiosity, the course and this text arose in a sense out of research carried out in significant measure by mathematicians at Stanford University (including their students) over the years. This includes Loewner, Karlin, and their students FitzGerald, Horn, Micchelli, and Pinkus. Less directly, there was also Katznelson, who had previously worked with Helson, Kahane, and Rudin, leading to Rudin's strengthening of Schoenberg's theorem. (Coincidentally, Pólya and Szegő, who made the original observation on entrywise preservers of positivity using the Schur product theorem, were again colleagues at Stanford.) On a personal note, the author's contributions to this area also have their origins in his time spent at Stanford University, collaborating with Alexander Belton, Dominique Guillot, Mihai Putinar, Bala Rajaratnam, and Terence Tao (though the collaboration with the last-named colleague was carried out almost entirely at IISc).

We now discuss the course, the notes that led to this text, and their mathematical contents. The notes were scribed by the students taking the course in Spring 2018 at IISc, followed by extensive "homogenization" by the author – and, in several chapters, addition of material. Each chapter was originally intended to cover the notes of roughly one 90-minute lecture, or occasionally two; that said, some material has subsequently been moved around for logical, mathematical, and expositional reasons. The notes, and the course itself, require an understanding of basic linear algebra and analysis, with a bit of measure theory as well. Beyond these basic topics, we have tried to keep these notes as self-contained as possible, with full proofs. To that end, we have included proofs of "preliminary" results, including:

(i) results of Schoenberg, Menger, von Neumann, Fréchet, and others connecting metric geometry and positive definite functions to matrix positivity;

(ii) results in Euclidean geometry (including on triangulation), Heron's formula for triangles, and linking Cayley–Menger matrices to simplicial volumes;

(iii) results of Boas–Widder and Bernstein on functions with positive differences;

(iv) Sierpiński's result: midconvexity and measurability imply continuity;

(v) an extension to normed linear spaces, of (a special case of) a classical result of Ostrowski on midconvexity and local boundedness implying continuity;

(vi) Descartes' rule of signs;

(vii) Fekete's result on TP matrices via positive contiguous minors;

(viii) Sylvester's criterion and the Schur product theorem for positive definite and semidefinite matrices (also, the Jacobi formula);

(ix) the Rayleigh–Ritz theorem;

(x) a special case of Weyl's inequality on eigenvalues;

(xi) the Cauchy–Binet formula and a continuous generalization;

(xii) the discreteness of zeros of real analytic functions (and a sketch of the continuity of roots of complex polynomials); and

(xiii) the equivalence of Cauchy's and Littlewood's definitions of Schur polynomials (and of the Jacobi–Trudi and von Nägelsbach–Kostka identities) via Lindström–Gessel–Viennot bijections.

As the reader will notice, the exposition herein has left out some proofs. These include proofs of theorems by Hamburger/Hausdorff/Stieltjes, Fubini, Tonelli, Cauchy, Montel, and Morera; a Schur positivity phenomenon for ratios of Schur polynomials; Lebesgue's dominated convergence theorem; as well as the closure of real analytic functions under composition. Most of these can be found in standard textbooks in mathematics. Nevertheless, as the previous and current paragraphs indicate, these notes cover many classical results by past experts and acquaint the reader with a variety of tools in analysis (especially the study of real functions) and in matrix theory – many of these tools are not found in more "traditional" courses on these subjects.

This text is broadly divided into three parts, each with detailed bibliographic notes (these are given at the end of the text). In Part I, the key objects of interest – positive semidefinite and Hankel TN matrices – are introduced via basic results and important classes of examples. We then begin the study of functions acting entrywise on such matrices and preserving the relevant notion of positivity. Here, we will mostly restrict ourselves to studying power functions that act on various sets of matrices of a *fixed* size. This is a long-studied question, including by Bhatia, Elsner, Fallat, FitzGerald, Hiai, Horn, Jain, Johnson, Sokal, and the author together with Guillot and Rajaratnam. An interesting highlight is a construction by Jain of individual (pairs of) matrices, which encode the entire set of entrywise powers preserving Loewner positivity, monotonicity, and convexity. We also obtain certain necessary conditions on general entrywise functions that preserve positivity, including multiplicative midconvexity and continuity. We explain some of the modern motivations to study entrywise preservers, and we end with some unsolved problems.

Part II deals with the foundational results on matrix positivity preservers. We begin with the early history – including related work by Menger, Fréchet, Bochner, and Schoenberg. We then classify the entrywise functions that preserve positive semidefiniteness (= positivity) in all dimensions, or total nonnegativity on Hankel matrices of *all* sizes. This is a celebrated result of Schoenberg – later strengthened by Rudin – which is a converse to the Schur product theorem, and we prove a stronger version by using a rank-constrained test set. The proof here is different from previous approaches of Schoenberg and Rudin, is essentially self-contained, and uses relatively less sophisticated machinery compared to the works of Schoenberg and Rudin. Moreover, it also proves a variant by Vasudeva for matrices with only positive entries, and it lends itself to a multivariate generalization (which will not be covered here). The starting point of these proofs is a necessary condition for entrywise preservers in a fixed dimension, proved by Loewner (and Horn) in the late 1960s. To this day, this result remains essentially the only known condition in a fixed dimension $n \geq 3$, and a proof of a (rank-constrained, as above) stronger version is also provided in these notes. In addition to techniques and ingredients introduced by the above authors, the text also borrows from the author's joint work with Belton, Guillot, and Putinar.

Following this development, this part of the text ends with several related results and follow-ups:

(i) Chapters 18 and 19 classify "dimension-free" preservers of Loewner positivity, monotonicity, and convexity (on kernels over infinite domains).

(ii) We next cover – in Chapter 20 – recent work by Vishwakarma on an "off-diagonal" variant of the positivity preserver problem.

(iii) Chapter 21 covers a result by Boas and Widder, which shows a converse "mean value theorem" for divided differences.

(iv) Finally, Chapter 22 explores the theme of Euclidean distance geometry, with a focus on some classical results by Menger.

The final Part III can be read directly following Chapter 5 and the statements of Theorems 11.2 and 12.1. Part III deals with entrywise functions preserving positivity in a fixed dimension. This is a challenging problem – it is still open in general, even for 3×3 matrices – and we restrict the discussion here to studying polynomial preservers. According to the Schur product theorem (1911), if the polynomial has all nonnegative coefficients, then it is at once a preserver; but, interestingly, until 2016 not a single other example was known in any fixed dimension $n \geq 3$. Very recently, this question has been answered to some degree of satisfaction by the author, in collaboration first with Belton, Guillot, and Putinar, and subsequently with Tao. The text ends by covering some of this recent progress, and it comes back full circle to Schur through symmetric function theory.

Also in this text, historical notes and further questions serve to acquaint the reader with past work(er)s, related areas, and possible avenues for future work – and can be accessed from the Index. See also the Exercises after each part of the text, and the Bibliographic Notes and References at the end of the text. The former include additional results not covered in the course or this text.

To conclude, thanks are owed to the scribes, as well as to Alexander Belton, Shabarish Chenakkod, Projesh Nath Choudhury, Julian R. D'Costa, Dominique Guillot, Prakhar Gupta, Roger A. Horn, Sarvesh Ravichandran Iyer, Poornendu Kumar, Frank Oertel, Aaradhya Pandey, Vamsi Pritham Pingali, Mihai Putinar, Shubham Rastogi, Aditya Guha Roy, Kartik Singh, G. V. Krishna Teja, Raghavendra Tripathi, Prateek Kumar Vishwakarma, and Pranjal Warade for helpful suggestions that improved the text. I am, of course, deeply indebted to my collaborators for their support and all of their research efforts in positivity – but also for many stimulating discussions, which helped shape my thinking about the field as a whole and the structure of this text in particular. I am also grateful to the University Grants Commission (UGC, Government of India), the Science and Engineering Research Board (SERB) and the Department of Science and Technology (DST) of the Government of India, the Infosys Foundation, and the Tata Trusts – for their support through a CAS-II grant, through a MATRICS grant and the Ramanujan and Swarnajayanti Fellowships, through a Young Investigator Award, and through their Travel Grants, respectively.

Finally, I thank my family for their constant support, encouragement, and patience during the writing of this book.

PART ONE

PRELIMINARIES: ENTRYWISE POWERS PRESERVING POSITIVITY IN A FIXED DIMENSION

1

The Cone of Positive Semidefinite Matrices

A *kernel* is a function $K : X \times Y \to \mathbb{R}$. Broadly speaking, the goal of this text is to understand:

Which functions $F : \mathbb{R} \to \mathbb{R}$, when applied to kernels that are positive semidefinite, preserve that notion?

To do so, we first study the test sets of such kernels K themselves, and then the post-composition operators F that preserve these test sets. We begin by understanding such kernels when the domains X, Y are finite, i.e., matrices.

In this text, we will assume familiarity with linear algebra and a first course in calculus/analysis. To set notation: an uppercase letter with a two-integer subscript (such as $A_{m \times n}$) represents a matrix with m rows and n columns. If m, n are clear from context or unimportant, then they will be omitted. Three examples of real matrices are $0_{m \times n}, 1_{m \times n}, \mathrm{Id}_{n \times n}$, which are the (rectangular) matrix consisting of all zeros, all ones, and the identity matrix, respectively. The entries of a matrix A will be denoted a_{ij}, a_{jk}, etc. Vectors are denoted by lowercase letters (occasionally in bold) and are columnar in nature. All matrices, unless specified otherwise, are real; and similarly, all functions, unless specified otherwise, are defined on – and take values in – $\mathbb{R}^m$ for some $m \geq 1$. As is standard, we let $\mathbb{C}, \mathbb{R}, \mathbb{Q}, \mathbb{Z}, \mathbb{N}$ denote the complex numbers, reals, rationals, integers, and positive integers respectively. Given $S \subset \mathbb{R}$, let $S^{\geq 0} := S \cap [0, \infty)$.

1.1 Preliminaries

We begin with several basic definitions.

Definition 1.1   A matrix $A_{n \times n}$ is said to be *symmetric* if $a_{jk} = a_{kj}$ for all $1 \leq j, k \leq n$. A real symmetric matrix $A_{n \times n}$ is said to be *positive semidefinite* if the real number $x^T A x$ is nonnegative for all $x \in \mathbb{R}^n$ – in other words, the quadratic form given by A is positive semidefinite. If, furthermore, $x^T A x > 0$ for all $x \neq 0$ then A is said to be *positive definite*. Denote the set of (real symmetric) positive semidefinite matrices by $\mathbb{P}_n$.

We state the spectral theorem for symmetric (i.e., self-adjoint) operators without proof.

Theorem 1.2 (Spectral theorem for symmetric matrices)   *For $A_{n \times n}$ a real symmetric matrix, $A = U^T D U$ for some orthogonal matrix U (i.e., $U^T U = $ Id) and real diagonal matrix D. D contains all the eigenvalues of A (counting multiplicities) along its diagonal.*

As a consequence, $A = \sum_{j=1}^{n} \lambda_j v_j v_j^T$, where each v_j is an eigenvector for A with *real* eigenvalue λ_j, and the v_j (which are the columns of U^T) form an orthonormal basis of $\mathbb{R}^n$.

We also have the following related results, stated here without proof: the spectral theorem for two commuting matrices, and the singular value decomposition.

Theorem 1.3 (Spectral theorem for commuting symmetric matrices)   *Let $A_{n \times n}$ and $B_{n \times n}$ be two commuting real symmetric matrices. Then A and B are simultaneously diagonalizable, i.e., for some common orthogonal matrix U, $A = U^T D_1 U$ and $B = U^T D_2 U$ for D_1 and D_2 diagonal matrices (whose diagonal entries comprise the eigenvalues of A, B respectively).*

Theorem 1.4 (Singular value decomposition)   *Every real matrix $A_{m \times n} \neq 0$ decomposes as $A = P_{m \times m} \begin{pmatrix} \Sigma_r & 0 \\ 0 & 0 \end{pmatrix}_{m \times n} Q_{n \times n}$, where P, Q are orthogonal and Σ_r is a diagonal matrix with positive eigenvalues. The entries of Σ_r are called the* singular values *of A and are the square roots of the nonzero eigenvalues of $A A^T$ (or $A^T A$).*

1.2 Criteria for Positive (Semi)Definiteness

We write down several equivalent criteria for positive (semi)definiteness. There are three initial criteria which are easy to prove, and a final criterion which requires separate treatment.

Theorem 1.5 (Criteria for positive (semi)definiteness)   *Given $A_{n \times n}$ a real symmetric matrix of rank $0 \leq r \leq n$, the following are equivalent:*

(1) *A is positive semidefinite (respectively, positive definite).*
(2) *All eigenvalues of A are nonnegative (respectively, positive).*
(3) *There exists a matrix $B \in \mathbb{R}^{r \times n}$ of rank r, such that $B^T B = A$. (In particular, if A is positive definite then B is square and nonsingular.)*

Proof   We prove only the positive semidefinite statements; minor changes show the corresponding positive definite variants. If (1) holds and λ is an eigenvalue – for an eigenvector x – then $x^T A x = \lambda \|x\|^2 \geq 0$. Hence, $\lambda \geq 0$, proving (2). Conversely, if (2) holds then by the spectral theorem, $A = \sum_j \lambda_j v_j v_j^T$ with all $\lambda_j \geq 0$, so A is positive semidefinite:

$$x^T A x = \sum_j \lambda_j x^T v_j v_j^T x = \sum_j \lambda_j (x^T v_j)^2 \geq 0, \qquad \forall x \in \mathbb{R}^n.$$

Next, if (1) holds then write $A = U^T D U$ by the spectral theorem; note that $D = U A U^T$ has the same rank as A. Since D has nonnegative diagonal entries d_{jj}, it has a square root $\sqrt{D}$, which is a diagonal matrix with diagonal entries $\sqrt{d_{jj}}$. Write $D = \begin{pmatrix} D'_{r \times r} & 0 \\ 0 & 0_{(n-r) \times (n-r)} \end{pmatrix}$, where D' is a diagonal matrix with positive diagonal entries. Correspondingly, write $U = \begin{pmatrix} P_{r \times r} & Q \\ R & S_{(n-r) \times (n-r)} \end{pmatrix}$. If we set $B := (\sqrt{D'}P \mid \sqrt{D'}Q)_{r \times n}$, then it is easily verified that

$$B^T B = \begin{pmatrix} P^T D' P & P^T D' Q \\ Q^T D' P & Q^T D' Q \end{pmatrix} = U^T D U = A.$$

Hence, (1) $\implies$ (3). Conversely, if (3) holds then $x^T A x = \|Bx\|^2 \geq 0$ for all $x \in \mathbb{R}^n$. Hence, A is positive semidefinite. Moreover, we claim that B and $B^T B$ have the same null space and hence the same rank. Indeed, if $Bx = 0$ then $B^T B x = 0$, while

$$B^T B x = 0 \implies x^T B^T B x = 0 \implies \|Bx\|^2 = 0 \implies Bx = 0. \quad \square$$

Corollary 1.6   *For any real symmetric matrix $A_{n \times n}$, the matrix $A - \lambda_{\min} \operatorname{Id}_{n \times n}$ is positive semidefinite, where $\lambda_{\min}$ denotes the smallest eigenvalue of A.*

We now state Sylvester's criterion for positive (semi)definiteness. (Incidentally, Sylvester is believed to have first introduced the use of "matrix" in mathematics, in the nineteenth century.) This requires some additional notation.

Definition 1.7   Given an integer $n \geq 1$, define $[n] := \{1, \ldots, n\}$. Now given a matrix $A_{m \times n}$ and subsets $J \subset [m]$, $K \subset [n]$, define $A_{J \times K}$ to be the submatrix of A with entries a_{jk} for $j \in J, k \in K$ (always considered to be arranged in increasing order in this text). If J, K have the same size then $\det A_{J \times K}$ is called a *minor* of A. If A is square and $J = K$ then $A_{J \times K}$ is called a *principal submatrix* of A, and $\det A_{J \times K}$ is a *principal minor*. The principal submatrix (and principal minor) are *leading* if $J = K = \{1, \ldots, m\}$ for some $1 \leq m \leq n$.

Theorem 1.8 (Sylvester's criterion)   *A symmetric matrix is positive semidefinite (respectively, positive definite) if and only if all its principal minors are nonnegative (respectively, positive).*

We will show Theorem 1.8 with the help of a few preliminary results.

Lemma 1.9   *If $A_{n \times n}$ is a positive semidefinite (respectively, positive definite) matrix, then so are all principal submatrices of A.*

Proof   Fix a subset $J \subset [n] = \{1, \ldots, n\}$ (so $B := A_{J \times J}$ is the corresponding principal submatrix of A), and let $x \in \mathbb{R}^{|J|}$. Define $x' \in \mathbb{R}^n$ to be the vector, such that $x'_j = x_j$ for all $j \in J$ and 0 otherwise. It is easy to see that $x^T B x = (x')^T A x'$. Hence, B is positive (semi)definite if A is.   □

As a corollary, all the principal minors of a positive semidefinite (positive definite) matrix are nonnegative (positive) since the corresponding principal submatrices have nonnegative (positive) eigenvalues and hence nonnegative (positive) determinants. So one direction of Sylvester's criterion holds trivially.

Lemma 1.10   *Sylvester's criterion is true for positive definite matrices.*

Proof   We induct on the dimension of the matrix A. Suppose $n = 1$. Then A is just an ordinary real number, so its only principal minor is A itself, and so the result is trivial.

Now, suppose the result is true for matrices of dimension $\leq n - 1$. We claim that A has at least $n - 1$ positive eigenvalues. To see this, let $\lambda_1, \lambda_2 \leq 0$ be eigenvalues of A. Let W be the $n - 1$ dimensional subspace of $\mathbb{R}^n$ with last entry 0. If v_j are orthogonal eigenvectors for λ_j, $j = 1, 2$, then the span of the v_j must intersect W nontrivially, since the sum of dimensions of these two subspaces of $\mathbb{R}^n$ exceeds n. Define $u := c_1 v_1 + c_2 v_2 \in W$; then $u^T A u > 0$ by Lemma 1.9. However,

$$u^T A u = \left(c_1 v_1^T + c_2 v_2^T\right) A (c_1 v_1 + c_2 v_2) = c_1^2 \lambda_1 ||v_1||^2 + c_2^2 \lambda_2 ||v_2||^2 \leq 0,$$

thereby giving a contradiction and proving the claim.

Now since the determinant of A is positive (it is the minor corresponding to A itself), it follows that all eigenvalues are positive, completing the proof.   □

We will now prove the Jacobi formula, an important result in its own right. A corollary of this result will be used, along with the previous result and the idea that positive semidefinite matrices can be expressed as entrywise limits of positive definite matrices, to prove Sylvester's criterion for all positive semidefinite matrices.

Theorem 1.11 (Jacobi formula)   *Let $A_t : \mathbb{R} \to \mathbb{R}^{n \times n}$ be a matrix-valued differentiable function. Then,*

$$\frac{d}{dt}(\det A_t) = \mathrm{tr}\left(\mathrm{adj}(A_t)\frac{dA_t}{dt} \right), \tag{1.1}$$

where $\mathrm{adj}(A_t)$ denotes the adjugate matrix of A_t.

Proof   The first step is to compute the differential of the determinant. We claim that for any $n \times n$ real matrices A, B,

$$d(\det)(A)(B) = \mathrm{tr}(\mathrm{adj}(A)B).$$

As a special case, at $A = \mathrm{Id}_{n \times n}$, the differential of the determinant is precisely the trace.

To show the claim, we need to compute the directional derivative

$$\lim_{\epsilon \to 0} \frac{\det(A + \epsilon B) - \det A}{\epsilon}.$$

The fraction is a polynomial in ϵ with vanishing constant term (e.g., set $\epsilon = 0$ to see this); and we need to compute the coefficient of the linear term. Expand $\det(A + \epsilon B)$ using the Laplace expansion as a sum over permutations $\sigma \in S_n$; now each individual summand $(-1)^\sigma \prod_{k=1}^{n}(a_{k\sigma(k)} + \epsilon b_{k\sigma(k)})$ splits as a sum of 2^n terms. (It may be illustrative to try and work out the $n = 3$ case by hand.) From these $2^n \cdot n!$ terms, choose the ones that are linear in ϵ. For each $1 \leq i$, $j \leq n$, there are precisely $(n - 1)!$ terms corresponding to ϵb_{ij}; and added together, they equal the (i, j)th cofactor C_{ij} of A – which equals $\mathrm{adj}(A)_{ji}$. Thus, the coefficient of ϵ is

$$d(\det)(A)(B) = \sum_{i,j=1}^{n} C_{ij}b_{ij},$$

and this is precisely $\mathrm{tr}(\mathrm{adj}(A)B)$, as claimed.

More generally, the above argument shows that if $B(\epsilon)$ is any family of matrices, with limit $B(0)$ as $\epsilon \to 0$, then

$$\lim_{\epsilon \to 0} \frac{\det(A + \epsilon B(\epsilon)) - \det A}{\epsilon} = \mathrm{tr}(\mathrm{adj}(A)B(0)). \tag{1.2}$$

Returning to the proof of the theorem, for $\epsilon \in \mathbb{R}$ small and $t \in \mathbb{R}$ we write

$$A_{t+\epsilon} = A_t + \epsilon B(\epsilon),$$

where $B(\epsilon) \to B(0) := \frac{dA_t}{dt}$ as $\epsilon \to 0$, by definition. Now compute using (1.2)

$$\frac{d}{dt}(\det A_t) = \lim_{\epsilon \to 0} \frac{\det(A_t + \epsilon B(\epsilon)) - \det A_t}{\epsilon} = \mathrm{tr}\left(\mathrm{adj}(A_t)\frac{dA_t}{dt}\right). \qquad \square$$

With these results at hand, we can finish the proof of Sylvester's criterion for positive semidefinite matrices.

Proof of Theorem 1.8   For positive definite matrices, the result was proved in Lemma 1.10. Now suppose $A_{n \times n}$ is positive semidefinite. One direction follows by the remarks preceding Lemma 1.10. We show the converse by induction on n, with an easy argument for $n = 1$ similar to the positive definite case.

Now suppose the result holds for matrices of dimension $\leq n - 1$ and let $A_{n \times n}$ have all principal minors nonnegative. Let B be any principal submatrix of A, and define $f(t) := \det(B + t \, \mathrm{Id}_{n \times n})$. Note that $f'(t) = \mathrm{tr}(\mathrm{adj}(B + t \, \mathrm{Id}_{n \times n}))$ by the Jacobi formula (1.1).

We claim that $f'(t) > 0 \; \forall t > 0$. Indeed, each diagonal entry of $\mathrm{adj}(B + t \, \mathrm{Id}_{n \times n})$ is a proper principal minor of $A + t \, \mathrm{Id}_{n \times n}$, which is positive definite since $x^T(A + t \, \mathrm{Id}_{n \times n})x = x^T Ax + t\|x\|^2$ for $x \in \mathbb{R}^n$. The claim now follows using Lemma 1.9 and the induction hypothesis.

The claim implies: $f(t) > f(0) = \det B \geq 0 \; \forall t > 0$. Thus, all principal minors of $A + tI$ are positive, and by Sylvester's criterion for positive definite matrices, $A + tI$ is positive definite for all $t > 0$. Now note that $x^T Ax = \lim_{t \to 0^+} x^T(A + t \, \mathrm{Id}_{n \times n})x$; therefore the nonnegativity of the right-hand side implies that of the left-hand side for all $x \in \mathbb{R}^n$, completing the proof. $\qquad \square$

1.3 Examples of Positive Semidefinite Matrices

We next discuss several examples of positive semidefinite matrices.

1.3.1 Gram Matrices

Definition 1.12   For any finite set of vectors $\mathbf{x}_1, \ldots, \mathbf{x}_n \in \mathbb{R}^m$, their *Gram matrix* is given by $\mathrm{Gram}((\mathbf{x}_j)_j) := (\langle \mathbf{x}_j, \mathbf{x}_k \rangle)_{1 \leq j, k \leq n}$.

A *correlation matrix* is a positive semidefinite matrix with ones on the diagonal.

In fact, we need not use $\mathbb{R}^m$ here; any inner product space/Hilbert space is sufficient.

Proposition 1.13   *Given a real symmetric matrix* $A_{n \times n}$, *it is positive semidefinite if and only if there exist an integer* $m > 0$ *and vectors* $\mathbf{x}_1, \ldots, \mathbf{x}_n \in \mathbb{R}^m$, *such that* $A = Gram((\mathbf{x}_j)_j)$.

As a special case, correlation matrices precisely correspond to those Gram matrices for which the $\mathbf{x}_j$ are unit vectors. We also remark that a "continuous" version of this result is given by a well-known result of Mercer [172].

Proof   If A is positive semidefinite, then by Theorem 1.5 we can write $A = B^T B$ for some matrix $B_{m \times n}$. It is now easy to check that A is the Gram matrix of the columns of B.

Conversely, if $A = Gram(\mathbf{x}_1, \ldots, \mathbf{x}_n)$ with all $\mathbf{x}_j \in \mathbb{R}^m$, then to show that A is positive semidefinite, we compute for any $\mathbf{u} = (u_1, \ldots, u_n)^T \in \mathbb{R}^n$

$$\mathbf{u}^T A \mathbf{u} = \sum_{j,k=1}^{n} u_j u_k \langle \mathbf{x}_j, \mathbf{x}_k \rangle = \left\| \sum_{j=1}^{n} u_j \mathbf{x}_j \right\|^2 \geq 0. \qquad \square$$

1.3.2 (Toeplitz) Cosine Matrices

Definition 1.14   A matrix $A = (a_{jk})$ is *Toeplitz* if a_{jk} depends only on $j - k$.

Lemma 1.15   *Let* $\theta_1, \ldots, \theta_n \in [0, 2\pi]$. *Then the matrix* $C := (\cos(\theta_j - \theta_k))_{j,k=1}^{n}$ *is positive semidefinite, with rank at most 2. In particular,* $\alpha \mathbf{1}_{n \times n} + \beta C$ *has rank at most 3 (for scalars* α, β), *and it is positive semidefinite if* α, $\beta \geq 0$.

Proof   Define the vectors $u, v \in \mathbb{R}^n$ via: $u^T = (\cos\theta_1, \ldots, \cos\theta_n)$, $v^T = (\sin\theta_1, \ldots, \sin\theta_n)$. Then $C = uu^T + vv^T$ via the identity $\cos(a - b) = \cos a \cos b + \sin a \sin b$, and clearly the rank of C is at most 2. (For instance, it can have rank 1 if the θ_j are equal.) As a consequence,

$$\alpha \mathbf{1}_{n \times n} + \beta C = \alpha \mathbf{1}_n \mathbf{1}_n^T + \beta uu^T + \beta vv^T$$

has rank at most 3; the final assertion is straightforward. $\qquad \square$

As a special case, if $\theta_1, \ldots, \theta_n$ are in arithmetic progression, i.e., $\theta_{j+1} - \theta_j = \theta \; \forall j$ for some θ, then we obtain a positive semidefinite *Toeplitz* matrix

$$C = \begin{pmatrix} 1 & \cos\theta & \cos 2\theta & \cdots & & & \\ \cos\theta & 1 & \cos\theta & \cos 2\theta & \cdots & & \\ \cos 2\theta & \cos\theta & 1 & \cos\theta & \cos 2\theta & \cdots & \\ & \cos 2\theta & \cos\theta & 1 & \cos\theta & \cos 2\theta & \cdots \\ \vdots & & \ddots & & \ddots & & \ddots \end{pmatrix}.$$

This family of Toeplitz matrices was used by Rudin in a 1959 paper [197] on entrywise positivity preservers; see Theorem 11.3 for his result.

1.3.3 Hankel Matrices

Definition 1.16   A matrix $A = (a_{jk})$ is *Hankel* if a_{jk} depends only on $j + k$.

Example 1.17 $\begin{pmatrix} 0 & 1 \\ 1 & 0 \end{pmatrix}$ is Hankel but not positive semidefinite.

Example 1.18   For $x \geq 0$, the matrix $\begin{pmatrix} 1 & x & x^2 \\ x & x^2 & x^3 \\ x^2 & x^3 & x^4 \end{pmatrix} = \begin{pmatrix} 1 \\ x \\ x^2 \end{pmatrix} \begin{pmatrix} 1 & x & x^2 \end{pmatrix}$ is Hankel and positive semidefinite of rank 1.

A more general perspective is as follows. Define

$$H_x := \begin{pmatrix} 1 & x & x^2 & \cdots \\ x & x^2 & x^3 & \cdots \\ x^2 & x^3 & x^4 & \cdots \\ \vdots & \vdots & \vdots & \ddots \end{pmatrix},$$

and let δ_x be the Dirac measure at $x \in \mathbb{R}$. The moments of this measure are given by

$$s_k(\delta_x) := \int_{\mathbb{R}} y^k \, d\delta_x(y) = x^k, \qquad k \geq 0.$$

Thus, H_x is the "moment matrix" of δ_x. More generally, given any nonnegative measure μ supported on $\mathbb{R}$, with all moments finite, the corresponding *Hankel moment matrix* is the bi-infinite "matrix" given by

$$H_\mu := \begin{pmatrix} s_0 & s_1 & s_2 & \cdots \\ s_1 & s_2 & s_3 & \cdots \\ s_2 & s_3 & s_4 & \cdots \\ \vdots & \vdots & \vdots & \ddots \end{pmatrix}, \qquad \text{where } s_k = s_k(\mu) := \int_{\mathbb{R}} y^k \, d\mu(y). \quad (1.3)$$

Lemma 1.19 *The matrix H_μ is positive semidefinite. In other words, every finite principal submatrix is positive semidefinite.*

Note, this is equivalent to every leading principal submatrix being positive semidefinite.

Proof Fix $n \geq 1$ and consider the finite principal (Hankel) submatrix H'_μ with the first n rows and columns. Let H'_{δ_x} be the Hankel matrix defined in a similar manner for the measure $\delta_x, x \in \mathbb{R}$. Now to show that H'_μ is positive semidefinite, we compute for any vector $\mathbf{u} \in \mathbb{R}^n$

$$\mathbf{u}^T H'_\mu \mathbf{u} = \int_{\mathbb{R}} \mathbf{u}^T H'_{\delta_x} \mathbf{u} \, d\mu(x) = \int_{\mathbb{R}} ((1, x, \ldots, x^{n-1}) \mathbf{u})^2 \, d\mu(x) \geq 0,$$

where the final equality holds because H'_{δ_x} has rank 1 and factorizes as in Example 1.18. (Note that the first equality holds because we are taking finite linear combinations of the integrals in the entries of H'_μ.) □

Remark 1.20 Lemma 1.19 is (the easier) half of a famous classical result by Hamburger. The harder converse result says that if a semi-infinite Hankel matrix is positive semidefinite, with (j, k)-entry s_{j+k} for $j, k \geq 0$, then there exists a nonnegative Borel measure on the real line whose kth moment is s_k for all $k \geq 0$. This theorem will be useful later; it was shown by Hamburger in 1920–21, when he extended the Stieltjes moment problem to the entire real line in the series of papers [112]. These works established the moment problem in its own right, as opposed to being a tool used to determine the convergence or divergence of continued fractions (as previously developed by Stieltjes – see Remark 3.9).

There is also a multivariate version of Lemma 1.19 which is no harder than the lemma, modulo notation:

Lemma 1.21 *Given a measure μ on $\mathbb{R}^d$ for some integer $d \geq 1$, we define its moments for tuples of nonnegative integers $\mathbf{n} = (n_1, \ldots, n_d)$ via*

$$s_{\mathbf{n}}(\mu) := \int_{\mathbb{R}^d} \mathbf{x}^{\mathbf{n}} \, d\mu(\mathbf{x}) = \int_{\mathbb{R}^d} \prod_{j=1}^{d} x_j^{n_j} \, d\mu,$$

if these integrals converge. (Here, $\mathbf{x}^{\mathbf{n}} := \prod_j x_j^{n_j}$.) Now suppose $\mu \geq 0$ on $\mathbb{R}^d$ and let $\Psi : (\mathbb{Z}^{\geq 0})^d \to \mathbb{Z}^{\geq 0}$ be any bijection, such that $\Psi(\mathbf{0}) = 0$ (although this restriction is not really required). Define the semi-infinite matrix

$$H_\mu := (a_{jk})_{j,k=0}^{\infty}, \qquad a_{jk} := s_{\Psi^{-1}(j)+\Psi^{-1}(k)},$$

where we assume that all moments of μ exist. Then H_μ is positive semidefinite.

Proof   Given a real vector $\mathbf{u} = (u_0, u_1, \ldots)^T$ with finitely many nonzero coordinates, we have

$$\mathbf{u}^T H_\mu \mathbf{u} = \sum_{j,k \geq 0} \int_{\mathbb{R}^d} u_j u_k \mathbf{x}^{\Psi^{-1}(j) + \Psi^{-1}(k)} \, d\mu(\mathbf{x})$$

$$= \int_{\mathbb{R}^d} ((1, \mathbf{x}^{\Psi^{-1}(1)}, \mathbf{x}^{\Psi^{-1}(2)}, \ldots) \mathbf{u})^2 \, d\mu(\mathbf{x}) \geq 0. \qquad \square$$

1.3.4 Matrices with Sparsity

Another family of positive semidefinite matrices involves matrices with a given zero pattern, i.e., structure of (non)zero entries. Such families are important in applications, as well as in combinatorial linear algebra, spectral graph theory, and graphical models/Markov random fields.

Definition 1.22   A graph $G = (V, E)$ is *simple* if the sets of vertices/nodes V and edges E are finite, and E contains no self-loops (v, v) or multi-edges. In this text, all graphs will be finite and simple. Given such a graph $G = (V, E)$, with node set $V = [n] = \{1, \ldots, n\}$, define

$$\mathbb{P}_G := \{A \in \mathbb{P}_n : a_{jk} = 0 \text{ if } j \neq k \text{ and } (j, k) \notin E\}, \tag{1.4}$$

where $\mathbb{P}_n$ comprises the (real symmetric) positive semidefinite matrices of dimension n.

Also, a subset $C \subset X$ of a real vector space X is *convex* if $\lambda v + (1 - \lambda) w \in C$ for all $v, w \in C$ and $\lambda \in [0, 1]$. If instead $\alpha C \subset C$ for all $\alpha \in (0, \infty)$, then we say C is a *cone*.

Remark 1.23   The set $\mathbb{P}_G$ is a natural mathematical generalization of the cone $\mathbb{P}_n$ (and shares several of its properties). In fact, two "extreme" special cases are: (i) G is the complete graph, in which case $\mathbb{P}_G$ is the full cone $\mathbb{P}_n$ for $n = |V|$; and (ii) G is the empty graph, in which case $\mathbb{P}_G$ is the cone of $|V| \times |V|$ diagonal matrices with nonnegative entries.

Akin to both of these cases, for all graphs G, the set $\mathbb{P}_G$ is in fact a closed convex cone.

Example 1.24   Let $G = \{\{v_1, v_2, v_3\}, \{(v_1, v_3), (v_2, v_3)\}\}$. The adjacency matrix is given by $A_G = \begin{pmatrix} 0 & 0 & 1 \\ 0 & 0 & 1 \\ 1 & 1 & 0 \end{pmatrix}$. This is not in $\mathbb{P}_G$ (but $A_G - \lambda_{\min}(A_G)$ $\mathrm{Id}_{3 \times 3} \in \mathbb{P}_G$, see Corollary 1.6).

Example 1.25   For any graph G with node set $[n]$, let D_G be the diagonal matrix with (j, j) entry the degree of node j, i.e., the number of edges adjacent to j. Then the *graph Laplacian*, defined to be $L_G := D_G - A_G$ (where A_G is the adjacency matrix), is in $\mathbb{P}_G$.

Example 1.26   An important class of examples of positive semidefinite matrices arises from the Hessian matrix of (suitably differentiable) functions. In particular, if the Hessian is positive definite at a point, then this is an isolated local minimum.

1.4 Schur Complements

We mention some more preliminary results here; these may be skipped for now but will get used in Lemma 5.5 below.

Definition 1.27   Given a matrix $M = \begin{pmatrix} P & Q \\ R & S \end{pmatrix}$, where P and S are square and S is nonsingular, the *Schur complement* of M with respect to S is given by $M/S := P - QS^{-1}R$.

Schur complements arise naturally in theory and applications. As an important example, suppose $X_1, \ldots, X_n$ and $Y_1, \ldots, Y_m$ are random variables with covariance matrix $\Sigma = \begin{pmatrix} A & B \\ B^T & C \end{pmatrix}$, with C nonsingular. Then the conditional covariance matrix of X given Y is $\mathrm{Cov}(X|Y) := A - BC^{-1}B^T = \Sigma/C$. That such a matrix is also positive semidefinite is implied by the following folklore result by Albert in 1969 [8].

Theorem 1.28   *Given a symmetric matrix* $\Sigma = \begin{pmatrix} A & B \\ B^T & C \end{pmatrix}$*, with C positive definite, the matrix Σ is positive (semi)definite if and only if the Schur complement Σ/C is thus.*

Proof   We first write down a more general matrix identity: for a nonsingular matrix C and a square matrix A, one uses a factorization shown by Schur in 1917 [217]:

$$\begin{pmatrix} A & B \\ B' & C \end{pmatrix} = \begin{pmatrix} \mathrm{Id} & BC^{-1} \\ 0 & \mathrm{Id} \end{pmatrix} \begin{pmatrix} A - BC^{-1}B' & 0 \\ 0 & C \end{pmatrix} \begin{pmatrix} \mathrm{Id} & 0 \\ C^{-1}B' & \mathrm{Id} \end{pmatrix}. \quad (1.5)$$

(Note, the identity matrices on the right have different sizes.) Now set $B' = B^T$; then $\Sigma = X^T Y X$, where $X = \begin{pmatrix} \mathrm{Id} & 0 \\ C^{-1}B^T & \mathrm{Id} \end{pmatrix}$ is nonsingular, and $Y = \begin{pmatrix} A - BC^{-1}B^T & 0 \\ 0 & C \end{pmatrix}$ is block diagonal (and real symmetric). The result is not hard to show from here. $\qquad\square$

Akin to Sylvester's criterion, the above characterization has a variant for when C is positive semidefinite; however, this is not as easy to prove, and requires a more flexible "inverse."

Definition 1.29 (Moore–Penrose inverse)   Given any real $m \times n$ matrix A, the *pseudo-inverse* or *Moore–Penrose inverse* of A is an $n \times m$ matrix $A^\dagger$ satisfying: (i) $AA^\dagger A = A$, (ii) $A^\dagger AA^\dagger = A^\dagger$, and (iii) $(AA^\dagger)_{m \times m}$, $(A^\dagger A)_{n \times n}$ are symmetric.

Lemma 1.30   *For every $A_{m \times n}$, the matrix $A^\dagger$ exists and is unique.*

Proof   By Theorem 1.4, write $A = P \begin{pmatrix} \Sigma_r & 0 \\ 0 & 0 \end{pmatrix} Q$, with $P_{m \times m}, Q_{n \times n}$ orthogonal, and Σ_r containing the nonzero singular values of A. It is easily verified that $Q^T \begin{pmatrix} \Sigma_r^{-1} & 0 \\ 0 & 0 \end{pmatrix}^T P^T$ works as a choice of $A^\dagger$. To show the uniqueness: if $A_1^\dagger, A_2^\dagger$ are both choices of Moore–Penrose inverse for a matrix $A_{m \times n}$, then first compute using the defining properties

$$AA_1^\dagger = (AA_2^\dagger A)A_1^\dagger = (AA_2^\dagger)^T (AA_1^\dagger)^T = (A_2^\dagger)^T (A^T (A_1^\dagger)^T A^T)$$
$$= (A_2^\dagger)^T A^T = (AA_2^\dagger)^T = AA_2^\dagger.$$

Similarly, $A_1^\dagger A = A_2^\dagger A$, so

$$A_1^\dagger = A_1^\dagger (AA_1^\dagger) = A_1^\dagger (AA_2^\dagger) = (A_1^\dagger A)A_2^\dagger = (A_2^\dagger A)A_2^\dagger = A_2^\dagger. \qquad\square$$

Example 1.31   Here are some examples of the Moore–Penrose inverse of square matrices.

(1) If $D = \mathrm{diag}(\lambda_1, \ldots, \lambda_r, 0, \ldots, 0)$, with all $\lambda_j \neq 0$, then

$$D^\dagger = \mathrm{diag}\left(\frac{1}{\lambda_1}, \ldots, \frac{1}{\lambda_r}, 0, \ldots, 0\right).$$

(2) If A is positive semidefinite, then $A = U^T DU$ where D is a diagonal matrix. It is easy to verify that $A^\dagger = U^T D^\dagger U$.
(3) If A is nonsingular then $A^\dagger = A^{-1}$.

We now mention the connection between the positivity of a matrix and its Schur complement with respect to a singular submatrix. First, note that the Schur complement is now defined in the expected way (here S, M are square) i.e., as follows:

$$M = \begin{pmatrix} P & Q \\ R & S \end{pmatrix} \quad \Longrightarrow \quad M/S := P - QS^\dagger R, \qquad (1.6)$$

Now the proof of the following result can be found in standard textbooks on matrix analysis.

Theorem 1.32 *Given a symmetric matrix* $\Sigma = \begin{pmatrix} A & B \\ B^T & C \end{pmatrix}$, *with* C *not necessarily invertible, the matrix* Σ *is positive semidefinite if and only if the following conditions hold:*

(1) C *is positive semidefinite.*
(2) *The Schur complement* Σ/C *is positive semidefinite.*
(3) $(\mathrm{Id} - CC^\dagger)B^T = 0.$

2

The Schur Product Theorem
and Nonzero Lower Bounds

2.1 The Schur Product Theorem

We now make some straightforward observations about the set $\mathbb{P}_n$. The first is that $\mathbb{P}_n$ is topologically closed, convex, and closed under scaling by positive multiples (a "cone").

Lemma 2.1 $\mathbb{P}_n$ *is a closed, convex cone in* $\mathbb{R}^{n \times n}$.

Proof   All of these properties are easily verified using the definition of positive semidefiniteness. □

If A and B are positive semidefinite matrices, then we expect the product AB to also be positive semidefinite. This is true if AB is symmetric.

Lemma 2.2   *For* $A, B \in \mathbb{P}_n$, *if* AB *is symmetric then* $AB \in \mathbb{P}_n$.

Proof   In fact, $AB = (AB)^T = B^T A^T = BA$, so A and B commute. Writing $A = U^T D_1 U$ and $B = U^T D_2 U$ as per the Spectral Theorem 1.3 for commuting matrices, we have

$$x^T (AB)x = x^T (U^T D_1 U \cdot U^T D_2 U)x = x^T U^T (D_1 D_2)U x$$
$$= \| \sqrt{D_1 D_2} U x \|^2 \geq 0.$$

Hence, $AB \in \mathbb{P}_n$. □

Note, however, that AB need not be symmetric even if A and B are symmetric. In this case, the matrix AB certainly cannot be positive semidefinite; however, it still satisfies one of the equivalent conditions for positive semidefiniteness (shown above for symmetric matrices), namely, having a nonnegative spectrum. We prove this with the help of another result, which shows the "tracial" property of the spectrum.

Lemma 2.3   *Given $A_{n \times m}, B_{m \times n}$, the nonzero eigenvalues of AB and BA (and their multiplicities) agree.*

(Here, "tracial" suggests that the expression for AB equals that for BA, as does the trace.)

Proof   Assume without loss of generality that $1 \leq m \leq n$. The result will follow if we can show that $\det(\lambda \operatorname{Id}_{n \times n} - AB) = \lambda^{n-m} \det(\lambda \operatorname{Id}_{m \times m} - BA)$ for all λ. In turn, this follows from the equivalence of characteristic polynomials of AB and BA up to a power of λ, which is why we must take the union of both spectra with zero. (In particular, the sought-for equivalence would also imply that the nonzero eigenvalues of AB and BA are equal up to multiplicity).

The proof finishes by considering the two following block matrix identities

$$\begin{pmatrix} \operatorname{Id}_{n \times n} & -A \\ 0 & \lambda \operatorname{Id}_{m \times m} \end{pmatrix} \begin{pmatrix} \lambda \operatorname{Id}_{n \times n} & A \\ B & \operatorname{Id}_{m \times m} \end{pmatrix} = \begin{pmatrix} \lambda \operatorname{Id}_{n \times n} - AB & 0 \\ \lambda B & \lambda \operatorname{Id}_{m \times m} \end{pmatrix},$$

$$\begin{pmatrix} \operatorname{Id}_{n \times n} & 0 \\ -B & \lambda \operatorname{Id}_{m \times m} \end{pmatrix} \begin{pmatrix} \lambda \operatorname{Id}_{n \times n} & A \\ B & \operatorname{Id}_{m \times m} \end{pmatrix} = \begin{pmatrix} \lambda \operatorname{Id}_{n \times n} & A \\ 0 & \lambda \operatorname{Id}_{m \times m} - BA \end{pmatrix}.$$

Note that the determinants on the two left-hand sides are equal. Now, equating the determinants on the right-hand sides and canceling λ^m shows the desired identity

$$\det(\lambda \operatorname{Id}_{n \times n} - AB) = \lambda^{n-m} \det(\lambda \operatorname{Id}_{m \times m} - BA) \tag{2.1}$$

for $\lambda \neq 0$. But since both sides here are polynomial (hence continuous) functions of λ, taking limits implies the identity for $\lambda = 0$ as well. (Alternately, AB is singular if $n > m$, which shows the identity for $\lambda = 0$.) $\square$

With Lemma 2.3 at hand, we can prove the following:

Proposition 2.4   *For $A, B \in \mathbb{P}_n$, AB has nonnegative eigenvalues.*

Proof   Let $X = \sqrt{A}$ and $Y = \sqrt{A}B$, where $A = U^T DU \implies \sqrt{A} = U^T \sqrt{D} U$. Then $XY = AB$ and $YX = \sqrt{A}B\sqrt{A}$. In fact, YX is (symmetric and) positive semidefinite, since

$$x^T Y X x = \left\| \sqrt{B} \sqrt{A} x \right\|^2, \quad \forall x \in \mathbb{R}^n.$$

It follows that YX has nonnegative eigenvalues, so the same holds by Lemma 2.3 for $XY = AB$, even if AB is not symmetric. $\square$

We next introduce a different multiplication operation on matrices (possibly rectangular, including row or column matrices), which features extensively in this text.

Definition 2.5   Given positive integers m, n, the *Schur product* of $A_{m \times n}$ and $B_{m \times n}$ is the matrix $C_{m \times n}$ with $c_{jk} = a_{jk} b_{jk}$ for $1 \leq j \leq m$, $1 \leq k \leq n$. We denote the Schur product by $\circ$ (to distinguish it from the conventional matrix product).

Lemma 2.6   *Given integers $m, n \geq 1$, $(\mathbb{R}^{m \times n}, +, \circ)$ is a commutative associative algebra.*

Proof   The easy proof is omitted. More formally, $\mathbb{R}^{m \times n}$ under coordinate-wise addition and multiplication is the direct sum (or direct product) of copies of $\mathbb{R}$ under these operations.   □

Remark 2.7   Schur products occur in a variety of settings in mathematics. These include the theory of Schur multipliers (introduced by Schur himself in 1911 [216], in the same paper where he shows the Schur product theorem 2.10), association schemes in combinatorics, characteristic functions in probability theory, and the weak minimum principle in partial differential equations. Another application is to products of integral equation kernels and the connection to Mercer's theorem (see [172]). Yet another, well-known result connects the functional calculus to this entrywise product: the Daletskii–Krein formula (1956) expresses the Fréchet derivative of $f(\cdot)$ (the usual "functional calculus") at a diagonal matrix A as the Schur product/multiplier against the Loewner matrix L_f of f (see Theorem 11.8). More precisely, let $(a, b) \subset \mathbb{R}$ be open, and $f \colon (a, b) \to \mathbb{R}$ be C^1. Choose scalars $a < x_1 < \cdots < x_k < b$ and let $A := \mathrm{diag}(x_1, \ldots, x_k)$. Then Daletskii and Krein [69] showed

$$(Df)(A)(C) := \frac{d}{d\lambda} f(A + \lambda C) \Big|_{\lambda = 0} = L_f(x_1, \ldots, x_k) \circ C,$$

$$\forall C = C^* \in \mathbb{C}^{k \times k}, \tag{2.2}$$

where L_f has (j, k) entry $\frac{f(x_j) - f(x_k)}{x_j - x_k}$ if $j \neq k$, and $f'(x_j)$ otherwise. A final appearance of the Schur product that we mention here is to trigonometric moments. Suppose $f_1, f_2 \colon \mathbb{R} \to \mathbb{R}$ are continuous and 2π-periodic, with Fourier coefficients/trigonometric moments

$$a_j^{(k)} := \int_0^{2\pi} e^{-ik\theta} f_j(\theta) \, d\theta, \qquad j = 1, 2, \, k \in \mathbb{Z}.$$

If one defines the convolution product of f_1, f_2 via

$$(f_1 * f_2)(\theta) := \int_0^{2\pi} f_1(t) f_2(\theta - t) \, dt,$$

then this function has corresponding kth Fourier coefficient $a_1^{(k)} a_2^{(k)}$. Thus, the bi-infinite Toeplitz matrix of Fourier coefficients for $f_1 * f_2$ equals the Schur product of the Toeplitz matrices $(a_1^{(p-q)})_{p,q \in \mathbb{Z}}$ and $(a_2^{(p-q)})_{p,q \in \mathbb{Z}}$.

Remark 2.8 The Schur product is also called the *entrywise product* or the *Hadamard product* in the literature; the latter is likely owing to the famous paper by Hadamard in 1899 [111], in which he shows (among other things) the Hadamard multiplication theorem. This relates the radii of convergence and singularities of two power series $\sum_{j \geq 0} a_j z^j$ and $\sum_{j \geq 0} b_j z^j$ with those of $\sum_{j \geq 0} a_j b_j z^j$. This "coefficientwise product" shows up in Maló's theorem [165], in the context of Pólya–Schur multipliers.

Before proceeding further, we define another product on matrices of any dimensions

Definition 2.9 Given matrices $A_{m \times n}, B_{p \times q}$, the *Kronecker product* of A and B, denoted $A \otimes B$ is the $mp \times nq$ block matrix, defined as

$$
A \otimes B =
\begin{pmatrix}
a_{11} B & a_{12} B & \cdots & a_{1n} B \\
a_{21} B & a_{22} B & \cdots & a_{2n} B \\
\vdots & \vdots & \ddots & \vdots \\
a_{m1} B & a_{m2} B & \cdots & a_{mn} B
\end{pmatrix}.
$$

While the Kronecker product (as defined) is asymmetric in its arguments, it is easily seen that the analogous matrix $B \otimes A$ is obtained from $A \otimes B$ by permuting its rows and columns.

The next result, by Schur in 1911 [216], is important later in this text. We provide four proofs.

Theorem 2.10 (Schur product theorem) $\mathbb{P}_n$ *is closed under* $\circ$.

Proof   Suppose $A, B \in \mathbb{P}_n$; we present four proofs that $A \circ B \in \mathbb{P}_n$.

(1) Let $A, B \in \mathbb{P}_n$ have eigenbases (λ_j, v_j) and (μ_k, w_k), respectively. Then,

$$
(A \otimes B)(v_j \otimes w_k) = \lambda_j \mu_k (v_j \otimes w_k), \qquad \forall 1 \leq j, k \leq n. \tag{2.3}
$$

It follows that the Kronecker product has spectrum $\{\lambda_j \mu_k\}$, and hence is positive (semi)definite if A, B are positive (semi)definite. Hence, every principal submatrix is also positive (semi)definite by Lemma 1.9. But now observe that the principal submatrix of $A \otimes B$ with entries $a_{jk} b_{jk}$ is precisely the Schur product $A \circ B$.

(2) By the spectral theorem and the bilinearity of the Schur product,

$$A = \sum_{j=1}^{n} \lambda_j v_j v_j^T, \quad B = \sum_{k=1}^{n} \mu_k w_k w_k^T$$

$$\implies \quad A \circ B = \sum_{j,k=1}^{n} \lambda_j \mu_k (v_j \circ w_k)(v_j \circ w_k)^T.$$

This is a nonnegative linear combination of rank-1 positive semidefinite matrices, hence lies in $\mathbb{P}_n$ by Lemma 2.1.

(3) This proof uses a clever computation. Given a commutative ring R, square matrices $A, B \in R^{n \times n}$ and vectors $u, v \in R^n$, we have

$$u^T (A \circ B) v = \operatorname{tr}\left(B^T D_u A D_v\right), \tag{2.4}$$

where D_v denotes the diagonal matrix with diagonal entries the coordinates of v (in the same order). Now if $R = \mathbb{R}$ and $A, B \in \mathbb{P}_n(\mathbb{R})$, then

$$v^T (A \circ B) v = \operatorname{tr}\left(A D_v B^T D_v\right) = \operatorname{tr}\left(A^{1/2} D_v B^T D_v A^{1/2}\right).$$

But $A^{1/2} D_v B D_v A^{1/2}$ is positive semidefinite, so its trace is nonnegative, as desired.

(4) Given $t > 0$, let X, Y be independent multivariate normal vectors centered at 0 and with covariance matrices $A + t \operatorname{Id}_{n \times n}, B + t \operatorname{Id}_{n \times n}$, respectively. (Note that these always exist.) The Schur product of X and Y is then a random vector with mean zero and covariance matrix $(A + t \operatorname{Id}_{n \times n}) \circ (B + t \operatorname{Id}_{n \times n})$. Now the result follows from the fact that covariance matrices are positive semidefinite, by letting $t \to 0^+$. □

Remark 2.11   The first of the above proofs also shows the *Schur product theorem for (complex) positive definite matrices:* If $A, B \in \mathbb{P}_n(\mathbb{C})$ are positive definite, then so is their Kronecker product $A \otimes B$, hence also its principal submatrix $A \circ B$, the Schur product.

2.2 Nonzero Lower Bounds for the Schur Product

Note that the Schur product theorem is "qualitative," in that it says $M \circ N \geq 0$ if $M, N \geq 0$. In the century after its formulation, tight quantitative nonzero lower bounds have been discovered. For instance, in their papers Fiedler and Markham (1995, [83]) and Reams (1999, [190]) showed

$$M \circ N \geq \lambda_{\min}(N)(M \circ \mathrm{Id}_{n \times n}),$$

$$M \circ N \geq \frac{1}{\mathbf{1}^T N^{-1} \mathbf{1}} M, \text{ if } \det(N) > 0. \tag{2.5}$$

Here we present a recent lower bound by Khare in 2021 [145], first obtained in a weaker form by Vybíral in 2020 [233]:

Theorem 2.12 *Fix integers $a, n \geq 1$ and nonzero matrices $A, B \in \mathbb{C}^{n \times a}$. Then we have the (rank ≤ 1) lower bound*

$$A A^* \circ B B^* \geq \frac{1}{\min(\mathrm{rk}\, A A^*, \, \mathrm{rk}\, B B^*)} d_{A B^T} d^*_{A B^T},$$

where given a square matrix $M = (m_{jk})$, the column vector $d_M := (m_{jj})$. Moreover, the coefficient $1/\min(\cdot, \cdot)$ is best possible.

We make several remarks here. First, if A, B are rank 1, then a stronger result holds: we get equality above (say unlike in (2.5), for instance). Second, reformulating the result in terms of $M = A A^*$, $N = B B^*$ says that $M \circ N$ is bounded below by many possible rank-1 submatrices, one for every (square) matrix decomposition $M = A A^*$, $N = B B^*$. Third, the result extends to Hilbert–Schmidt operators, in which case it again provides nonzero positive lower bounds – and in a more general form even on $\mathbb{P}_n(\mathbb{C})$; see the paper by Khare in 2021 [145]. Finally, specializing the result to the positive semidefinite square roots $A = \sqrt{M}$ and $B = \sqrt{N}$ yields a novel connection between the matrix functional calculus and entrywise operations on matrices

$$M \circ N \geq \frac{1}{\min(\mathrm{rk}(M), \, \mathrm{rk}(N))} d_{\sqrt{M}\sqrt{N}} d^*_{\sqrt{M}\sqrt{N}}, \quad \forall M, N \in \mathbb{P}_n(\mathbb{C}), \, n \geq 1.$$

Proof We write down the proof as it serves to illustrate another important tool in matrix analysis: the *trace form* on matrix space $\mathbb{C}^{a \times a}$. Compute using (2.4):

$$u^*(A A^* \circ B B^*)u = \mathrm{tr}\left(\overline{B} B^T D_{\overline{u}} A A^* D_u\right) = \mathrm{tr}(T^* T), \quad \text{where} \quad T := A^* D_u \overline{B}.$$

Use the inner product on $\mathbb{C}^{a \times a}$, given by $\langle X, Y \rangle := \mathrm{tr}(X^* Y)$. Define the projection operator

$$P := \mathrm{proj}_{(\ker A)^\perp}|_{\mathrm{im}(B^T)};$$

thus, $P \in \mathbb{C}^{a \times a}$. Now compute

$$\langle P, P \rangle \leq \min\left(\dim(\ker A)^\perp, \, \dim \mathrm{im}\left(B^T\right)\right) = \min(\mathrm{rk}(A^*), \, \mathrm{rk}(B^*))$$
$$= \min(\mathrm{rk}(A A^*), \, \mathrm{rk}(B B^*)).$$

Here we use that $A A^*$ and A^* have the same null space, since

$$A^* x = 0 \implies A A^* x = 0 \implies \|A^* x\|^2 = 0 \implies A^* x = 0,$$

and hence the same rank. Now using the Cauchy–Schwarz inequality (for this tracial inner product) and the above computations, we have

$$u^*(AA^* \circ BB^*)u = \langle T, T \rangle \geq \frac{|\langle T, P \rangle|^2}{\langle P, P \rangle} = \frac{\left| \operatorname{tr}\left(APB^T D_{\bar{u}} \right) \right|^2}{\langle P, P \rangle} = \frac{|u^* d_{APB^T}|^2}{\langle P, P \rangle}$$

$$\geq \frac{1}{\min(\operatorname{rk}(AA^*), \operatorname{rk}(BB^*))} u^* d_{APB^T} d^*_{APB^T} u.$$

As this holds for all vectors u, the result follows because by the choice of P, we have $APB^T = AB^T$.

Finally, to see the tightness of the lower bound, choose integers $1 \leq r, s \leq a$ with $r, s \leq n$, and complex block diagonal matrices

$$A_{n \times a} := \begin{pmatrix} D_{r \times r} & 0 \\ 0 & 0 \end{pmatrix}, \quad B_{n \times a} := \begin{pmatrix} D'_{s \times s} & 0 \\ 0 & 0 \end{pmatrix},$$

with both D, D' nonsingular. Now $P = \operatorname{Id}_{\min(r,s)} \oplus \mathbf{0}_{a-\min(r,s)}$, and the inequality is indeed tight, as can be verified using the Cauchy–Schwarz identity. □

Remark 2.13   Vybíral also provided a simpler lower bound for Schur products for square matrix decompositions: if $A, B \in \mathbb{C}^{n \times n}$, then

$$AA^* \circ BB^* \geq (A \circ B)(A \circ B)^*,$$

where both sides are positive semidefinite. Indeed, if v_j, w_k denote the columns of A, B respectively, then $AA^* = \sum_j v_j v_j^*$ and $BB^* = \sum_k w_k w_k^*$, so

$$AA^* \circ BB^* = \sum_{j,k=1}^{n} (v_j v_j^*) \circ (w_k w_k^*) \geq \sum_{j=1}^{n} (v_j \circ w_j)(v_j^* \circ w_j^*)$$

$$= (A \circ B)(A \circ B)^*.$$

3
Totally Positive (*TP*) and Totally Nonnegative (*TN*) Matrices

We now study a different notion of positivity – the class of totally positive (TP) and totally nonnegative (TN) matrices (or kernels). In this text, we will primarily focus on Hankel totally nonnegative matrices, as their preservers are closely related to the preservers of positive semidefinite matrices.

Definition 3.1   Given an integer $p \geq 1$, we say a matrix is *totally positive (totally nonnegative) of order* p, denoted TP_p (TN_p), if all its $1 \times 1, 2 \times 2, \ldots$, and $p \times p$ minors are positive (nonnegative). We will also abuse notation and write $A \in TP_p$ ($A \in TN_p$) if A is TP_p (TN_p). A matrix is *totally positive (TP)* (resp. *totally nonnegative (TN)*) if A is TP_p (resp. TN_p) for all $p \geq 1$.

Remark 3.2   In classical works, as well as the books by Karlin and Pinkus, totally nonnegative and totally positive matrices were referred to, respectively, as *totally positive* and *strictly totally positive matrices*.

Here are some distinctions between totally positive/nonnegative matrices and positive (semi)definite ones:

- For TP/TN matrices we consider all minors, not just the principal ones.
- As a consequence of considering the 1×1 minors, it follows that the entries of *TP* (*TN*) matrices are all positive (nonnegative).
- TP/TN matrices need not be symmetric or even square, unlike positive semidefinite matrices.

Example 3.3   The matrix $\begin{pmatrix} 1 & 2 \\ 3 & 16 \end{pmatrix}$ is *TP*, while the matrix $\begin{pmatrix} 1 & 0 \\ 0 & 1 \end{pmatrix}$ is *TN* negative but not *TP*.

TP/TN matrices and kernels have featured in the mathematics literature in a variety of classical and modern topics. A few of these topics are now listed, as well as a few of the experts who have worked/written on them:

- Interacting particle systems, mechanics, and physics (Gantmacher and Krein), see [93, 94] and follow-up papers, as well as [54] and the references therein.
- Analysis (Aissen, Edrei, Pólya, Schoenberg, Whitney, Hirschman and Widder), see [3, 4, 74, 75, 76, 123, 124, 208, 209, 210, 211, 212, 213, 214, 215]. (See also Loewner [159] and Belton et al. and Khare [22, 24], and [144].)
- Differential equations – see, e.g., Loewner [159], Schwarz [218], the book of Karlin [136], and numerous follow-ups.
- Probability and statistics (Efron, Karlin, Kim, McGregor, Pitman, Proschan, Rinott), see, e.g., [77, 135, 136, 137, 138, 139, 140, 147, 184] and numerous follow-ups.
- Matrix theory and applications (including Ando, Cryer, Fallat, Garloff, Johnson, Pinkus, Sokal, Wagner), see, e.g., [9, 65, 68, 80, 81, 95, 96, 183, 234]. See also the conference proceedings [97].
- Gabor analysis (Gröchenig, Romero, Stöckler), see [101, 102] and related references.
- Interpolation theory and splines – see, e.g., numerous works by Schoenberg (e.g., with Curry, Whitney), de Boor, Karlin, Micchelli, Ziegler (e.g., [48, 66, 67, 141, 173, 207, 215]).
- Combinatorics (including Brenti, Gessel and Viennot, Karlin and McGregor, Lindström, Skandera, Sturmfels), see, e.g., [51, 52, 98, 138, 157, 222, 228] and follow-up works, for example, [236].
- Representation theory, flag varieties, and canonical bases (Goodearl et al., Lusztig, Postnikov, Rietsch), for example, [100, 162, 163, 188, 192, 193] and follow-up papers.
- Cluster algebras (Berenstein, Fomin, Zelevinsky), see, for example, [26, 27, 28, 86, 87, 88] and numerous follow-up papers.
- Quadratic algebras, Witt vector theory (Borger, Davydov, Grinberg, Hô Hai, Skryabin), see [49, 50, 70, 125, 223].
- Integrable systems (Kodama and Williams), see, e.g., [149, 150].

A very important, and widely used, property of TN_p matrices and kernels K is their *variation diminishing property*. Roughly speaking, if a vector u (or a function f on an interval J) has finitely many sign changes in its values, say s, then the vector Ku (or the function $\int_J K(u,x)\, dx$) has at most $\min(p,s)$ sign changes. In 1930, Schoenberg [200] showed that if K is a *TN* matrix, then

$S^-(Kx) \le S^-(x) \; \forall x \in \mathbb{R}^n$ (we define $S^-(\cdot)$ presently). A characterization of such matrices was then shown by Motzkin in his thesis:

Theorem 3.4 (Motzkin, 1936, [174])   *The following are equivalent for a matrix $K \in \mathbb{R}^{m \times n}$:*

(1) *K is variation diminishing: $S^-(Kx) \le S^-(x) \; \forall x \in \mathbb{R}^n$. Here, $S^-(x)$ for a vector x denotes the number of changes in sign, after removing all zero entries in x.*

(2) *Let K have rank r. Then K should not have two minors of equal size $< r$ but opposite signs; and K should not have two minors of equal size $= r$ but opposite signs if these minors come from the same rows or columns of K.*

In this chapter and Chapter 4, we discuss examples of *TN* and *TP* matrices. We begin by showing that the (positive semidefinite) Toeplitz cosine matrices and Hankel moment matrices considered above are in fact totally nonnegative.

Example 3.5 (Toeplitz cosine matrices)   We claim that the following matrices are *TN*:

$$C(\theta) := (\cos(j - k)\theta)_{j,k=1}^n, \quad \text{where } \theta \in \left[0, \tfrac{\pi}{2(n-1)}\right].$$

Indeed, all 1×1 minors are nonnegative, and as discussed above, $C(\theta)$ has rank at most 2, and so all 3×3 and larger minors vanish. It remains to consider all 2×2 minors. Now a 2×2 submatrix of $C(\theta)$ is of the form $C' = \begin{pmatrix} C_{ab} & C_{ac} \\ C_{db} & C_{dc} \end{pmatrix}$, where $1 \le a < d \le n$, $1 \le b < c \le n$, and C_{ab} denotes the matrix entry $\cos((a - b)\theta)$. Writing a,b,c,d in place of $a\theta, b\theta, c\theta, d\theta$ in the next computation for ease of exposition, the corresponding minor is

$$\begin{aligned}
\det C' &= \frac{1}{2}\{2\cos(a - b)\cos(d - c) - 2\cos(a - c)\cos(d - b)\} \\
&= \frac{1}{2}\{\cos(a - b + d - c) + \cos(a - b - d + c) \\
&\qquad - \cos(a - c + d - b) - \cos(a - c - d + b)\} \\
&= \frac{1}{2}\{\cos(a - b - d + c) - \cos(a - c - d + b)\} \\
&= \frac{1}{2}(-2)\sin(a - d)\sin(c - b).
\end{aligned}$$

Thus, $\det C' = \sin(d\theta - a\theta)\sin(c\theta - b\theta)$, which is nonnegative because $a < d$, $b < c$, and $\theta \in \left[0, \tfrac{\pi}{2(n-1)}\right]$. This shows that $C(\theta)$ is *TN*.

3.1 Total Nonnegativity of Hankel Matrices

Akin to the preceding example (which had been introduced in Section 1.3), we show that the Hankel moment matrices H_μ (also introduced previously) are not only positive semidefinite, but more strongly, *TN*. Just as the angle θ was restricted in the preceding example of Toeplitz cosine matrices to ensure the entries are nonnegative, we restrict the support of the measure to $[0, \infty)$, which guarantees that the entries are nonnegative.

To achieve these goals, we prove the following result, which is crucial in relating positive semidefinite matrices and their preservers, to Hankel *TN* matrices and their preservers:

Theorem 3.6   *If* $1 \le p \le n$ *are integers and* $A_{n \times n}$ *is a real Hankel matrix, then A is TP$_p$ (TN$_p$) if and only if all contiguous principal submatrices of both A and* $A^{(1)}$*, of order* $\le p$*, are positive (semi)definite, Here* $A^{(1)}$ *is obtained from A by removing the first row and last column. By a contiguous submatrix (or minor) we mean (the determinant of) a square submatrix corresponding to successive rows and to successive columns.*

In particular, A is TP (TN) if and only if $A, A^{(1)}$ *are positive (semi)definite.*

From this theorem, we derive two consequences, both of which are useful later. The first follows from the fact that all contiguous submatrices of a Hankel matrix are Hankel, hence symmetric:

Corollary 3.7   *For all integers* $1 \le p \le n$*, the set of Hankel TN$_p$ $n \times n$ matrices is a closed, convex cone, further closed under taking Schur products.*

The second corollary of Theorem 3.6 provides a large class of examples of such Hankel *TN* matrices:

Corollary 3.8   *Suppose* μ *is a nonnegative measure supported in* $[0, \infty)$*, with all moments finite. Then* H_μ *is TN.*

The proofs are left as easy exercises; the second proof uses Lemma 1.19.

Remark 3.9   Akin to Lemma 1.19 and the remark following its proof, Corollary 3.8 is also the easy half of a well-known classical result on moment problems – this time, by Stieltjes. The harder converse of Stieltjes' result says (in particular) that if a semi-infinite Hankel matrix H is TN, with (j, k)-entry $s_{j+k} \ge 0$ for $j, k \ge 0$, then there exists a nonnegative Borel measure μ on $\mathbb{R}$ with support in $[0, \infty)$, whose kth moment is s_k for all $k \ge 0$. By Theorem 3.6, this is equivalent to both H as well as $H^{(1)}$ being positive semidefinite, where $H^{(1)}$ is obtained by truncating the first row (or the first column) of H.

In the 1890s, Stieltjes was working on continued fractions and divergent series, following Euler, Laguerre, Hermite, and others. One result that is relevant here is that Stieltjes produced a nonzero function $\varphi \colon [0, \infty) \to \mathbb{R}$, such that $\int_0^\infty x^k \varphi(x)\, dx = 0$ for all $k = 0, 1, \ldots$ – an indeterminate moment problem. (The work in this setting also led him to develop the Stieltjes integral; see, e.g., [148] for a detailed historical account.) This marks the beginning of his exploration of the moment problem, which he resolved in his well-known memoir [227].

A curious follow-up, by Boas in 1939, is that if one replaces the nonnegativity of the Borel measure μ by the hypothesis of being of the form $d\alpha(t)$ on $[0, \infty)$ with α of bounded variation, then this recovers *all* real sequences! See [42].

The remainder of this chapter is devoted to showing Theorem 3.6. The proof uses a sequence of lemmas shown by Gantmacher [92], Gantmacher and Krein [93], Fekete [82], Schoenberg [213], and others. The first of these lemmas may be (morally) attributed to Laplace.

Lemma 3.10   *Let $r \geq 1$ be an integer and $U = (u_{jk})$ an $(r+2) \times (r+1)$ matrix. Given subsets $[a,b], [c,d] \subset (0, \infty)$, let $U_{[a,b] \times [c,d]}$ denote the submatrix of U with entries u_{jk}, such that j, k are integers and $a \leq j \leq b$, $c \leq k \leq d$. Then*

$$\det U_{[1,r] \cup \{r+2\} \times [1, r+1]} \cdot \det U_{[2, r+1] \times [1, r]}$$

$$= \det U_{[2, r+2] \times [1, r+1]} \cdot \det U_{[1, r] \times [1, r]} \qquad (3.1)$$

$$+ \det U_{[1, r+1] \times [1, r+1]} \cdot \det U_{[2, r] \cup \{r+2\} \times [1, r]}.$$

Note that in each of the three products of determinants, the second factor in the subscript for the first (respectively second) determinant terms is the same: $[1, r+1]$ (respectively $[1, r]$).

To give a feel for the result, the special case of $r = 1$ asserts that

$$u_{11} \begin{vmatrix} u_{21} & u_{22} \\ u_{31} & u_{32} \end{vmatrix} - u_{21} \begin{vmatrix} u_{11} & u_{12} \\ u_{31} & u_{32} \end{vmatrix} + u_{31} \begin{vmatrix} u_{11} & u_{12} \\ u_{21} & u_{22} \end{vmatrix} = 0.$$

But this is precisely the Laplace expansion along the third column of the singular matrix

$$\det \begin{pmatrix} u_{11} & u_{12} & u_{11} \\ u_{21} & u_{22} & u_{21} \\ u_{31} & u_{32} & u_{31} \end{pmatrix} = 0.$$

Proof   Consider the $(2r + 1) \times (2r + 1)$ block matrix of the form

$$
M = \left(
\begin{array}{ccc|c}
\mathbf{b}^T & u_{1,r+1} & | & \mathbf{b}^T \\
A & \mathbf{a} & | & A \\
\mathbf{c}^T & u_{r+1,r+1} & | & \mathbf{c}^T \\
\mathbf{d}^T & u_{r+2,r+1} & | & \mathbf{d}^T \\
\hline
A & \mathbf{a} & | & 0_{(r-1)\times r}
\end{array}
\right)
$$

$$
= \left(
\begin{array}{cc}
(u_{jk})_{j\in[1,r+2],\, k\in[1,r+1]} & (u_{jk})_{j\in[1,r+2],\, k\in[1,r]} \\
(u_{jk})_{j\in[2,r],\, k\in[1,r+1]} & 0_{(r-1)\times r}
\end{array}
\right);
$$

that is, where

$$
\mathbf{a} = (u_{2,r+1}, \ldots, u_{r,r+1})^T, \quad \mathbf{b} = (u_{1,1}, \ldots, u_{1,r})^T, \quad \mathbf{c} = (u_{r+1,1}, \ldots, u_{r+1,r})^T,
$$

$$
\mathbf{d} = (u_{r+2,1}, \ldots, u_{r+2,r})^T, \quad A = (u_{jk})_{2\le j\le r,\, 1\le k\le r}.
$$

Notice that M is a square matrix whose first $r + 2$ rows have column space of dimension at most $r + 1$; hence $\det M = 0$. Now we compute $\det M$ using the (generalized) Laplace expansion by complementary minors: choose all possible $(r + 1)$-tuples of rows from the first $r + 1$ columns to obtain a submatrix $M'_{(r+1)}$, and deleting these rows and columns from M yields the complementary $r \times r$ submatrix $M''_{(r)}$ from the final r columns. The generalized Laplace expansion says that if one multiplies $\det M'_{(r+1)} \cdot \det M''_{(r)}$ by $(-1)^{\Sigma}$, with Σ the sum of the row numbers in $M'_{(r+1)}$, then summing over all such products (running over subsets of rows) yields $\det M$ – which vanishes for this particular matrix M.

Now in the given matrix, to avoid obtaining zero terms, the rows in $M'_{(r+1)}$ must include all entries from the final $r - 1$ rows (and the first $r + 1$ columns). But then it, moreover, cannot include entries from the rows of M labelled $2, \ldots, r$; and it must include two of the remaining three rows (and entries from only the first $r + 1$ columns).

Thus, we obtain three product terms that sum to: $\det M = 0$. Upon carefully examining the terms and computing the companion signs (by row permutations), we obtain (3.1).   □

The next two results are classical facts about *TP* matrices, first shown by Fekete in his correspondence with Pólya [82].

Lemma 3.11   *Given integers $m \ge n \ge 1$ and a real matrix $A_{m\times n}$, such that*

(1) *all $(n - 1) \times (n - 1)$ minors $\det A_{J\times[1,n-1]} > 0$ for $J \subset [1,m]$ of size $n - 1$, and*

(2) *all* $n \times n$ *minors* $\det A_{[j+1, j+n] \times [1,n]} > 0$ *for* $0 \le j \le m - n$,

we have that all $n \times n$ *minors of A are positive.*

Proof Define the *gap*, or "index" of a subset of integers $J = \{j_1 < j_2 < \cdots < j_n\} \subset [1, m]$, to be $g_J := j_n - j_1 - (n - 1)$. Thus, the gap is zero if and only if J consists of successive integers, and in general it counts precisely the number of integers between j_1 and j_n that are missing from J.

We claim that $\det A_{J \times [1,n]} > 0$ for $|J| = n$, by induction on the gap $g_J \ge 0$; the base case $g_J = 0$ is given as hypothesis. For the induction step, suppose j^0 is a missing index (or row number) in $J = \{j_1 < \cdots < j_n\}$. By suitably specializing the identity (3.1), we obtain

$$
\det A_{(j_1, \dots, j_n) \times [1,n]} \cdot \det A_{(j_2, \dots, j_{n-1}, j^0) \times [1, n-1]}
$$
$$
= \det A_{(j_1, \dots, j_{n-1}, j^0) \times [1,n]} \cdot \det A_{(j_2, \dots, j_n) \times [1, n-1]} \qquad (3.2)
$$
$$
- \det A_{(j_2, \dots, j_n, j^0) \times [1,n]} \cdot \det A_{(j_1, \dots, j_{n-1}) \times [1, n-1]}.
$$

Consider the six factors in serial order. The first, fourth, and sixth factors have indices listed in increasing order, while the other three factors have j^0 listed at the end, so their indices are not listed in increasing order. For each of the six factors, the number of "bubble sorts" required to rearrange indices in increasing order (by moving j^0 down the list) equals the number of row exchanges in the corresponding determinants; label these numbers $n_1, \dots, n_6$. Thus, $n_1 = n_4 = n_6 = 0$ as above, while $n_2 = n_3$ (since $j_1 < j^0 < j_n$), and $|n_2 - n_5| = 1$. Now multiply Equation (3.2) by $(-1)^{n_2}$, and divide both sides by

$$
c_0 := (-1)^{n_2} \det A_{(j_2, \dots, j_{n-1}, j^0) \times [1, n-1]} > 0.
$$

Using the given hypotheses as well as the induction step (since all terms involving j^0 have a gap equal to $g_J - 1$), it follows that

$$
\det A_{(j_1, \dots, j_n) \times [1,n]}
$$
$$
= c_0^{-1} \Big((-1)^{n_2} \det A_{(j_1, \dots, j_{n-1}, j^0) \times [1,n]} \cdot \det A_{(j_2, \dots, j_n) \times [1, n-1]}
$$
$$
+ (-1)^{n_2+1} \det A_{(j_2, \dots, j_n, j^0) \times [1,n]} \cdot \det A_{(j_1, \dots, j_{n-1}) \times [1, n-1]} \Big)
$$
$$
> 0.
$$

This completes the induction step, and with it, the proof. □

We can now state and prove another 1912 result by Fekete [82] for *TP* matrices – extended to *TP*$_p$ matrices by Schoenberg in 1955 [213]:

Lemma 3.12 (Fekete–Schoenberg lemma) *Suppose* $m, n \ge p \ge 1$ *are integers, and* $A \in \mathbb{R}^{m \times n}$ *is a matrix, all of whose contiguous minors of order at most p are positive. Then A is TP*$_p$.

Notice that the analogous statement for TN_p is false, e.g., $p = 2$ and
$$A = \begin{pmatrix} 1 & 0 & 2 \\ 1 & 0 & 1 \end{pmatrix}.$$

Proof   We show that for any integer $s \in [1, p]$, every $s \times s$ minor of A is positive. The proof is by induction on s (and running over all real matrices satisfying the hypotheses); note that the base case of $s = 1$ is immediate from the assumptions. For the induction step, suppose
$$2 \le s = |J| = |K| \le p, \qquad J \subset \mathbb{Z} \cap [1, m], \ K \subset \mathbb{Z} \cap [1, n].$$

First fix a subset K that consists of consecutive rows, i.e., has gap $g_K = 0$ (as in the proof of Lemma 3.11). Let B denote the submatrix $A_{[1,m] \times K}$. Then all $s \times s$ minors of B are positive, by Lemma 3.11. In particular, it follows for all J that all $s \times s$ minors $\det A_{J \times K'}$ are positive, whenever $K' \subset [1, n]$ has size s and gap $g_{K'} = 0$. Now apply Lemma 3.11 to the matrix $B := (A_{J \times [1,n]})^T$ to obtain: $\det(A_{J,K})^T > 0$ for (possibly nonconsecutive subsets) K. This concludes the proof.   □

The final ingredient required to prove Theorem 3.6 is the following result:

Lemma 3.13   *If $A_{n \times n}$ is a Hankel matrix, then every contiguous minor of A (see Lemma 3.12) is a contiguous principal minor of A or of $A^{(1)}$.*

Recall that $A^{(1)}$ was defined in Theorem 3.6.

Proof   Let the first row (respectively, last column) of A contain the entries $s_0, s_1, \dots, s_{n-1}$ (respectively, $s_{n-1}, s_n, \dots, s_{2n-2}$). Then every contiguous minor is the determinant of a submatrix of the form
$$M = \begin{pmatrix} s_j & \cdots & s_{j+m} \\ \vdots & \ddots & \vdots \\ s_{j+m} & \cdots & s_{j+2m} \end{pmatrix}, \qquad 0 \le j \le j + m \le n - 1.$$

It is now immediate that if j is even (respectively odd), then M is a contiguous principal submatrix of A (respectively $A^{(1)}$).   □

With these results at hand, we conclude by proving the above theorem.

Proof of Theorem 3.6   If the Hankel matrix A is TN_p (TP_p), then all contiguous minors of A of order $\le p$ are nonnegative (positive), proving one implication. Conversely, suppose all contiguous principal minors of A and $A^{(1)}$ of order $\le p$ are positive. By Lemma 3.13, this implies every contiguous minor of A of order $\le p$ is positive. By the Fekete–Schoenberg Lemma 3.12, A is TP_p as desired.

Finally, suppose all contiguous principal minors of $A, A^{(1)}$ or size $\leq p$ are nonnegative. It follows by Lemma 3.13 that every contiguous square submatrix of A of order $\leq p$ is positive semidefinite. Also choose and fix an $n \times n$ Hankel *TP* matrix B (note by (4.3) or Lemma 4.10 that such matrices exist for all $n \geq 1$). Applying Lemma 3.13, $B, B^{(1)}$ are positive definite, hence so is every contiguous square submatrix of B. Now for $\epsilon > 0$, it follows (by Sylvester's criterion, Theorem 1.8) that every contiguous principal minor of $A + \epsilon B$ of size $\leq p$ is positive. Again applying Lemma 3.12, the Hankel matrix $A + \epsilon B$ is necessarily TP_p, and taking $\epsilon \to 0^+$ finishes the proof.

The final statement is the special case $p = n$. $\qquad\square$

4

Totally Positive Matrices – Generalized Vandermonde and Hankel Moment Matrices

Chapter 3 discussed examples (Toeplitz, Hankel) of totally nonnegative (TN) matrices. These examples consisted of symmetric matrices.

- We will now look at some examples of nonsymmetric matrices that are totally positive (TP). We then prove the Cauchy–Binet formula, which will lead to the construction of additional examples of symmetric TP matrices.
- Let $H_\mu := (s_{j+k}(\mu))_{j,k \geq 0}$ denote the moment matrix associated to a nonnegative measure μ supported on $[0, \infty)$. We have already seen that this matrix is Hankel and positive semidefinite – in fact, TN. We will show in this chapter that H_μ is in fact TP in many cases. The proof will use a continuous generalization of the Cauchy–Binet formula.

4.1 Generalized Vandermonde Matrices

A *generalized Vandermonde matrix* is a matrix $(x_j^{\alpha_k})_{j,k=1}^n$, where $x_j > 0$ and $\alpha_j \in \mathbb{R}$ for all j. If the x_j are pairwise distinct, as are the α_k, then the corresponding generalized Vandermonde matrix is nonsingular. In fact, a stronger result holds:

Theorem 4.1   *If $0 < x_1 < \cdots < x_m$ and $\alpha_1 < \cdots < \alpha_n$ are real numbers, then the generalized Vandermonde matrix $V_{m \times n} := (x_j^{\alpha_k})$ is TP.*

As an illustration, consider the special case $m = n$ and $\alpha_k = k - 1$, which recovers the usual Vandermonde matrix

$$V = \begin{pmatrix} 1 & x_1 & \cdots & x_1^{n-1} \\ 1 & x_2 & \cdots & x_2^{n-1} \\ \vdots & \vdots & \ddots & \vdots \\ 1 & x_n & \cdots & x_n^{n-1} \end{pmatrix}.$$

This matrix has determinant $\prod_{1 \leq j < k \leq n}(x_k - x_j) > 0$. Thus, if $0 < x_1 < x_2 < \cdots < x_n$ then $\det V > 0$. However, note this is not enough to prove that the matrix is *TP*. Thus, more work is required to prove total positivity, even for usual Vandermonde matrices. The following 1637 result by Descartes [71] will help in the proof. (This preliminary result is also the beginning of a mathematical journey that led to both total positivity and to the variation diminishing property, via works of Laguerre [153] and Fekete [82] in 1883 and 1912, respectively.)

Lemma 4.2 (Descartes' rule of signs, weaker version)   *Fix pairwise distinct real numbers $\alpha_1, \alpha_2, \ldots, \alpha_n \in \mathbb{R}$ and n scalars $c_1, c_2, \ldots, c_n \in \mathbb{R}$, such that not all scalars are 0. Then the function $f(x) := \sum_{k=1}^{n} c_k x^{\alpha_k}$ can have at most $(n-1)$ distinct positive roots.*

The following proof is due to Laguerre (1883); that said, the trick of multiplying by a faster-decaying function and applying Rolle's theorem was previously employed by Poulain in 1867 to prove the Hermite–Poulain theorem [189].

Proof   By induction on n. For $n = 1$, clearly $f(x)$ has no positive root. For the induction step, without loss of generality we may assume $\alpha_1 < \alpha_2 < \cdots < \alpha_n$ and that all c_j are nonzero. If f has n distinct positive roots, then so does the function

$$g(x) := x^{-\alpha_1} f(x) = c_1 + \sum_{k=2}^{n} c_k x^{\alpha_k - \alpha_1}.$$

But then Rolle's theorem implies that $g'(x) = \sum_{k=2}^{n} c_k(\alpha_k - \alpha_1)x^{\alpha_k - \alpha_1 - 1}$ has $(n-1)$ distinct positive roots. This contradicts the induction hypothesis, completing the proof.   □

With this result at hand, we can prove that generalized Vandermonde matrices are *TP*.

Proof of Theorem 4.1   As any submatrix of V is also a generalized Vandermonde matrix, it suffices to show that the determinant of V is positive when $m = n$.

We first claim that $\det V \neq 0$. Indeed, suppose for contradiction that V is singular. Then there is a nonzero vector $\mathbf{c} = (c_1, c_2, \ldots, c_n)^T$, such that $V\mathbf{c} = 0$. But then there exist n distinct positive numbers $x_1, x_2, \ldots, x_n$, such that $\sum_{k=1}^{n} c_k x_j^{\alpha_k} = 0$, which contradicts Lemma 4.2 for $f(x) = \sum_{k=1}^{n} c_k x^{\alpha_k}$. Thus, the claim follows.

We now prove the theorem via a homotopy argument. Consider a (continuous) path $\gamma : [0,1] \to \mathbb{R}^n$ going from $\gamma(0) = (0,1,\ldots,n-1)$ to $\gamma(1) = (\alpha_1, \alpha_2, \ldots, \alpha_n)$, such that at each timepoint $t \in [0,1]$, the coordinates of $\gamma(t)$ are in increasing order. It is possible to choose such a path; indeed, the straight line geodesic path is one such path.

Now let $W(t) := \det(x_j^{\gamma_k(t)})_{j,k=1}^n$. Then $W : [0,1] \to \mathbb{R}$ is a continuous map that never vanishes. Since $[0,1]$ is connected and $W(0) > 0$ (see remarks above), it follows that $W(1) = \det V > 0$.   □

Remark 4.3   If we have $0 < x_n < x_{n-1} < \cdots < x_1$ and $\alpha_n < \alpha_{n-1} < \cdots < \alpha_1$, then observe that the corresponding generalized Vandermonde matrix $V' := \left(x_j^{\alpha_k}\right)_{j,k=1}^n$ is also *TP*. Indeed, once again we only need to show $\det V' > 0$, and this follows from applying the same permutation to the rows and to the columns of V' to reduce it to the situation in Theorem 4.1 (since then the determinant does not change in sign).

4.2 The Cauchy–Binet Formula

The following is a recipe to construct new examples of TP/TN matrices from known ones.

Proposition 4.4   *If $A_{m \times n}, B_{n \times k}$ are both TN, then so is the matrix $(AB)_{m \times k}$. This assertion is also valid upon replacing "TN" by "TP," provided $n \geq \max\{m,k\}$.*

To prove this proposition, we require the following important result.

Theorem 4.5 (Cauchy–Binet formula)   *Given matrices $A_{m \times n}$ and $B_{n \times m}$, we have*

$$\det(AB)_{m \times m} = \sum_{J \subset [n] \ of \ size \ m} \det(A_{[m] \times J^\uparrow}) \det(B_{J^\uparrow \times [m]}), \qquad (4.1)$$

where $J^\uparrow$ reiterates the fact that the elements of J are arranged in increasing order.

For example, if $m = n$, this theorem just reiterates the fact that the determinant map is multiplicative on square matrices. If $m > n$, the theorem says that determinants of singular matrices are zero. If $m = 1$, we obtain the inner product of a row and column vector.

Proof   Notice that

$$
\det(AB) = \det
\begin{pmatrix}
\sum\limits_{j_1=1}^{n} a_{1j_1} b_{j_11} & \cdots & \sum\limits_{j_m=1}^{n} a_{1j_m} b_{j_mm} \\
\vdots & \ddots & \vdots \\
\sum\limits_{j_1=1}^{n} a_{mj_1} b_{j_11} & \cdots & \sum\limits_{j_m=1}^{n} a_{mj_m} b_{j_mm}
\end{pmatrix}_{m \times m} .
$$

By the multilinearity of the determinant, expanding $\det(AB)$ as a sum over all j_l yields

$$
\det(AB) = \sum_{(j_1, j_2, \ldots, j_m) \in [n]^m} b_{j_11} b_{j_22} \ldots b_{j_mm} \cdot \det
\begin{pmatrix}
a_{1j_1} & \cdots & a_{1j_m} \\
\vdots & \ddots & \vdots \\
a_{mj_1} & \cdots & a_{mj_m}
\end{pmatrix} .
$$

The determinant in the summand vanishes if $j_k = j_m$ for any $k \neq m$. Therefore,

$$
\det(AB) = \sum_{\substack{(j_1, j_2, \ldots, j_m) \in [n]^m, \\ \text{all } j_l \text{ are distinct}}} b_{j_11} b_{j_22} \ldots b_{j_mm} \cdot \det
\begin{pmatrix}
a_{1j_1} & \cdots & a_{1j_m} \\
\vdots & \ddots & \vdots \\
a_{mj_1} & \cdots & a_{mj_m}
\end{pmatrix}
$$

$$
= \sum_{\substack{(j_1, j_2, \ldots, j_m) \in [n]^m, \\ \text{all } j_l \text{ are distinct}}} b_{j_11} b_{j_22} \ldots b_{j_mm} \cdot \det A_{[m] \times (j_1, j_2, \ldots, j_m)} .
$$

We split this sum into two subsummations. One part runs over all collections of indices, while the other runs over all possible orderings – that is, permutations – of each fixed collection of indices. Thus, for each ordering $\mathbf{j} = (j_1, \ldots, j_m)$ of $J = \{j_1, \ldots, j_m\}$, there exists a unique permutation $\sigma_{\mathbf{j}} \in S_m$, such that $(j_1, \ldots, j_m) = \sigma_{\mathbf{j}}(J^\uparrow)$. Now,

$$
\det(AB)
$$

$$
= \sum_{J = \{j_1, j_2, \ldots, j_m\} \subset [n], \text{ all } j_l \text{ distinct}} \sum_{\sigma = \sigma_{\mathbf{j}} \in S_m} b_{j_11} b_{j_22} \ldots b_{j_mm} (-1)^{\sigma_{\mathbf{j}}} \det A_{[m] \times J^\uparrow}
$$

$$
= \sum_{J \subset [n] \text{ of size } m} \det(A_{[m] \times J^\uparrow}) \det \left(B_{J^\uparrow \times [m]} \right) . \qquad \square
$$

Proof of Proposition 4.4   Suppose two matrices $A_{m \times n}$ and $B_{n \times k}$ are both *TN*. Let $I \subset [m]$ and $K \subset [k]$ be index subsets of the same size; we are to show

$\det(AB)_{I\times K}$ is nonnegative. Define matrices, $A' := A_{I\times[n]}$ and $B' := B_{[n]\times K}$. Now it is easy to show that $(AB)_{I\times K} = A'B'$. In particular, $\det(AB)_{I\times K} = \det(A'B')$. Hence, the Cauchy–Binet theorem implies

$$\det(AB)_{I\times K} = \sum_{\substack{J\subset[n],\\ |J|=|K|=|I|}} \det A'_{I\times J\uparrow} \det B'_{J\uparrow\times K}$$

$$= \sum_{\substack{J\subset[n],\\ |J|=|K|=|I|}} \det A_{I\times J\uparrow} \det B_{J\uparrow\times K} \geq 0. \qquad (4.2)$$

It follows that AB is *TN* if both A and B are *TN*. For the corresponding *TP*-version, the above proof works as long as the sums in the preceding equation are always over nonempty sets; but this happens whenever $n \geq \max\{m,k\}$. □

Remark 4.6   This proof shows that Proposition 4.4 holds upon replacing *TN/TP* by TP_p for any $1 \leq p \leq n$. (E.g. the condition TP_n coincides with TP_{n-1} if $\min\{m,k\} < n$.)

4.3 Generalized Cauchy–Binet Formula and Totally Positive Moment Matrices

We showed in Section 4.1 that generalized Vandermonde matrices are examples of *TP* but nonsymmetric matrices. Using these, we can construct additional examples of *TP* symmetric matrices: let $V = (x_j^{\alpha_k})_{j,k=1}^n$ be a be a generalized Vandermonde matrix with $0 < x_1 < x_2 < \cdots < x_n$ and $\alpha_1 < \alpha_2 < \cdots < \alpha_n$. Then Proposition 4.4 implies that the symmetric matrices $V^T V$ and $V V^T$ are *TP*.

For instance, if we take $n = 3$ and $\alpha_k = k - 1$, then

$$V = \begin{pmatrix} 1 & x_1 & x_1^2 \\ 1 & x_2 & x_2^2 \\ 1 & x_3 & x_3^2 \end{pmatrix}, \quad V^T V = \begin{pmatrix} 3 & \sum_{j=1}^3 x_j & \sum_{j=1}^3 x_j^2 \\ \sum_{j=1}^3 x_j & \sum_{j=1}^3 x_j^2 & \sum_{j=1}^3 x_j^3 \\ \sum_{j=1}^3 x_j^2 & \sum_{j=1}^3 x_j^3 & \sum_{j=1}^3 x_j^4 \end{pmatrix}.$$

$$(4.3)$$

This is clearly the Hankel moment matrix H_μ for the counting measure on the set $\{x_1, x_2, x_3\}$. Moreover, $V^T V$ is (symmetric and) *TP* by the Cauchy–Binet formula. More generally, for all increasing α_k which are in arithmetic progression, the matrix $V^T V$ (defined similarly as above) is a *TP* Hankel moment matrix for some nonnegative measure on $[0, \infty)$ – more precisely, supported on $\left\{ x_1^{\alpha_2 - \alpha_1}, x_2^{\alpha_2 - \alpha_1}, \ldots, x_n^{\alpha_2 - \alpha_1} \right\}$.

The following discussion aims to show (among other things) that the moment matrices H_μ defined in (1.3) are *TP* for "most" nonnegative measures μ supported in $[0, \infty)$. We begin with by studying functions that are *TP* or *TN*.

Definition 4.7   Let $X, Y \subset \mathbb{R}$ and $K : X \times Y \to \mathbb{R}$ be a function. Given $p \in \mathbb{N}$, we say $K(x, y)$ is a *TN/TP kernel of order p* (denoted TN_p or TP_p) if for any integer $1 \le n \le p$ and elements

$$x_1 < x_2 < \cdots < x_n \in X, \qquad y_1 < y_2 < \cdots < y_n \in Y,$$

we have $\det K(x_j, y_k)_{j,k=1}^n$ is nonnegative (positive). Similarly, we say that the kernel $K : X \times Y \to \mathbb{R}$ is *TN/TP* if K is TN_p (or TP_p) for all $p \ge 1$.

Here are a few examples of *TP* kernels on $\mathbb{R} \times \mathbb{R}$.

Example 4.8   The kernel $K(x, y) = e^{xy}$ is *TP*, with $X = Y = \mathbb{R}$. Indeed, choosing real numbers $x_1 < x_2 < \cdots < x_n$ and $y_1 < y_2 < \cdots < y_n$, the matrix $(e^{x_j y_k})_{j,k=1}^n = ((e^{x_j})^{y_k})_{j,k=1}^n$ is a generalized Vandermonde matrix, hence *TP*, so its determinant is positive.

Lemma 4.9 (Pólya)   *For all $\sigma > 0$, the Gaussian kernel $T_{G_\sigma} : \mathbb{R} \times \mathbb{R} \to \mathbb{R}$ given by*

$$T_{G_\sigma}(x, y) := e^{-\sigma(x-y)^2}$$

is TP.

Note that the function $f(x) = e^{-\sigma x^2}$ is such that the *TP* kernel $T_{G_\sigma}(x, y)$ can be rewritten as $f(x - y)$. Such (integrable) functions on the real line are known as *Pólya frequency functions*.

Proof   Given real numbers $x_1 < x_2 < \cdots < x_n$ and $y_1 < y_2 < \cdots < y_n$, we have

$$\left(T_{G_\sigma}(x_j, y_k) \right)_{j,k=1}^n$$
$$= \left(e^{-\sigma x_j^2} e^{2\sigma x_j y_k} e^{-\sigma y_k^2} \right)_{j,k=1}^n$$
$$= \operatorname{diag} \left(e^{-\sigma x_j^2} \right)_{j=1}^n \begin{pmatrix} e^{2\sigma x_1 y_1} & \cdots & e^{2\sigma x_1 y_n} \\ \vdots & \ddots & \vdots \\ e^{2\sigma x_n y_1} & \cdots & e^{2\sigma x_n y_n} \end{pmatrix} \operatorname{diag} \left(e^{-\sigma y_k^2} \right)_{k=1}^n,$$

and this has positive determinant by Example 4.8. $\square$

In a similar vein, we have the following:

Lemma 4.10   *For all $\sigma > 0$, the kernel $H'_\sigma : \mathbb{R} \times \mathbb{R} \to \mathbb{R}$, given by*

$$H'_\sigma(x, y) := e^{\sigma(x+y)^2}$$

is TP. In particular, the kernels T_{G_σ} (from Lemma 4.9) and H'_σ provide examples of TP Toeplitz and Hankel matrices, respectively.

Proof   The proof of the total positivity of H'_σ is similar to that of T_{G_σ} above, and hence left as an exercise. To obtain *TP* Toeplitz and Hankel matrices, akin to Example 3.5, we choose any arithmetic progression $x_1, \ldots, x_n$ of finite length and consider the matrices with (j,k)th entry $T_{G_\sigma}(x_j, x_k)$ and $H'_\sigma(x_j, x_k)$, respectively.   □

We next generalize the Cauchy–Binet formula to TP/TN *kernels*. Suppose $X, Y, Z \subset \mathbb{R}$ and μ is a nonnegative Borel measure on Y. Let $K(x, y)$ and $L(y, z)$ be "nice" functions (i.e., Borel measurable with respect to Y), and assume the following function is well defined:

$$M : X \times Z \to \mathbb{R}, \qquad M(x, z) := \int_Y K(x, y) L(y, z) d\mu(y). \qquad (4.4)$$

For example, consider $K(x, y) = e^{xy}$ and $L(y, z) = e^{yz}$. Take $X = Z = \{\alpha_1, \alpha_2, \ldots, \alpha_n\}$ and $Y = \{\log(x_1), \log(x_2), \ldots, \log(x_n)\}$, such that $0 < x_1 < x_2 < \cdots < x_n$ and $\alpha_1 < \alpha_2 < \cdots < \alpha_n$. Finally, suppose μ denotes the counting measure on Y. Then $M(\alpha_i, \alpha_k) = \sum_{j=1}^n x_j^{\alpha_i} x_j^{\alpha_k}$. So $(M(\alpha_i, \alpha_k))_{i,k=1}^n = V^T V$, where $V = \left(x_j^{\alpha_k}\right)_{j,k=1}^n$ is a generalized Vandermonde matrix.

In this "discrete" example (i.e., where the support of μ is a discrete set), $\det M$ is shown to be positive using the total positivity of V, V^T and the Cauchy–Binet formula. In fact, this phenomenon extends to the more general setting above as follows:

Exercise 4.11 (Pólya–Szegő, Basic Composition formula, aka Generalized Cauchy–Binet formula)   Suppose $X, Y, Z \subset \mathbb{R}$ and $K(x, y), L(y, z), M(x, z)$ are as above. Then using an argument similar to the above proof of the Cauchy–Binet formula show that

$$\det \begin{pmatrix} M(x_1, z_1) & \cdots & M(x_1, z_m) \\ \vdots & \ddots & \vdots \\ M(x_m, z_1) & \cdots & M(x_m, z_m) \end{pmatrix}$$

$$= \int \cdots \int_{y_1 < y_2 < \cdots < y_m \text{ in } Y} \det(K(x_i, y_j))_{i,j=1}^m \cdot \det(L(y_j, z_k))_{j,k=1}^m \prod_{j=1}^m d\mu(y_j).$$

$$(4.5)$$

Remark 4.12   In the right-hand side of Equation (4.5), we may also integrate over the region $y_1 \leq \cdots \leq y_m$ in Y, since matrices with equal rows or columns are singular.

From (4.5) and Remark 4.6 we obtain the following consequence:

Corollary 4.13   *If the kernels K and L are both TN_p for some integer $p > 0$ (or even TN), then so is M, where M was defined in (4.4). If instead, K and L are TP_p kernels, where $p \leq |\mathrm{supp}(\mu)|$, then so is M.*

We will apply this result to the moment matrices H_μ defined in (1.3). We begin more generally: suppose $Y \subset \mathbb{R}$ and $u : Y \to (0, \infty)$ is a positive and strictly increasing function, all of whose moments exist with respect to some nonnegative measure μ

$$\int_Y u(y)^n \, d\mu(y) < \infty, \qquad \forall n \geq 0.$$

Then we claim:

Proposition 4.14   *The kernel $M : \mathbb{R} \times \mathbb{R} \to \mathbb{R}$, given by*

$$M(n, m) := \int_Y u(y)^{n+m} \, d\mu(y)$$

is TN as well as $TP_{|Y_+|}$, where $Y_+ := \mathrm{supp}(\mu) \subset Y$ is finite or infinite.

Proof   To show M is $TP_{|Y_+|}$, the first claim is that the kernel $K(n, y) := u(y)^n$ is TP on $\mathbb{R} \times Y$. Indeed, we can rewrite $K(n, y) = e^{n \log(u(y))}$. Now given increasing tuples of elements $n_j \in \mathbb{R}$ and $y_k \in Y$, the matrix $K(n_j, y_k)$ is TP, by the total positivity of e^{xy} (see Example 4.8).

Similarly, $L(y, m) := u(y)^m$ is also TP on $Y \times \mathbb{R}$. The result now follows by Corollary 4.13. That M is TN follows from the same arguments, via Remark 4.12.   □

This result implies the total positivity of the Hankel moment matrices (1.3). Indeed, setting $u(y) = y$ on domains $Y \subset [0, \infty)$, we obtain:

Corollary 4.15   *Suppose $Y \subset [0, \infty)$ and μ is a nonnegative measure on Y with infinite support. Then the moment matrix H_μ is TP (of all orders).*

5

Entrywise Powers Preserving Positivity in a Fixed Dimension

In the remainder of this text, we discuss operations that preserve the positive semidefiniteness of matrices and kernels, when applied via composition operators on positive kernels. With the exception of Chapters 18 and 19, in this text we deal with kernels on finite domains – aka matrices – which translates to the functions being applied *entrywise* to various classes of matrices. The remainder of this part of the text discusses the important special case of entrywise *powers* preserving positivity on matrices; to understand some of the modern and classical motivations behind this study, we refer the reader to Sections 7.1 and 11.1, respectively.

We begin with some preliminary definitions.

Definition 5.1 Given a subset $I \subset \mathbb{R}$, define $\mathbb{P}_n(I) := \mathbb{P}_n \cap I^{n \times n}$ to be the set of $n \times n$ positive semidefinite matrices, all of whose entries are in I.

A function $f \colon I \to \mathbb{R}$ acts *entrywise* on vectors/matrices with entries in I via: $A = (a_{jk}) \mapsto f[A] := (f(a_{jk}))$. We say f is *Loewner positive on* $\mathbb{P}_n(I)$ if $f[A] \in \mathbb{P}_n$ whenever $A \in \mathbb{P}_n(I)$.

Note that the entrywise operator $f[-]$ differs from the usual holomorphic calculus (except when acting on diagonal matrices by functions that vanish at the origin).

Remark 5.2 The entrywise calculus was initiated by Schur in the same paper in 1911 [216] where he proved the Schur product theorem. Schur defined $f[A]$ – but using different notation – and proved the first result involving entrywise maps (see, e.g., [73, p. cxii] for additional commentary).

We fix the following **notation** for future use. If $f(x) = x^\alpha$ for some $\alpha \geq 0$ and $I \subset [0, \infty)$, then we write $A^{\circ \alpha}$ for $f[A]$, where A is any vector or matrix. By convention we shall take $0^0 = 1$ whenever required, so that $A^{\circ 0}$ is the matrix $\mathbf{1}$ of all ones, and this is positive semidefinite whenever A is square.

At this point, one can ask the following question: *Which entrywise power functions preserve positive semidefiniteness on* $n \times n$ *matrices?* (We will also study in Chapters 12 and 13, the case of general functions.) This question was considered by Loewner [160, 161] while studying the Green's function of the unit circle and schlicht/univalent functions (on a separate note, the coefficients of such functions feature in the Bieberbach conjecture). The question – i.e., which entrywise powers preserve positivity – was eventually answered in 1977 by two of Loewner's students, C.H. FitzGerald and R.A. Horn [84]:

Theorem 5.3 (FitzGerald–Horn)   *Given an integer* $n \geq 2$ *and a scalar* $\alpha \in \mathbb{R}$, $f(x) = x^\alpha$ *preserves positive semidefiniteness on* $\mathbb{P}_n((0, \infty))$ *if and only if* $\alpha \in \mathbb{Z}^{\geq 0} \cup [n - 2, \infty)$.

Remark 5.4   We will in fact show that if α is not in this set, there exists a rank-2 Hankel *TN* matrix $A_{n \times n}$, such that $A^{\circ \alpha} \notin \mathbb{P}_n$. (In fact, it is the (partial) moment matrix of a nonnegative measure on two points.) Also notice that Theorem 5.3 holds for entrywise powers applied to $\mathbb{P}_n([0, \infty))$, since as we show, $\alpha < 0$ never works while $\alpha = 0$ always does work by convention; and for $\alpha > 0$ the power x^α is continuous on $[0, \infty)$, and we use the density of $\mathbb{P}_n((0, \infty))$ in $\mathbb{P}_n([0, \infty))$.

The "phase transition" at $n - 2$ in Theorem 5.3 is a remarkable and oft-repeated phenomenon in entrywise calculus (we will see additional examples of such events in Chapter 8). The value $n - 2$ is called the *critical exponent* for the given problem of preserving positivity. (Loewner's aforementioned papers [160, 161] are perhaps the first time that the term *critical exponent* was used in this context).

To prove Theorem 5.3, we require a preliminary lemma, also by FitzGerald and Horn. Recall the preliminaries in Section 1.4.

Lemma 5.5   *Given a matrix* $A \in \mathbb{P}_n(\mathbb{R})$ *with last column* ζ, *the matrix* $A - a_{nn}^\dagger \zeta \zeta^T$ *is positive semidefinite with last row and column zero.*

Here, $a_{nn}^\dagger$ denotes the Moore–Penrose inverse of the 1×1 matrix (a_{nn}).

Proof   If $a_{nn} = 0$, then $\zeta = 0$ by positive semidefiniteness and $a_{nn}^\dagger = 0$ as well. The result follows. Now suppose $a_{nn} > 0$ and write $A = \begin{pmatrix} B & \omega \\ \omega^T & a_{nn} \end{pmatrix}$. Then a straightforward computation shows that

$$A - a_{nn}^{-1} \zeta \zeta^T = \begin{pmatrix} B - \dfrac{\omega \omega^T}{a_{nn}} & 0 \\ 0 & 0 \end{pmatrix}.$$

Notice that $B - \frac{\omega \omega^T}{a_{nn}}$ is the Schur complement of A with respect to $a_{nn} > 0$.
Now since A is positive semidefinite, so is $B - \frac{\omega \omega^T}{a_{nn}}$ by Theorem 1.28.   □

Proof of Theorem 5.3   Notice that x^α preserves positivity on $\mathbb{P}_n((0, \infty))$ for
all $\alpha \in \mathbb{Z}^{\geq 0}$, by the Schur product theorem 2.10. Now we prove by induction
on $n \geq 2$ that if $\alpha \geq n - 2$, then x^α preserves positivity on $\mathbb{P}_n((0, \infty))$. If
$n = 2$ and $A = \begin{pmatrix} a & b \\ b & c \end{pmatrix} \in \mathbb{P}_2((0, \infty))$, then $ac \geq b^2 \implies (ac)^\alpha \geq b^{2\alpha}$ for
all $\alpha \geq 0$. It follows that $A^{\circ \alpha} \in \mathbb{P}_2((0, \infty))$, proving the base case.

For the induction step, assume that the result holds for $n - 1 \geq 2$. Suppose
$\alpha \geq n - 2$ and $A \in \mathbb{P}_n((0, \infty))$; thus, $a_{nn} > 0$. Consider the following elementary definite integral

$$x^\alpha - y^\alpha = \alpha(x - y) \int_0^1 (\lambda x + (1 - \lambda)y)^{\alpha - 1} \, d\lambda. \tag{5.1}$$

Let ζ denote the final column of A; applying (5.1) entrywise to x, an entry of
A, and y, the corresponding entry of $B := \frac{\zeta \zeta^T}{a_{nn}}$ yields

$$A^{\circ \alpha} - B^{\circ \alpha} = \alpha \int_0^1 (A - B) \circ (\lambda A + (1 - \lambda)B)^{\circ(\alpha - 1)} \, d\lambda. \tag{5.2}$$

By the induction hypothesis, the leading principal $(n - 1) \times (n - 1)$ submatrix
of the matrix $(\lambda A + (1 - \lambda)B)^{\circ(\alpha - 1)}$ is positive semidefinite (even though the
entire matrix need not be so). By Lemma 5.5, $A - B$ is positive semidefinite
and has last row and column zero. It follows by the Schur product theorem
that the integrand on the right-hand side is positive semidefinite. Since $B^{\circ \alpha}$ is
a rank-1 positive semidefinite matrix (this is easy to verify), it follows that $A^{\circ \alpha}$
is also positive. This concludes one direction of the proof.

To prove the other half, suppose $\alpha \notin \mathbb{Z}^{\geq 0} \cup [n - 2, \infty)$; now consider H_μ
where $\mu = \delta_1 + \epsilon \delta_x$ for $\epsilon, x > 0$, $x \neq 1$. Note that $(H_\mu)_{jk} = s_{j+k}(\mu) =
1 + \epsilon x^{j+k}$; and as shown previously, H_μ is positive semidefinite of rank 2.

First, suppose $\alpha < 0$. Then consider the leading principal 2×2 submatrix
of $H_\mu^{\circ \alpha}$

$$B := \begin{pmatrix} (1 + \epsilon)^\alpha & (1 + \epsilon x)^\alpha \\ (1 + \epsilon x)^\alpha & (1 + \epsilon x^2)^\alpha \end{pmatrix}.$$

We claim that $\det B < 0$, which shows $H_\mu^{\circ \alpha}$ is not positive semidefinite.
Indeed, note that

$$(1 + \epsilon)(1 + \epsilon x^2) - (1 + \epsilon x)^2 = \epsilon(x - 1)^2 > 0,$$

so $\det B = (1 + \epsilon)^\alpha (1 + \epsilon x^2)^\alpha - (1 + \epsilon x)^{2\alpha} < 0$ because $\alpha < 0$.

Next, suppose that $\alpha \in (0, n-2) \setminus \mathbb{N}$. Given $x > 0$, for small ϵ we know by the binomial theorem that

$$(1 + \epsilon x)^\alpha = 1 + \sum_{k \geq 1} \binom{\alpha}{k} \epsilon^k x^k, \quad \text{where} \quad \binom{\alpha}{k} = \frac{\alpha(\alpha - 1) \cdots (\alpha - k + 1)}{k!}.$$

We will produce $u \in \mathbb{R}^n$, such that $u^T H_\mu^{\circ \alpha} u < 0$; note this shows that $H_\mu^{\circ \alpha} \notin \mathbb{P}_n$.

Starting with the matrix $H_\mu = \mathbf{1}\mathbf{1}^T + \epsilon v v^T$ where $v = (1, x, \ldots, x^{n-1})^T$, we obtain

$$H_\mu^{\circ \alpha} = \mathbf{1}\mathbf{1}^T + \sum_{k=1}^{\lfloor \alpha \rfloor + 2} \epsilon^k \binom{\alpha}{k} (v^{\circ k})(v^{\circ k})^T + o(\epsilon^{\lfloor \alpha \rfloor + 2}), \tag{5.3}$$

where $o(\epsilon^{\lfloor \alpha \rfloor + 2})$ is a matrix, such that the quotient of any entry by $\epsilon^{\lfloor \alpha \rfloor + 2}$ goes to zero as $\epsilon \to 0^+$.

Note that the first term and the sum together contain at most n terms. Since the corresponding vectors $\mathbf{1}, v, v^{\circ 2}, \ldots, v^{\circ(\lfloor \alpha \rfloor + 2)}$ are linearly independent (by considering the – possibly partial – usual Vandermonde matrix formed by them), there exists a vector $u \in \mathbb{R}^n$ satisfying

$$u^T \mathbf{1} = u^T v = u^T v^{\circ 2} = \cdots = u^T v^{\circ(\lfloor \alpha \rfloor + 1)} = 0, \qquad u^T v^{\circ(\lfloor \alpha \rfloor + 2)} = 1.$$

Substituting these into the above computation, we obtain

$$u^T H_\mu^{\circ \alpha} u = \epsilon^{\lfloor \alpha \rfloor + 2} \binom{\alpha}{\lfloor \alpha \rfloor + 2} + u^T \cdot o(\epsilon^{\lfloor \alpha \rfloor + 2}) \cdot u.$$

Since $\binom{\alpha}{\lfloor \alpha \rfloor + 2}$ is negative if α is not an integer, it follows that

$$\lim_{\epsilon \to 0^+} \frac{u^T H_\mu^{\circ \alpha} u}{\epsilon^{\lfloor \alpha \rfloor + 2}} < 0.$$

Hence, one can choose a small $\epsilon > 0$, such that $u^T H_\mu^{\circ \alpha} u < 0$. It follows for this ϵ that $H_\mu^{\circ \alpha}$ is not positive semidefinite.   □

Remark 5.6   As the above proof reveals, the following are equivalent for $n \geq 2$ and $\alpha \in \mathbb{R}$:

(1) The entrywise map x^α preserves positivity on $\mathbb{P}_n((0, \infty))$ (or $\mathbb{P}_n([0, \infty))$).
(2) $\alpha \in \mathbb{Z}^{\geq 0} \cup [n-2, \infty)$.
(3) The entrywise map x^α preserves positivity on the (leading principal $n \times n$ truncations of) Hankel moment matrices of nonnegative measures supported on $\{1, x\}$, for any fixed $x > 0$, $x \neq 1$.

The use of the Hankel moment matrix counterexample $\mathbf{1}\mathbf{1}^T + \epsilon v v^T$ for $v = (1, x, \ldots, x^{n-1})^T$ and small $\epsilon > 0$ was not due to FitzGerald and Horn – who used $v = (1, 2, \ldots, n)^T$ instead – but due to Fallat et al. [81]. In fact, the above proof can be made to work if one uses any vector v with distinct positive real coordinates, and small enough $\epsilon > 0$.

As these remarks show, to isolate the entrywise powers preserving positivity on $\mathbb{P}_n((0, \infty))$, it suffices to consider a much smaller family – namely, the one-parameter family of truncated moment matrices of the measures $\delta_1 + \epsilon \delta_x$ – or the one-parameter family $\mathbf{1}_{n \times n} + \epsilon v v^T$, where $v = (x_1, \ldots, x_n)^T$ for pairwise distinct $x_j > 0$. In fact, a stronger result is true. In her 2017 paper [131], Jain was able to eliminate the dependence on ϵ:

Theorem 5.7 (Jain)   *Suppose $n > 0$ is an integer, and $x_1, x_2, \ldots, x_n$ are pairwise distinct positive real numbers. Let $C := (1 + x_j x_k)_{j,k=1}^n$. Then $C^{\circ \alpha}$ is positive semidefinite if and only if $\alpha \in \mathbb{Z}^{\geq 0} \cup [n - 2, \infty)$.*

In other words, this result identifies a multiparameter family of matrices, *each one of which* encodes the positivity preserving powers in the original result of FitzGerald–Horn.

We defer the proof of Theorem 5.7 (in fact, a stronger form shown by Jain in 2020, [132]) to Chapter 9. We then use this stronger variant to prove the corresponding result for entrywise powers preserving other Loewner properties: monotonicity (again shown by Jain in 2020), and hence convexity, both with the same multiparameter family of matrices.

The next result is an application of Theorem 5.3 to classify the entrywise powers that preserve positive definiteness:

Corollary 5.8   *Given an integer $n \geq 2$ and a scalar $\alpha \in \mathbb{R}$, the following are equivalent:*

(1) *The entrywise αth power preserves positive definiteness for $n \times n$ matrices with positive entries.*
(2) *The entrywise αth power preserves positive definiteness for $n \times n$ Hankel matrices with positive entries.*
(3) $\alpha \in \mathbb{Z}^{\geq 0} \cup [n - 2, \infty)$.

Proof   Clearly, (1) $\implies$ (2). Next, if α is not in $\mathbb{Z}^{\geq 0} \cup [n - 2, \infty)$, then given $1 \neq x \in (0, \infty)$, there exists $\epsilon > 0$, such that $A^{\circ \alpha}$ has a negative principal minor, where $A := (1 + \epsilon x^{j+k})_{j,k=0}^{n-1}$ is Hankel. Now perturb A by $\delta H_1'$ for small enough $\delta > 0$, where $H_1' := (e^{(j+k)^2})_{j,k=0}^{n-1}$ is a Hankel "principal submatrix" drawn from the kernel in Lemma 4.10. By Theorem 3.6, $A + \delta H_1'$ is TP Hankel for all $\delta > 0$, hence positive definite; and for small enough $\delta > 0$, its αth power also has a negative principal minor. This shows the contrapositive to (2) $\implies$ (3).

Finally, suppose (3) holds. Since the Schur product is a principal submatrix of the Kronecker product, it follows from the first proof of Theorem 2.10 that positive integer powers entrywise preserve positive definiteness. Now suppose $\alpha \geq n - 2$ and $A_{n \times n}$ is positive definite. Then all eigenvalues of A are positive, so there exists $\epsilon > 0$, such that $A - \epsilon \operatorname{Id}_{n \times n} \in \mathbb{P}_n([0, \infty))$. Now we have

$$A^{\circ \alpha} = (A - \epsilon \operatorname{Id})^{\circ \alpha} + \operatorname{diag}(a_{jj}^{\alpha} - (a_{jj} - \epsilon)^{\alpha})_{j=1}^n,$$

and the first term on the right-hand side is in $\mathbb{P}_n$ by Theorem 5.3, so $A^{\circ \alpha}$ is positive definite. $\square$

We conclude by highlighting the power and applicability of the "integration trick" (5.2) of FitzGerald and Horn. First, it in fact applies to general functions, not just to powers. The following observation (by Khare and Tao [146]) will be useful below:

Theorem 5.9 (Extension principle) *Let $0 < \rho \leq \infty$ and $I = (0, \rho)$, $(-\rho, \rho)$. Fix an integer $n \geq 2$ and a continuously differentiable function $h \colon I \to \mathbb{R}$. If $h[-]$ preserves positivity on rank-1 matrices in $\mathbb{P}_n(I)$ and $h'[-]$ preserves positivity on $\mathbb{P}_{n-1}(I)$, then $h[-]$ preserves positivity on $\mathbb{P}_n(I)$.*

The proof is exactly as before, but now using the more general integral identity

$$h(x) - h(y) = \int_y^x h'(t)\, dt = \int_0^1 (x - y) h'(\lambda x + (1 - \lambda) y)\, d\lambda.$$

Second, this integration trick is even more powerful, in that it further applies to classify the entrywise powers that preserve other properties of $\mathbb{P}_n$, including monotonicity and superadditivity. See Chapter 8 for details on these properties, their power-preservers, and their further application to the distinguished subcones $\mathbb{P}_G$ for noncomplete graphs G.

6

Midconvex Implies Continuity, and 2 × 2 Preservers

We now proceed toward the more general question of classifying *functions* which preserve positivity, when applied entrywise to matrices of a fixed size or of all sizes. The first observation along these lines is that every such preserver is necessarily continuous on $(0, \infty)$. The continuity follows from a variant of a 1929 result by Ostrowski [180] on midconvex functions on normed linear spaces, and we begin by proving this result.

6.1 Midconvex Functions and Continuity

Definition 6.1   Given a convex subset U of a real vector space, a function $f : U \to \mathbb{R}$ is said to be *midconvex* if

$$f\left(\frac{x+y}{2}\right) \leq \frac{f(x) + f(y)}{2}, \quad \forall x, y \in U;$$

and f is *convex* if $f(\lambda x + (1 - \lambda)y) \leq \lambda f(x) + (1 - \lambda)f(y)$ for all $x, y \in U$ and $\lambda \in (0, 1)$.

Notice that convex functions are automatically midconvex. The converse need not be true in general. However, if a midconvex function is continuous, then it is easy to see that it is also convex. Thus, a natural question for midconvex functions is to find sufficient conditions under which they are continuous. We now discuss two such conditions, both classical results. The first condition is mild: f is locally bounded, *on one neighborhood of one point*.

Theorem 6.2   *Let $\mathbb{B}$ be a normed linear space (over $\mathbb{R}$) and let U be a convex open subset. Suppose $f : U \to \mathbb{R}$ is midconvex and f is bounded above in an open neighborhood of a single point $x_0 \in U$. Then f is continuous on U, and hence convex.*

46

This generalizes to normed linear spaces a special case of a result by Ostrowski [180], who showed in 1929 the same conclusion, but over $\mathbb{B} = \mathbb{R}$ and assuming f is bounded above in a measurable subset.

The proof requires the following useful observation:

Lemma 6.3 *If $f : U \to \mathbb{R}$ is midconvex, then f is rationally convex, i.e., $f(\lambda x + (1 - \lambda)y) \le \lambda f(x) + (1 - \lambda)f(y)$ for all $x, y \in U$ and $\lambda \in (0, 1) \cap \mathbb{Q}$.*

Proof   Inductively using midconvexity, it follows that

$$f\left(\frac{x_1 + x_2 + \cdots + x_{2^n}}{2^n}\right) \le \frac{f(x_1) + \cdots + f(x_{2^n})}{2^n},$$

$$\forall n \in \mathbb{N}, \ x_1, \ldots, x_{2^n} \in U.$$

Now suppose $\lambda = \frac{p}{q} \in (0, 1)$, where $p, q > 0$ are integers and $2^{n-1} \le q < 2^n$ for some $n \in \mathbb{N}$. Let $x_1, \ldots, x_q \in U$ and define $\overline{x} = \frac{1}{q}(x_1 + \cdots + x_q)$. Setting $x_{q+1} = \cdots = x_{2^n} = \overline{x}$, we obtain

$$f\left(\frac{x_1 + \cdots + x_q + (2^n - q)\overline{x}}{2^n}\right) \le \frac{f(x_1) + \cdots + f(x_q) + (2^n - q)f(\overline{x})}{2^n}$$

$$\implies \quad 2^n f(\overline{x}) \le f(x_1) + \cdots + f(x_q) + (2^n - q)f(\overline{x})$$

$$\implies \quad q f(\overline{x}) \le f(x_1) + \cdots + f(x_q)$$

$$\implies \quad f\left(\frac{x_1 + \cdots + x_q}{q}\right) \le \frac{f(x_1) + \cdots + f(x_q)}{q}$$

In this inequality, set $x_1 = \cdots = x_p = x$ and $x_{p+1} = \cdots = x_q = y$ to complete the proof

$$f(\lambda x + (1 - \lambda)y) \le \lambda f(x) + (1 - \lambda)f(y). \qquad \square$$

With Lemma 6.3 at hand, we can prove the theorem above.

Proof of Theorem 6.2   We may assume without loss of generality that $x_0 = 0 \in U \subset \mathbb{B}$, and also that $f(x_0) = f(\mathbf{0}) = 0$.

We claim that f is continuous at $\mathbf{0}$, where f was assumed to be bounded above in an open neighborhood of $\mathbf{0}$. Write this as: $f(B(\mathbf{0}, r)) < M$ for some $r, M > 0$, where $B(x, r) \subset \mathbb{B}$ denotes the open ball of radius r centered at $x \in \mathbb{B}$. Now given $\epsilon \in (0, 1) \cap \mathbb{Q}$ rational and $x \in B(\mathbf{0}, \epsilon r)$, we compute using Lemma 6.3

$$x = \epsilon\left(\frac{x}{\epsilon}\right) + (1 - \epsilon)\mathbf{0} \quad \implies \quad f(x) \le \epsilon f\left(\frac{x}{\epsilon}\right) + 0 < \epsilon M.$$

Moreover,

$$0 = \left(\frac{\epsilon}{1+\epsilon}\right)\left(\frac{-x}{\epsilon}\right) + \frac{x}{1+\epsilon},$$

so applying Lemma 6.3 once again, we obtain

$$0 \leq \left(\frac{\epsilon}{1+\epsilon}\right) f\left(\frac{-x}{\epsilon}\right) + \frac{f(x)}{1+\epsilon} < \frac{\epsilon M}{1+\epsilon} + \frac{f(x)}{1+\epsilon} \quad\Longrightarrow\quad f(x) > -\epsilon M.$$

Therefore, we have $x \in B(\mathbf{0}, \epsilon r) \implies |f(x)| < \epsilon M$.

Now given $\epsilon > 0$, choose $0 < \epsilon' < \min(M, \epsilon)$, such that ϵ'/M is rational, and set $\delta := r\epsilon'/M$. Then $\delta < r$, so $|f(x)| < \delta M/r = \epsilon' < \epsilon$ whenever $x \in B(\mathbf{0}, \delta)$. Hence, f is continuous at x_0.

We have shown that if f is bounded above in some open neighborhood of $x_0 \in U$, then f is continuous at x_0. To finish the proof, we claim that for all $y \in U$, f is bounded above on some open neighborhood of y. This would show that f is continuous on U, which combined with midconvexity implies convexity.

To show the claim, choose a rational $\rho > 1$, such that $\rho y \in U$ (this is possible as U is open), and set $U_y := B(y, (1 - 1/\rho)r)$. Note that $U_y \subset U$ since for every $v \in U_y$ there exists $x \in B(\mathbf{0}, r)$, such that

$$v = y + (1 - 1/\rho)x = \frac{1}{\rho}(\rho y) + \left(1 - \frac{1}{\rho}\right)x.$$

Thus, v is a convex combination of $\rho y \in U$ and $x \in B(\mathbf{0}, r) \subset U$. Hence, $U_y \subset U$; in turn,

$$f(v) \leq \frac{1}{\rho} f(\rho y) + \left(1 - \frac{1}{\rho}\right) f(x) \leq \frac{f(\rho y)}{\rho} + \left(1 - \frac{1}{\rho}\right) M, \qquad \forall v \in U_y$$

by Lemma 6.3. Since the right-hand side is independent of $v \in U_y$, the above claim follows. Hence, by the first claim, f is indeed continuous at every point in U. $\qquad\square$

The second sufficient condition is that f is Lebesgue measurable. Its sufficiency was proved a decade before Ostrowski's result, independently by Blumberg in 1919 [40] and Sierpińsky in 1920 [220]. However, the following proof goes via Theorem 6.2:

Theorem 6.4   *If $I \subset \mathbb{R}$ is an open interval and $f : I \to \mathbb{R}$ is Lebesgue measurable and midconvex, then f is continuous, hence convex.*

Proof   Suppose f is not continuous at a point $x_0 \in I$. Fix $c > 0$, such that $(x_0 - 2c, x_0 + 2c) \subset I$. By Theorem 6.2, f is unbounded on $(x_0 - c, x_0 + c)$.

Now let $B_n := \{x \in I : f(x) > n\}$ for $n \geq 1$; note this is Lebesgue measurable. Choose $u_n \in B_n \cap (x_0 - c, x_0 + c)$ and $\lambda \in [0,1]$; then by midconvexity,

$$n < f(u_n) = f\left[\frac{u_n + \lambda c}{2} + \frac{u_n - \lambda c}{2}\right] \leq \frac{1}{2}(f(u_n + \lambda c) + f(u_n - \lambda c)).$$

Thus, B_n contains at least one of the points $u_n \pm \lambda c \in I$, i.e., one of $\pm \lambda c$ lies in $B_n - u_n$. We now **claim** that each B_n has Lebesgue measure $\mu(B_n) \geq c$. Assuming this claim,

$$c \leq \lim_{n \to \infty} \mu(B_n) = \mu\left(\cap_{n \geq 1} B_n\right),$$

since the B_n are a nested family of subsets. But then $S := \cap_{n \geq 1} B_n$ is nonempty, so for any $v \in S$, we have $f(v) > n$ for all n, which produces the desired contradiction.

Thus, it remains to show the above claim. Fix $n \geq 1$ and note from above that $M_n := B_n - u_n$ is a Lebesgue measurable set, such that for every $\lambda \in [0,1]$, at least one of λc, $- \lambda c$ lies in M_n. Define the measurable sets $A_1 := M_n \cap [-c, 0]$ and $A_2 := M_n \cap [0, c]$, so that $-A_1 \cup A_2 = [0, c]$. This implies

$$c \leq \mu(-A_1) + \mu(A_2) = \mu(A_1) + \mu(A_2) = \mu(A_1 \cup A_2) \leq \mu(M_n). \qquad \square$$

6.2 Functions Preserving Positivity on 2 × 2 Matrices

With Theorem 6.2 at hand, it is possible to classify all entrywise functions that preserve positive semidefiniteness on 2×2 matrices. To state this result, we need the following notion:

Definition 6.5   Suppose $I \subset [0, \infty)$ is an interval. A function $f : I \to [0, \infty)$ is *multiplicatively midconvex* on I if and only if $f(\sqrt{xy}) \leq \sqrt{f(x)f(y)}$ for all $x, y \in I$.

Remark 6.6   A straightforward computation yields that if $f : I \to \mathbb{R}$ is always positive and $0 \notin I$, then f is multiplicatively midconvex on I if and only if the auxiliary function $g(y) := \log f(e^y)$ is midconvex on $\log(I)$.

We now prove the classification promised above; this is also crucially used in Part II of this book.

Theorem 6.7   *Suppose* $I = [0, \infty)$ *and* $I^+ := I \setminus \{0\}$. *A function* $f : I \to \mathbb{R}$ *satisfies* $\begin{pmatrix} f(a) & f(b) \\ f(b) & f(c) \end{pmatrix}$ *is positive semidefinite whenever* $a, b, c \in I$ *and*

$\begin{pmatrix} a & b \\ b & c \end{pmatrix}$ *is so, if and only if f is nonnegative, nondecreasing, and multiplicatively midconvex on I. In particular,*

(1) $f|_{I^+}$ *is never zero or always zero.*
(2) $f|_{I^+}$ *is continuous.*

The same results hold if $I = [0, \infty)$ *is replaced by* $I = (0, \infty), [0, \rho),$ *or* $(0, \rho)$ *for* $0 < \rho < \infty$.

This result was essentially proved by H.L. Vasudeva [230], under some reformulation. Note that all of the matrices above are clearly Hankel. This result will therefore also play an important role when we classify the entrywise preservers of positive semidefiniteness on low-rank Hankel matrices (see Theorems 12.1 and 14.1).

Proof Let *I* be any of the domains mentioned in the theorem. We begin by showing the equivalence. Given a positive semidefinite matrix

$$\begin{pmatrix} a & b \\ b & c \end{pmatrix}, \qquad a, b, c \in I, \ 0 \le b \le \sqrt{ac},$$

we compute using the nonnegativity, monotonicity, and multiplicative midconvexity respectively:

$$0 \le f(b) \le f(\sqrt{ac}) \le \sqrt{f(a)f(c)}.$$

It follows that $\begin{pmatrix} f(a) & f(b) \\ f(b) & f(c) \end{pmatrix}$ is positive semidefinite.

Conversely, if $f[-] \colon \mathbb{P}_2(I) \to \mathbb{P}_2$, then apply $f[-]$ entrywise to the matrices

$$\begin{pmatrix} a & b \\ b & a \end{pmatrix}, \qquad \begin{pmatrix} a & \sqrt{ac} \\ \sqrt{ac} & c \end{pmatrix}, \qquad a, b, c \in I, \tag{6.1}$$

with $0 \le b \le a$. From the hypotheses, it successively (and respectively) follows that f is nonnegative, nondecreasing, and multiplicatively midconvex. This proves the equivalence.

As a brief digression, we remark that the test matrices $\begin{pmatrix} a & b \\ b & a \end{pmatrix}$ can be replaced by

$$\begin{pmatrix} a & b \\ b & b \end{pmatrix}, \qquad \begin{pmatrix} a & \sqrt{ac} \\ \sqrt{ac} & c \end{pmatrix}, \qquad a, b, c \in I, \tag{6.2}$$

with $0 \leq b \leq a$, to conclude as above that f is nondecreasing and multiplicatively midconvex on I. Indeed, we obtain $f(a), f(b) \geq 0$, and either $f(b) = 0 \leq f(a)$ or $0 < f(b)^2 \leq f(b)f(a)$, leading to the same conclusion.

We now show the two final assertions (1) and (2) in the theorem, again on I^+ for any of the domains I above; in other words, $I^+ = (0, \rho)$ for $0 < \rho \leq \infty$. For (1), suppose $f(x) = 0$ for some $x \in I^+$. Since f is nonnegative and nondecreasing on I^+, it follows that $f \equiv 0$ on $(0, x)$. Now claim that $f(y) = 0$ if $y > x$, $y \in I^+ = (0, \rho)$. Indeed, choose a large enough $n > 0$, such that $y \sqrt[n]{y/x} < \rho$. Set $\zeta := \sqrt[n]{y/x} > 1$ and consider the following rank-1 matrices in $\mathbb{P}_2(I^+)$

$$A_1 := \begin{pmatrix} x & x\zeta \\ x\zeta & x\zeta^2 \end{pmatrix}, \quad A_2 := \begin{pmatrix} x\zeta & x\zeta^2 \\ x\zeta^2 & x\zeta^3 \end{pmatrix}, \quad \ldots, \quad A_n := \begin{pmatrix} x\zeta^{n-1} & x\zeta^n \\ x\zeta^n & x\zeta^{n+1} \end{pmatrix}.$$

The inequalities $\det f[A_k] \geq 0$, $1 \leq k \leq n$ yield

$$0 \leq f(x\zeta^k) \leq \sqrt{f(x\zeta^{k-1})f(x\zeta^{k+1})}, \quad k = 1, 2, \ldots, n.$$

From this inequality for $k = 1$, it follows that $f(x\zeta) = 0$. Similarly, these inequalities inductively yield: $f(x\zeta^k) = 0$ for all $1 \leq k \leq n$. In particular, we have $f(y) = f(x\zeta^n) = 0$. This shows that $f \equiv 0$ on I^+, as claimed.

We provide two proofs of (2). If $f \equiv 0$ on I^+, then f is continuous on I^+. Otherwise by (1), f is strictly positive on $(0, \rho) = I^+$. Now the "classical" proof uses the above "Ostrowski-result": define the function $g: \log I^+ := (-\infty, \log \rho) \to \mathbb{R}$ via

$$g(y) := \log f(e^y), \quad y < \log \rho.$$

By the assumptions on f and the observation in Remark 6.6, g is midconvex and nondecreasing on $(-\infty, \log \rho)$. In particular, g is bounded above on compact sets. Now apply Theorem 6.2 to deduce that g is continuous. It follows that f is continuous on $(0, \rho)$.

A more recent, shorter proof is by Hiai [119] in 2009: given f as above which is strictly positive and nondecreasing on $(0, \rho)$, fix $t \in (0, \rho)$ and let $0 < \epsilon < \min(t/5, (\rho - t)/4)$. Then $0 < t + \epsilon \leq \sqrt{(t + 4\epsilon)(t - \epsilon)} < \rho$, so

$$f(t + \epsilon) \leq f\left(\sqrt{(t + 4\epsilon)(t - \epsilon)}\right) \leq \sqrt{f(t + 4\epsilon)f(t - \epsilon)}.$$

Now letting $\epsilon \to 0^+$, this implies $f(t^+) \leq f(t^-)$, so

$$0 < f(t) \leq f(t^+) \leq f(t^-) \leq f(t), \quad \forall t \in (0, \rho).$$

Since $t \in (0, \rho)$ was arbitrary, this shows f is continuous as claimed. $\square$

Remark 6.8   From the proof – see (6.1) – it follows that the assumptions may be further weakened to not work with all symmetric 2×2 positive semidefinite matrices, but with only the rank-1 symmetric and the Toeplitz symmetric 2×2 positive semidefinite matrices.

As a variant, we now show that if $f[-]$ preserves positive definiteness on 2×2 matrices, then f is continuous. More strongly, we have:

Lemma 6.9   *Suppose $p \in (1, \infty)$, and $f : (0, \infty) \to \mathbb{R}$. The following are equivalent:*

(1) *$f[-]$ preserves positive definiteness on 2×2 matrices (with positive entries).*
(2) *$f[-]$ preserves positive definiteness on the symmetric TP Hankel matrices*

$$\begin{pmatrix} a & b \\ b & c \end{pmatrix}, \qquad a, c > 0, \ \sqrt{ac/p} < b < \sqrt{ac}.$$

(3) *f is positive, increasing, and multiplicatively midconvex on $(0, \infty)$.*

In particular, f is continuous.

Proof   Clearly, (1) $\implies$ (2), and that (3) $\implies$ (1) is left to the reader as it is similar to the proof of Theorem 6.7. We now assume (2) and show (3). The first step is to claim that f is positive and strictly increasing on $(0, \infty)$. Suppose $0 < x < y < \infty$. Choose $n > \log_p (y/x)$, and define the increasing sequence

$$x_0 = x, \quad x_1 = x(y/x)^{1/n}, \quad x_2 = x(y/x)^{2/n}, \quad \ldots, \quad x_n = y.$$

Now the matrix $\begin{pmatrix} x_{k+1} & x_k \\ x_k & x_k \end{pmatrix}$ is in the given test set, by choice of n, so applying $f[-]$ and taking determinants, we have

$$f(x_k), \ f(x_{k+1}), \ f(x_k)(f(x_{k+1}) - f(x_k)) > 0, \qquad 0 \leq k \leq n-1.$$

It follows that f is positive on $(0, \infty)$, hence also strictly increasing, since $f(x) = f(x_0) < f(x_1) < \cdots < f(x_n) = f(y)$.

We next show continuity, proceeding indirectly. From above, $f : (0, \infty) \to (0, \infty)$ has at most countably many discontinuities, and they are all jump discontinuities. Let $f(x^+) := \lim_{y \to x^+} f(y)$, for $x > 0$. Then $f(x^+) \geq f(x) \ \forall x$, and $f(x^+)$ coincides with $f(x)$ at all points of right continuity and has the same jumps as f. Thus, it suffices to show that $f(x^+)$ is continuous (since this implies f is also continuous).

Now given $0 < x < y < \infty$, apply $f[-]$ to the matrices

$$M(x,y,\epsilon) := \begin{pmatrix} x + \epsilon & \sqrt{xy} + \epsilon \\ \sqrt{xy} + \epsilon & y + \epsilon \end{pmatrix}, \qquad x, y, \epsilon > 0,$$

where $\epsilon > 0$ is small enough that $(x + \epsilon)(y + \epsilon) < p(\sqrt{xy} + \epsilon)^2$. Then an easy verification shows that $M(x,y,\epsilon)$ is in the given test set. It follows that $\det f[M(x,y,\epsilon)] > 0$, i.e., $f(x+\epsilon)f(y+\epsilon) > f(\sqrt{xy}+\epsilon)^2$. Taking $\epsilon \to 0^+$, we obtain

$$f(x^+)f(y^+) \geq f\big(\sqrt{xy}^+\big)^2, \qquad \forall x, y > 0.$$

Thus, $f(x^+)$ is positive, nondecreasing and multiplicatively midconvex on $(0, \infty)$. From the proof of Theorem 6.7(2), we conclude that $f(x^+)$ is continuous on $(0, \infty)$, so $f(x) = f(x^+)$ is also continuous and multiplicatively midconvex on $(0, \infty)$. $\square$

6.3 Totally Nonnegative Hankel Matrices – Entrywise Preservers

We have seen that the powers x^α that entrywise preserve positive semidefiniteness on $\mathbb{P}_n((0, \infty))$ (for fixed $n \geq 2$) are $\mathbb{Z}^{\geq 0} \cup [n - 2, \infty)$. In contrast, one can show (see, e.g., the 2020 preprint [22] by Belton et al.) that if the entrywise map $f[-]$ preserves the $m \times n$ TP/TN matrices for $m, n \geq 4$, then f is either constant on $(0, \infty)$ (and $f(0)$ equals either this constant or zero) or $f(x) = f(1)x$ for all x.

This discrepancy is also supported by the fact that $\mathbb{P}_n$ is closed under the Schur (or entrywise) product, but already the 3×3 TN matrices are not. For example, $A := \begin{pmatrix} 1 & 1 & 1 \\ 1 & 1 & 1 \\ 0 & 1 & 1 \end{pmatrix}$, $B := A^T = \begin{pmatrix} 1 & 1 & 0 \\ 1 & 1 & 1 \\ 1 & 1 & 1 \end{pmatrix}$ are both TN, but

$A \circ B = \begin{pmatrix} 1 & 1 & 0 \\ 1 & 1 & 1 \\ 0 & 1 & 1 \end{pmatrix}$ has determinant -1, and hence cannot be TN.

Thus, a more refined (albeit technical) question would be to isolate and work with a class of TN matrices that is a closed, convex cone, and which is further closed under the Schur product. In fact, such a class has already been discussed earlier: the family of *Hankel TN* matrices (see Corollary 3.7). With those results in mind, and for future use, we introduce the following notation:

Definition 6.10   Given $n \geq 1$, let HTN_n denote the $n \times n$ Hankel *TN* matrices.

We also study in this text the notion of entrywise preservers of TN on HTN_n for a fixed n and for all n – see Remark 5.6, Corollary 27.7, and Corollary 14.4. This study turns out to be remarkably similar (and related) to the study of positivity preservers on $\mathbb{P}_n$ – which is not surprising, given Theorem 3.6. For now, we work in the setting under current consideration: entrywise power-preservers.

Theorem 6.11   *For $n \geq 2$ and $\alpha \in \mathbb{R}$, x^α entrywise preserves TN on HTN_n, if and only if $\alpha \in \mathbb{Z}^{\geq 0} \cup [n - 2, \infty)$.*

In other words, x^α preserves total nonnegativity on HTN_n if and only if it preserves positive semidefiniteness on $\mathbb{P}_n([0, \infty))$.

Proof   If $\alpha \in \mathbb{Z}^{\geq 0} \cup [n - 2, \infty)$, then we use Theorem 3.6 together with Theorem 5.3. Conversely, suppose $\alpha \in (0, n - 2) \setminus \mathbb{Z}$. We have previously shown that the moment matrix $H := \left(1 + \epsilon x^{j+k-2}\right)_{j,k=1}^n$ lies in HTN_n for x, $\epsilon > 0$; but if $x \neq 1$ and $\epsilon > 0$ is small, then $H^{\circ \alpha} \notin \mathbb{P}_n$, as shown in the proof of Theorem 5.3. (Alternately, this holds for all $\epsilon > 0$ by Theorem 5.7.) It follows that $H^{\circ \alpha} \notin TN_n$. $\square$

7

Entrywise Preservers of Positivity on Matrices with Zero Patterns

In this chapter and Chapter 8, we continue to study entrywise functions preserving positive semidefiniteness – henceforth termed *positivity* – in a fixed dimension, by refining the test set of positive semidefinite matrices. The plan for these two chapters is as follows:

(1) We begin by recalling the test set $\mathbb{P}_G([0, \infty))$ associated to any graph G, and discussing some of the modern-day motivations in studying entrywise functions (including powers) that preserve positivity.

(2) We then prove some results on general entrywise functions preserving positivity on $\mathbb{P}_G$ for arbitrary noncomplete graphs. (The case of complete graphs is the subject of the remainder of the text.) As a consequence, the powers – in fact, the *functions* – preserving $\mathbb{P}_G([0, \infty))$ where G is any tree (or collection of trees) are completely classified.

(3) We show how the integration trick of FitzGerald and Horn (see the discussion around Equations (5.1) and (5.2)) extends to help classify the entrywise powers preserving other Loewner properties, including monotonicity, and in turn, superadditivity.

(4) Using these results, we classify the powers preserving $\mathbb{P}_G$ for G the *almost complete graph* (i.e., the complete graph minus any one edge).

(5) We then state some recent results on powers preserving $\mathbb{P}_G$ for other G (all chordal graphs; cycles), and conclude with some questions for general graphs G, which arise naturally from these results.

7.1 Modern Motivations: Graphical Models and Covariance Estimation

As we discuss in Chapter 11, the question of which functions preserve positivity when applied entrywise has a long history, having been studied for the best part of a century within the analysis literature. For now, we explain why this question has attracted renewed attention owing to its importance in high-dimensional covariance estimation.

In modern-day scientific applications, one of the most important challenges involves understanding complex multivariate structures and dependencies. Such questions naturally arise in various domains: understanding the interactions of financial instruments, studying markers of climate parameters to understand climate patterns, and modeling gene-gene associations in cancer and cardiovascular disease, to name a few. In such applications, one works with very large random vectors $X \in \mathbb{R}^p$, and a fundamental measure of dependency that is commonly used (given a sample of vectors) is the covariance matrix (or correlation matrix) and its inverse. Unlike traditional regimes, where the sample size n exceeds the dimension of the problem p (i.e., the number of random variables in the model), the reverse holds in these modern applications (and others): $n \ll p$. This is due to the high cost of making, storing, and working with observations, for instance; but moreover, an immediate consequence is that the corresponding covariance matrix built out of the samples $x_1, \ldots, x_n \in \mathbb{R}^p$

$$\widehat{\Sigma} := \frac{1}{n-1} \sum_{j=1}^{n} (x_j - \overline{x})(x_j - \overline{x})^T,$$

is highly singular. (Its rank is bounded above by the sample size $n \ll p$.) This makes $\widehat{\Sigma}$ a poor estimator of the true underlying covariance matrix Σ.

A second shortcoming of the sample covariance matrix concerns zero patterns. In the underlying model, there is often additional domain-specific knowledge which leads to *sparsity*. In other words, certain pairs of variables are known to be independent, or conditionally independent given other variables. For instance, in probability theory one has the notion of a *Markov random field*, or graphical model, in which the nodes of a graph represent random variables and the edges the dependency structure between them. Or, in the aforementioned climate-related application – specifically, temperature reconstruction – the temperature at one location is assumed to not influence that at another (perhaps far away) location, at least when conditioned on the neighboring points. Such (conditional) independences are reflected in zero entries in the associated (inverse) covariance matrix. In fact, in the aforementioned applications, several models assume most of the entries ($\sim 90\%$ or more) to be zero.

However, in the observed sample covariance matrix, there is almost always some noise, as a result of which very few entries are zero. This is another reason why sample covariance is a poor estimator in modern applications.

For such reasons, it is common for statistical practitioners to *regularize* estimators like the sample covariance matrix, to improve its properties for a given application. Popular state-of-the-art methods involve inducing sparsity – i.e., zero entries – via convex optimization techniques that impose an ℓ^1-penalty (since ℓ^0-penalties are not amenable to such techniques). While these methods induce sparsity and are statistically consistent as $n, p \to \infty$, they are iterative, hence involve computationally expensive optimization. In particular, they are not scalable to ultra-high-dimensional data, say for $p \sim 100,000$ or $500,000$, as one often encounters in the aforementioned situations.

A recent promising alternative is to apply entrywise functions on the entries of sample covariance matrices (see, e.g., [10, 38, 78, 115, 116, 155, 196, 243] and numerous follow-up papers). For example, the hard and soft thresholding functions set very small entries to zero (operating under the assumption that these often come from noise, and do not represent the most important associations). Another popular family of functions used in applications consists of entrywise powers. Indeed, powering up the entries provides an effective way in applications to separate signal from noise.

Note that these entrywise operations do not suffer from the same drawback of scalability, since they operate directly on the entries of the matrix, and do not involve optimization. The key question now is to understand *when such entrywise operations preserve positivity*. Indeed, the regularized matrix that these operations yield must serve as a proxy for the sample covariance matrix in further statistical analyses, and hence is required to be positive semidefinite.

It is thus crucial to understand when these entrywise operations preserve positivity – and in a fixed dimension, since in a given application one knows the dimension of the problem. Note that while the motivation here comes from downstream applications, the heart of the issue is very much a mathematical question involving analysis on the cone $\mathbb{P}_n$.

With these motivations, the current and next two chapters deal with entrywise powers preserving positive semidefiniteness in a fixed dimension; progress on these questions impacts applied fields. At the same time, the question of when entrywise powers and functions preserve positivity, has been studied in mathematics literature for almost a century. Thus (looking slightly ahead), in Parts II and III of this text, we return to the mathematical advances, both classical and recent. This includes proving some of the celebrated characterization results in this area – by Schoenberg, Rudin, Loewner/Horn, and Vasudeva – using fairly accessible mathematical machinery.

7.2 Entrywise Preservers of Positivity on $\mathbb{P}_G$ for Noncomplete Graphs

We continue with the theme of entrywise powers and functions preserving positivity in a fixed dimension, now under additional sparsity constraints – i.e., on $\mathbb{P}_G$ for a fixed graph G. In this chapter, we obtain certain necessary conditions on general functions preserving positivity on $\mathbb{P}_G$.

As we will see in Section 11.1, the functions preserving positive semidefiniteness on $\mathbb{P}_n$ for all n for all integers $n \geq 1$ can be classified, and they are precisely the power series with nonnegative coefficients

$$f(x) = \sum_{k=0}^{\infty} c_k x^k, \qquad \text{with } c_k \geq 0 \ \forall \, k.$$

This is a celebrated result of Schoenberg and Rudin (see Theorems 11.2 and 11.3). However, the situation is markedly different for entrywise preservers of $\mathbb{P}_n$ for a *fixed* dimension $n \geq 1$:

- For $n = 1$, clearly any $f : [0, \infty) \to [0, \infty)$ works.
- For $n = 2$, the entrywise preservers of positivity on $\mathbb{P}_2((0, \infty))$ have been classified by Vasudeva in 1979: see Theorem 6.7.
- For $n \geq 3$, the problem remains open to date.

Given the open (and challenging!) nature of the problem in fixed dimension, efforts along this direction have tended to work on refinements of the problem: either restricting the class of entrywise functions (to, e.g., power functions, or polynomials as we study in Part III), or restricting the class of matrices: to Toeplitz matrices (by Rudin), or Hankel *TN* matrices, or to matrices with rank bounded above (by Schoenberg, Rudin, Loewner and Horn, and subsequent authors), or to matrices with a given sparsity pattern – i.e., $\mathbb{P}_G$ for fixed G. We focus on this last approach in this chapter and Chapter 8.

Given a (finite simple) graph $G = (V, E)$, with $V = [n] = \{1, \ldots, n\}$ for some $n \geq 1$, and a subset $0 \in I \subset \mathbb{R}$, the subset $\mathbb{P}_G(I)$ is defined to be

$$\mathbb{P}_G(I) := \{A \in \mathbb{P}_n(I) \ : \ a_{jk} = 0 \text{ if } j \neq k \text{ and } (j, k) \notin E\}. \qquad (7.1)$$

For example, when $G = A_3$ (the path graph on three nodes),

$$\mathbb{P}_G = \left\{ \begin{pmatrix} a & b & e \\ d & b & 0 \\ e & 0 & c \end{pmatrix} \in \mathbb{P}_3 \right\},$$

and when $G = K_n$ (the complete graph on n vertices), we have $\mathbb{P}_G(I) = \mathbb{P}_n(I)$.

We now study the entrywise preservers of $\mathbb{P}_G$ for a graph G. To begin, we extend the notion of entrywise functions to $\mathbb{P}_G$, by acting only on the "unconstrained" entries:

Definition 7.1 Let $0 \in I \subset \mathbb{R}$. Given a graph G with vertex set $[n]$ and $f : I \to \mathbb{R}$, define $f_G[-] : \mathbb{P}_G(I) \to \mathbb{R}^{n \times n}$ via

$$(f_G[A])_{jk} := \begin{cases} 0, & \text{if } j \neq k, \ (j,k) \notin E, \\ f(a_{jk}), & \text{otherwise.} \end{cases}$$

Here are some straightforward observations on entrywise preservers of the cone $\mathbb{P}_G([0, \infty))$:

(1) When G is the empty graph, i.e., $G = (V, \emptyset)$, the functions f, such that $f_G[-]$ preserves $\mathbb{P}_G$ are precisely the functions sending $[0, \infty)$ to itself.

(2) When G is the disjoint union of a positive number of disconnected copies of K_2 and isolated nodes, $\mathbb{P}_G$ consists of block diagonal matrices of the form $\oplus_{j=1}^{k} A_j$, where the A_j are either 2×2 or 1×1 matrices (blocks), corresponding to copies of K_2 or isolated points respectively, and $\oplus$ denotes a block diagonal matrix of the form

$$\begin{pmatrix} A_1 & & & \\ & A_2 & & \\ & & \ddots & \\ & & & A_k \end{pmatrix}.$$

(The remaining entries are zero.) By assumption, at least one of the A_j must be a 2×2 block. For such graphs, we conclude by Theorem 6.7 that $f_G[-] : \mathbb{P}_G([0, \infty)) \to \mathbb{P}_G([0, \infty))$ if and only if f is nonnegative, nondecreasing, multiplicatively midconvex, and $0 \leq f(0) \leq \lim_{x \to 0^+} f(x)$.

(3) More generally, if G is a disconnected union of graphs: $G = \bigsqcup_{j \in J} G_j$, then $f_G[-] : \mathbb{P}_G([0, \infty)) \to \mathbb{P}_G([0, \infty))$ if and only if the entrywise map $f_{G_j}[-]$ preserves $\mathbb{P}_{G_j}([0, \infty))$ for all j.

In light of these examples, we shall henceforth consider only connected, noncomplete graphs G, and the functions f, such that $f_G[-]$ preserves the cone $\mathbb{P}_G([0, \infty))$. We begin with the following necessary conditions:

Proposition 7.2   *Let $I = [0, \infty)$ and G be a connected, noncomplete graph. Suppose $f : I \to \mathbb{R}$ is such that $f_G[-] : \mathbb{P}_G(I) \to \mathbb{P}_G(I)$. Then the following statements hold:*

(1) $f(0) = 0$.
(2) *f is continuous on I (and not just on $(0, \infty)$).*
(3) *f is superadditive on I, i.e., $f(x + y) \geq f(x) + f(y) \; \forall \, x, y \geq 0$.*

Remark 7.3   In particular, $f_G[-] = f[-]$ for (non)complete graphs G. Thus, following the proof of Proposition 7.2, we use $f[-]$ in the sequel.

Proof   Clearly, $f : I \to I$. Assume that G has at least three nodes, since for connected graphs with two nodes, the proposition is vacuous. A small observation – made by Horn [126], if not earlier – reveals that there exist three nodes, which we may relabel as 1, 2, and 3 without loss of generality, such that 2 and 3 are adjacent to 1 but not to each other. Since $\mathbb{P}_2(I) \hookrightarrow \mathbb{P}_G(I)$ via

$$\begin{pmatrix} a & b \\ b & c \end{pmatrix} \mapsto \begin{pmatrix} a & b \\ b & c \end{pmatrix} \oplus 0_{(|V|-2) \times (|V|-2)},$$

it follows from Theorem 6.7 that $f|_{(0, \infty)}$ is nonnegative, nondecreasing, and multiplicatively midconvex; moreover, $f|_{(0, \infty)}$ is continuous and is identically zero or never zero.

To prove (1), define

$$B(\alpha, \beta) := \begin{pmatrix} \alpha + \beta & \alpha & \beta \\ \alpha & \alpha & 0 \\ \beta & 0 & \beta \end{pmatrix}, \qquad \alpha, \beta \geq 0.$$

Note that $B(\alpha, \beta) \oplus 0_{(|V|-3) \times (|V|-3)} \in \mathbb{P}_G(I)$. Hence, $f_G[B(\alpha, \beta) \oplus 0] \in \mathbb{P}_G(I)$, from which we obtain

$$f_G[B(\alpha, \beta)] = \begin{pmatrix} f(\alpha + \beta) & f(\alpha) & f(\beta) \\ f(\alpha) & f(\alpha) & 0 \\ f(\beta) & 0 & f(\beta) \end{pmatrix} \in \mathbb{P}_3(I), \qquad \forall \alpha, \beta \geq 0. \quad (7.2)$$

For $\alpha = \beta = 0$, (7.2) yields that $\det f_G[B(0, 0)] = -f(0)^3 \geq 0$. But since f is nonnegative, it follows that $f(0) = 0$, proving (1).

Now if $f|_{(0, \infty)} \equiv 0$, the remaining assertions immediately follow. Thus, we assume in the sequel that $f|_{(0, \infty)}$ is always positive.

To prove (2), let $\alpha = \beta > 0$. Then (7.2) gives

$$\det f_G[B(\alpha,\alpha)] \geq 0 \quad \Longrightarrow \quad f(\alpha)^2(f(2\alpha) - 2f(\alpha)) \geq 0$$
$$\Longrightarrow \quad f(2\alpha) - 2f(\alpha) \geq 0.$$

Taking the limit as $t \to 0^+$, we obtain $-f(0^+) \geq 0$. Since f is nonnegative, $f(0^+) = 0 = f(0)$, so f is continuous at 0. The continuity of f on I now follows from the above discussion.

Finally, to prove (3), let $\alpha, \beta > 0$. Invoking (7.2) and again starting with $\det f_G[B(\alpha,\beta)] \geq 0$, we obtain

$$f(\alpha)f(\beta)(f(\alpha + \beta) - f(\alpha) - f(\beta)) \geq 0$$
$$\Longrightarrow \quad f(\alpha + \beta) \geq f(\alpha) + f(\beta).$$

This shows that f is superadditive on $(0, \infty)$; since $f(0) = 0$, we obtain superadditivity on all of I. $\qquad\qquad\square$

Proposition 7.2 is the key step in classifying *all entrywise functions* preserving positivity on $\mathbb{P}_G$ for every tree G. In fact, apart from the case of $\mathbb{P}_2 = \mathbb{P}_{K_2}$, this is perhaps the only known case (i.e., family of individual graphs) for which a complete classification of the entrywise preservers of $\mathbb{P}_G$ is available – and proved in the next result.

Recall that a tree is a connected graph in which there is a unique path between any two vertices; equivalently, where the number of edges is one less than the number of nodes; or also where there are no cycle subgraphs. For example, the graph A_3 considered above (with $V = \{1, 2, 3\}$ and $E = \{(1, 2), (1, 3)\}$) is a tree.

Theorem 7.4 *Suppose $I = [0, \infty)$ and a function $f : I \to I$. Let G be a tree on at least three vertices. Then the following are equivalent:*

(1) $f[-] : \mathbb{P}_G(I) \to \mathbb{P}_G(I)$.
(2) $f[-] : \mathbb{P}_T(I) \to \mathbb{P}_T(I)$ *for all trees T.*
(3) $f[-] : \mathbb{P}_{A_3}(I) \to \mathbb{P}_{A_3}(I)$.
(4) f *is multiplicatively midconvex and superadditive on I.*

Proof Note that G contains three vertices on which the induced subgraph is A_3 (consider any induced connected subgraph on three vertices). By padding $\mathbb{P}_{A_3}$ by zeros to embed inside $\mathbb{P}_{|G|}$, we obtain (1) $\Longrightarrow$ (3). Moreover, that (2) $\Longrightarrow$ (1) is clear.

To prove that (3) $\Longrightarrow$ (4), note that $K_2 \hookrightarrow A_3$. Hence, f is multiplicatively midconvex on $(0, \infty)$ by Theorem 6.7. By Proposition 7.2, $f(0) = 0$ and f is superadditive on I. In particular, f is also multiplicatively midconvex on all of I.

Finally, we show that (4) $\implies$ (2) by induction on n for all trees T with at least $n \geq 2$ nodes. For the case $n = 2$ by Theorem 6.7, it suffices to show that f is nondecreasing. Given $\gamma \geq \alpha \geq 0$, by superadditivity we have

$$f(\gamma) \geq f(\alpha) + f(\gamma - \alpha) \geq f(\alpha),$$

proving the result.

For the induction step, suppose that (2) holds for all trees on n nodes and let $G' = (V, E)$ be a tree on $n + 1$ nodes. Without loss of generality, let $V = [n+1] = \{1, \ldots, n+1\}$, such that node $n+1$ is adjacent only to node n. (Note: there always exists such a node in every tree.) Let G be the induced subgraph on the subset $[n]$ of vertices. Then, any $A \in \mathbb{P}_{G'}(I)$ can be written as

$$A = \begin{pmatrix} B & b e_n \\ b e_n^T & c \end{pmatrix}_{(n+1) \times (n+1)}, \quad \text{where } b \in \mathbb{R}, \ c \in I, \ B \in \mathbb{P}_G(I),$$

and $e_n := ((0, 0, \ldots, 0, 1)_{1 \times n})^T$ is a standard basis vector. Since f is nonnegative and superadditive, $f(0) = f(0 + 0) \geq 2f(0) \geq 0$, hence $f(0) = 0$. If $f \equiv 0$, we are done. Thus, we assume that $f \not\equiv 0$, so $f|_{(0,\infty)}$ is positive by Theorem 6.7.

If $c = 0$, then $b_{nn}c - b^2 \geq 0$ implies

$$b = 0, \quad \text{and so} \quad f[A] = \begin{pmatrix} f[B] & 0_{n \times 1} \\ 0_{1 \times n} & 0 \end{pmatrix} \in \mathbb{P}_G \oplus 0_{1 \times 1}$$

by the induction hypothesis. Otherwise, $c > 0$, hence $f(c) > 0$. From the properties of Schur complements (Theorem 1.28) we obtain that A is positive semidefinite if and only if $B - \frac{b^2}{c} E_{nn}$ is positive semidefinite, where E_{nn} is the elementary $n \times n$ matrix with (j, k) entry $\delta_{j,n}\delta_{k,n}$; and similarly, $f[A]$ is positive semidefinite if and only if $f[B] - \frac{f(b)^2}{f(c)} E_{nn}$ is positive semidefinite.

By the induction hypothesis, we have that $f[B - \frac{b^2}{c} E_{nn}]$ is positive semidefinite. Thus, it suffices to prove that $f[B] - \frac{f(b)^2}{f(c)} E_{nn} - f[B - \frac{b^2}{c} E_{nn}]$ is positive semidefinite. Now compute

$$f[B] - \frac{f(b)^2}{f(c)} E_{nn} - f\left[B - \frac{b^2}{c} E_{nn} \right] = \alpha E_{nn},$$

$$\text{where } \alpha = f(b_{nn}) - \frac{f(b)^2}{f(c)} - f\left(b_{nn} - \frac{b^2}{c} \right).$$

Therefore, it suffices to show that $\alpha \geq 0$. But by superadditivity, we have

$$
\alpha = f(b_{nn}) - \frac{f(b)^2}{f(c)} - f\left(b_{nn} - \frac{b^2}{c}\right)
$$
$$
= f\left(b_{nn} - \frac{b^2}{c} + \frac{b^2}{c}\right) - \frac{f(b)^2}{f(c)} - f\left(b_{nn} - \frac{b^2}{c}\right)
$$
$$
\geq f\left(b_{nn} - \frac{b^2}{c}\right) + f\left(\frac{b^2}{c}\right) - \frac{f(b)^2}{f(c)} - f\left(b_{nn} - \frac{b^2}{c}\right)
$$
$$
= f\left(\frac{b^2}{c}\right) - \frac{f(b)^2}{f(c)}.
$$

Moreover, by multiplicative midconvexity, we obtain that $f\left(\frac{b^2}{c}\right) f(c) \geq f(b)^2$. Hence, $\alpha \geq 0$ and $f[A]$ is positive semidefinite as desired. □

An immediate consequence is the complete classification of entrywise powers preserving positivity on $\mathbb{P}_T([0,\infty))$ for T a tree.

Corollary 7.5 $f(x) = x^\alpha$ preserves $\mathbb{P}_T([0,\infty))$ for a tree on at least three nodes if and only if $\alpha \geq 1$.

The proof follows from the observation that x^α is superadditive on $[0,\infty)$ if and only if $\alpha \geq 1$.

8

Entrywise Powers Preserving Positivity, Monotonicity, and Superadditivity

We next study the set of entrywise powers preserving positivity on matrices with zero patterns. Recall the closed convex cone $\mathbb{P}_G([0,\infty))$ studied in Chapter 7, for a (finite simple) graph G. Now define

$$\mathcal{H}_G := \{\alpha \geq 0 : A^{\circ\alpha} \in \mathbb{P}_G([0,\infty)) \; \forall A \in \mathbb{P}_G([0,\infty))\}, \tag{8.1}$$

with the convention that $0^0 := 1$. Thus, $\mathcal{H}_G$ is the set of entrywise, or Hadamard, powers that preserve positivity on $\mathbb{P}_G$.

Observe that if $G \subset H$ are graphs, then $\mathcal{H}_G \supset \mathcal{H}_H$. In particular, by the FitzGerald–Horn classification in Theorem 5.3,

$$\mathcal{H}_G \supset \mathcal{H}_{K_n} = \mathbb{Z}^{\geq 0} \cup [n - 2, \infty), \tag{8.2}$$

whenever G has n vertices. Specifically, there is always a point $\beta \geq 0$ beyond which every real power preserves positivity on $\mathbb{P}_G$. We are interested in the smallest such point, which leads us to the next definition (following the FitzGerald–Horn theorem 5.3 in the special case $G = K_n$):

Definition 8.1 The *critical exponent of a graph* G is

$$\alpha_G := \min\{\beta \geq 0 : \alpha \in \mathcal{H}_G \; \forall \alpha \geq \beta\}.$$

Example 8.2 We saw earlier that if G is a tree (but not a disjoint union of copies of K_2), then $\alpha_G = 1$; and FitzGerald and Horn [84] showed that $\alpha_{K_n} = n - 2$ for all $n \geq 2$.

In this chapter we are interested in closed-form expressions for α_G and $\mathcal{H}_G$. Not only is this a natural mathematical refinement of Theorem 5.3, but as discussed in Section 7.1, this moreover impacts applied fields, providing modern motivation to study the question. Somewhat remarkably, the above examples were the only known cases until very recently.

On a more mathematical note, we are also interested in understanding a combinatorial interpretation of the critical exponent α_G. This is a graph invariant that arises out of positivity; it is natural to ask if it is related to previously known (combinatorial) graph invariants, and more broadly, how it relates to the geometry of the graph.

We explain in this chapter that there is a uniform answer for a large family of graphs, which includes complete graphs, trees, split graphs, banded graphs, cycles, and other classes; and moreover, there are no known counterexamples to this answer. Before stating the results, we remark that the question of computing $\mathcal{H}_G, \alpha_G$ for a given graph is easy to formulate, and one can carry out easy numerical simulations by running (software code) over large sets of matrices in $\mathbb{P}_G$ (possibly chosen randomly), to better understand which powers preserve $\mathbb{P}_G$. This naturally leads to accessible research problems for various classes of graphs: say triangle-free graphs, or graphs with small numbers of vertices. For instance, there is a graph on five vertices for which the critical exponent is not known!

Now on to the known results. We begin by computing the critical exponent α_G – and $\mathcal{H}_G$, more generally – for a family of graphs that turns out to be crucial in understanding several other families (split, Apollonian, banded, and in fact all chordal graphs)

Definition 8.3   The *almost complete graph* $K_n^{(1)}$ is the complete graph on n nodes, with one edge missing.

We will choose a specific labeling of the nodes in $K_n^{(1)}$; note this does not affect the set $\mathcal{H}_G$ or the threshold α_G. Specifically, we set the $(1,n)$ and $(n,1)$ entries to be zero, so that $\mathbb{P}_{K_n^{(1)}}$ consists of matrices in $\mathbb{P}_n$ of the form

$$\begin{pmatrix} & \cdots & 0 \\ \vdots & \ddots & \vdots \\ 0 & \cdots & \end{pmatrix}.$$ Our goal is to prove:

Theorem 8.4   *For all $n \geq 2$, we have $\mathcal{H}_{K_n^{(1)}} = \mathcal{H}_{K_n} = \mathbb{Z}^{\geq 0} \cup [n-2, \infty)$.*

8.1 Other Loewner Properties

In order to prove Theorem 8.4, we need to understand the powers that preserve superadditivity on $n \times n$ matrices under the positive semidefinite ordering. We now define this notion, as well as a related notion of monotonicity.

Definition 8.5   Let $I \subset \mathbb{R}$ and $n \in \mathbb{N}$. A function $f : I \to \mathbb{R}$ is said to be

(1) *Loewner monotone* on $\mathbb{P}_n(I)$ if we have $A \geq B \geq 0_{n \times n} \implies$ $f[A] \geq f[B]$.
(2) *Loewner superadditive* on $\mathbb{P}_n(I)$ if $f[A + B] \geq f[A] + f[B]$ for all $A, B \in \mathbb{P}_n(I)$.

In these definitions, we are using the *Loewner ordering* (or positive semidefinite ordering) on $n \times n$ matrices: $A \geq B$ if $A - B \in \mathbb{P}_n(\mathbb{R})$.

Remark 8.6   A few comments to clarify these definitions are in order. First, if $n = 1$, then these notions both reduce to their usual counterparts for real functions defined on $[0, \infty)$. Second, if $f(0) \geq 0$, then Loewner monotonicity implies Loewner positivity. Third, a Loewner monotone function differs from – in fact is the entrywise analogue of – the more commonly studied *operator monotone* functions, which have the same property but for the functional calculus: $A \geq B \geq 0 \implies f(A) \geq f(B) \geq 0$.

Note that if $n = 1$ and f is differentiable, then f is monotonically increasing if and only if f' is positive. The following result generalizes this fact to powers acting entrywise on $\mathbb{P}_n$ and classifies the Loewner monotone powers.

Theorem 8.7 (FitzGerald–Horn)   *Given an integer $n \geq 2$ and a scalar $\alpha \in \mathbb{R}$, the power x^α is Loewner monotone on $\mathbb{P}_n([0, \infty))$ if and only if $\alpha \in \mathbb{Z}^{\geq 0} \cup [n - 1, \infty)$. In particular, the critical exponent for Loewner monotonicity on $\mathbb{P}_n$ is $n - 1$.*

We will see in Chapter 9 a strengthening of Theorem 8.7 by using individual matrices from a multiparameter family, in the spirit of Jain's theorem 5.7 for Loewner positive powers.

Proof   The proof strategy is similar to that of Theorem 5.3: use the Schur product theorem for nonnegative integer powers, perform induction on n for the powers above the critical exponent, and employ (the same) rank-2 Hankel moment matrix counterexample for the remaining powers. First, if $\alpha \in \mathbb{N}$ and $0 \leq B \leq A$, then repeated application of the Schur product theorem yields

$$0_{n \times n} \leq B^{\circ \alpha} \leq B^{\circ(\alpha-1)} \circ A \leq B^{\circ(\alpha-2)} \circ A^{\circ 2} \leq \cdots \leq A^{\circ \alpha}.$$

Now, suppose $\alpha \geq n - 1$. We prove that x^α is Loewner monotone on $\mathbb{P}_n$ by induction on n; the base case of $n = 1$ is clear. For the induction step, if $\alpha \geq n - 1$, then recall the integration trick (5.2) of FitzGerald and Horn

$$A^{\circ \alpha} - B^{\circ \alpha} = \alpha \int_0^1 (A - B) \circ (\lambda A + (1 - \lambda)B)^{\circ(\alpha-1)} \, d\lambda.$$

Since $\alpha - 1 \geq n - 2$, the matrix $(\lambda A + (1-\lambda)B)^{\circ(\alpha-1)}$ is positive semidefinite by Theorem 5.3, and thus, $A^{\circ\alpha} - B^{\circ\alpha} \in \mathbb{P}_n$. Therefore, $A^{\circ\alpha} \geq B^{\circ\alpha}$, and we are done by induction.

Finally, to show that the threshold $\alpha = n - 1$ is sharp, suppose $\alpha \in (0, n-1) \setminus \mathbb{N}$ (we leave the case of $\alpha < 0$ as an easy exercise). Consider again the Hankel moment matrices

$$A(\epsilon) := H_\mu \text{ for } \mu = \delta_1 + \epsilon\delta_x, \qquad B := A(0) = \mathbf{1}_{n \times n},$$

where $x, \epsilon > 0$, $x \neq 1$, and H_μ is understood to denote the leading principal $n \times n$ submatrix of the Hankel moment matrix for μ. Clearly, $A(\epsilon) \geq B \geq 0_{n \times n}$. As above, let $v = (1, x, \dots, x^{n-1})^T$, so that $A(\epsilon) = \mathbf{1}\mathbf{1}^T + \epsilon vv^T$. Choose a vector $u \in \mathbb{R}^n$ that is orthogonal to $v, v^{\circ 2}, \dots, v^{\circ(\lfloor\alpha\rfloor+1)}$, and $u^T v^{\circ(\lfloor\alpha\rfloor+2)} = 1$. (Note, this is possible since the vectors $v, v^{\circ 2}, \dots, v^{\circ(\lfloor\alpha\rfloor+2)}$ are linearly independent, forming the columns of a possibly partial generalized Vandermonde matrix.)

We claim that $u^T(A(\epsilon)^{\circ\alpha} - B^{\circ\alpha})u < 0$ for small $\epsilon > 0$, which will show that x^α is not Loewner monotone on $\mathbb{P}_n([0,\infty))$. Indeed, compute using the binomial series for $(1+x)^\alpha$

$$u^T A(\epsilon)^{\circ\alpha} u - u^T B u = u^T(\mathbf{1}\mathbf{1}^T + \epsilon vv^T)^{\circ\alpha} u - u^T \mathbf{1}\mathbf{1}^T u$$

$$= u^T \cdot \sum_{k=1}^{\lfloor\alpha\rfloor+2} \binom{\alpha}{k} \epsilon^k v^{\circ k}(v^{\circ k})^T \cdot u + u^T \cdot o(\epsilon^{\lfloor\alpha\rfloor+2}) \cdot u$$

$$= \binom{\alpha}{\lfloor\alpha\rfloor+2} \epsilon^{\lfloor\alpha\rfloor+2} + u^T o(\epsilon^{\lfloor\alpha\rfloor+2})u$$

$$= \epsilon^{\lfloor\alpha\rfloor+2}\left(\binom{\alpha}{\lfloor\alpha\rfloor+2} + u^T \cdot o(1) \cdot u\right),$$

and this is negative for small $\epsilon > 0$. (Here, $o(\cdot)$ always denotes a matrix, as in (5.3)) □

Theorem 8.7 is now used to classify the powers preserving Loewner superadditivity. Note that if $n = 1$, then x^α is superadditive on $\mathbb{P}_n([0,\infty)) = [0,\infty)$ if and only if $\alpha \geq n = 1$. The following result generalizes this to all integers $n \geq 1$:

Theorem 8.8 (Guillot et al. [104])   *Given an integer $n \geq 1$ and a scalar $\alpha \in \mathbb{R}$, the power x^α is Loewner superadditive on $\mathbb{P}_n([0,\infty))$ if and only if $\alpha \in \mathbb{N} \cup [n,\infty)$. Moreover, for each $\alpha \in (0,n) \setminus \mathbb{N}$ and $x \in (0,1)$, for $\epsilon > 0$ small enough the matrix*

$$\left(\mathbf{1}\mathbf{1}^T + \epsilon vv^T\right)^{\circ\alpha} - \mathbf{1}\mathbf{1}^T - \left(\epsilon vv^T\right)^{\circ\alpha}$$

is not positive semidefinite, where $v = (1, x, \ldots, x^{n-1})^T$. *In particular, the critical exponent for Loewner superadditivity on* $\mathbb{P}_n$ *is* n.

Thus, once again the same rank-2 Hankel moment matrices provide the desired counterexamples, for noninteger powers α below the critical exponent.

Proof   As above, we leave the proof of the case $\alpha < 0$ or $n = 1$ to the reader. Next, if $\alpha = 0$, then superadditivity fails, since we always get $-\mathbf{1}\mathbf{1}^T$ from the superadditivity condition, and this is not positive semidefinite.

Thus, henceforth $\alpha > 0$ and $n \geq 2$. If α is an integer, then by the binomial theorem and the Schur product theorem,

$$(A + B)^{\circ \alpha} = \sum_{k=0}^{\alpha} \binom{\alpha}{k} A^{\circ k} \circ B^{\circ(\alpha - k)} \geq A^{\circ \alpha} + B^{\circ \alpha}, \qquad \forall A, B \in \mathbb{P}_n.$$

Next, if $\alpha \geq n$ and $A, B \in \mathbb{P}_n([0, \infty))$, then $x^{\alpha - 1}$ preserves Loewner monotonicity on $\mathbb{P}_n$, by Theorem 8.7. Again, use the integration trick (5.2) to compute

$$(A + B)^{\circ \alpha} - A^{\circ \alpha} = \alpha \int_0^1 B \circ (\lambda(A + B) + (1 - \lambda)A)^{\circ(\alpha - 1)} \, d\lambda$$

$$\geq \alpha \int_0^1 B \circ (\lambda B)^{\circ(\alpha - 1)} \, d\lambda = B^{\circ \alpha}.$$

The final case is if $\alpha \in (0, n) \setminus \mathbb{Z}$. As above, we fix $x > 0$, $x \neq 1$, and define

$$v := (1, x, \ldots, x^{n-1})^T, \qquad A(\epsilon) := \epsilon v v^T \ (\epsilon > 0), \qquad B := A(0) = \mathbf{1}_{n \times n}.$$

Clearly, $A(\epsilon), B \geq 0_{n \times n}$. Now, since $\alpha \in (0, n)$, the vectors $v, v^{\circ 2}, \ldots, v^{\circ \lfloor \alpha \rfloor}$, $v^{\circ \alpha}$ are linearly independent (since the matrix with these columns is part of a generalized Vandermonde matrix). Thus, we may choose $u \in \mathbb{R}^n$ that is orthogonal to $v, \ldots, v^{\circ \lfloor \alpha \rfloor}$ (if $\alpha \in (0, 1)$, this is vacuous) and such that $u^T v^{\circ \alpha} = 1$. Now compute as in the previous proof, using the binomial theorem

$$(A(\epsilon) + B)^{\circ \alpha} - A(\epsilon)^{\circ \alpha} - B^{\circ \alpha}$$

$$= \sum_{k=1}^{\lfloor \alpha \rfloor} \binom{\alpha}{k} \epsilon^k v^{\circ k} (v^{\circ k})^T - \epsilon^\alpha v^{\circ \alpha} (v^{\circ \alpha})^T + o(\epsilon^\alpha);$$

the point here is that the last term shrinks at least as fast as $\epsilon^{\lfloor \alpha \rfloor + 1}$. Hence, by the choice of u,

$$u^T \left((A(\epsilon) + B)^{\circ \alpha} - A(\epsilon)^{\circ \alpha} - B^{\circ \alpha} \right) u = -\epsilon^\alpha + u^T \cdot o(\epsilon^\alpha) \cdot u,$$

and this is negative for small $\epsilon > 0$. Hence, x^α is not Loewner superadditive even on rank-1 matrices in $\mathbb{P}_n((0, 1])$. $\square$

Remark 8.9 The above proofs of Theorems 8.7 and 8.8 apply for arbitrary $v = (v_1, \ldots, v_n)^T$, consisting of pairwise distinct positive real scalars.

8.2 Entrywise Powers Preserving $\mathbb{P}_G$

We now apply the above results to compute the set of entrywise powers preserving positivity on $\mathbb{P}_{K_n^{(1)}}$ (the almost complete graph).

Proof of Theorem 8.4 The result is straightforward for $n = 2$, so we assume henceforth that $n \geq 3$. It suffices to show that $\mathcal{H}_{K_n^{(1)}} \subset \mathbb{Z}^{\geq 0} \cup [n-2, \infty)$, since the reverse inclusion follows from Theorem 5.3 via (8.2). Fix $x > 0$, $x \neq 1$, and define

$$v := (1, x, \ldots, x^{n-3})^T \in \mathbb{R}^{n-2},$$

$$A(\epsilon) := \begin{pmatrix} 1 & \mathbf{1}^T & 0 \\ \mathbf{1} & \mathbf{11}^T + \epsilon v v^T & \sqrt{\epsilon} v \\ 0 & \sqrt{\epsilon} v^T & 1 \end{pmatrix}_{n \times n}, \quad \epsilon > 0.$$

Note that if $p, q > 0$ are scalars, $\mathbf{a}, \mathbf{b} \in \mathbb{R}^{n-2}$ are vectors, and B is an $(n-2) \times (n-2)$ matrix, then using Schur complements (Theorem 1.28),

$$\begin{pmatrix} p & \mathbf{a}^T & 0 \\ \mathbf{a} & B & \mathbf{b} \\ 0 & \mathbf{b}^T & q \end{pmatrix} \in \mathbb{P}_n \iff \begin{pmatrix} p & \mathbf{a}^T \\ \mathbf{a} & B - q^{-1} \mathbf{b} \mathbf{b}^T \end{pmatrix} \in \mathbb{P}_{n-1} \tag{8.3}$$

$$\iff B - p^{-1} \mathbf{a} \mathbf{a}^T - q^{-1} \mathbf{b} \mathbf{b}^T \in \mathbb{P}_{n-2}.$$

Applying this to the matrices $A(\epsilon)$ and $A(\epsilon)^{\circ \alpha}$, we obtain: $A(\epsilon) \in \mathbb{P}_n$, and

$$A(\epsilon)^{\circ \alpha} \in \mathbb{P}_n \iff (\mathbf{11}^T + \epsilon v v^T)^{\circ \alpha} - (\mathbf{11}^T)^{\circ \alpha} - (\epsilon v v^T)^{\circ \alpha} \in \mathbb{P}_{n-2}.$$

For small $\epsilon > 0$, Theorem 8.8 now shows that $\alpha \in \mathbb{Z}^{\geq 0} \cup [n-2, \infty)$, as desired. □

In the remainder of this chapter, we present what is known about the critical exponents α_G and power-preserver sets $\mathcal{H}_G$ for various graphs. We do not provide proofs below, instead referring the reader to the 2016 paper by Guillot et al. [106].

The first family of graphs is that of *chordal graphs*, and it subsumes not only complete graphs, trees, and almost complete graphs (for all of which we have computed $\mathcal{H}_G, \alpha_G$ with full proofs above), but also other graphs including split, banded, and Apollonian graphs, which we shall now discuss.

Definition 8.10   A graph is *chordal* if it has no induced cycle of length ≥ 4.

Chordal graphs are important in many fields. They are also known as triangulated graphs, decomposable graphs, and rigid circuit graphs. They occur in spectral graph theory, but also in network theory, optimization, and Gaussian graphical models. Chordal graphs play a fundamental role in areas including maximum likelihood estimation in Markov random fields, perfect Gaussian elimination, and the matrix completion problem.

The following is the main result of the aforementioned 2016 paper [106], and it computes $\mathcal{H}_G$ for every chordal graph:

Theorem 8.11   *Let G be a chordal graph with $n \geq 2$ nodes and at least one edge. Let r denote the largest integer, such that K_r or $K_r^{(1)} \subset G$. Then $\mathcal{H}_G = \mathbb{Z}^{\geq 0} \cup [r - 2, \infty)$.*

The point of the theorem is that the study of powers preserving positivity reduces solely to the geometry of the graph and can be understood combinatorially rather than through matrix analysis (given the theorem). While we do not prove this result here, we remark that the proof crucially uses Theorem 8.4 and the "clique-tree decomposition" of a chordal graph.

As applications of Theorem 8.11, we mention several examples of chordal graphs G and their critical exponents α_G; by the preceding theorem, the only powers below α_G that preserve positivity on $\mathbb{P}_G$ are the nonnegative integers.

(1) The complete and almost complete graph on n vertices are chordal and have critical exponent $n - 2$.

(2) Trees are chordal and have critical exponent 1.

(3) Let C_n denote a cycle graph (for $n \geq 4$), which is clearly not chordal. Any minimal planar triangulation G of C_n is chordal, and one can check that $\alpha_G = 2$ regardless of the size of the original cycle graph or the locations of the additional chords drawn.

(4) A banded graph with bandwidth $d > 0$ is a graph with vertex set $[n] = \{1, \dots, n\}$ and edges $(j, j + x)$ for $x \in \{-d, -d + 1, \dots, d - 1, d\}$, such that $1 \leq j, j + x \leq n$. Such graphs are chordal, and one checks (combinatorially) that $\alpha_G = \min(d, n - 2)$ if $n > d$.

(5) A split graph consists of a clique $C \subset V$ and an independent (i.e., pairwise disconnected) set $V \setminus C$, whose nodes are connected to various nodes of C. Split graphs are an important class of chordal graphs, because it can be shown that the proportion of (connected) chordal graphs with n nodes that are split graphs grows to 1 as $n \to \infty$. Theorem 8.11 implies that for a split graph G,

$$\alpha_G = \max(|C| - 2, \max \deg(V \setminus C)).$$

(6) Apollonian graphs are constructed as follows: start with a triangle as the first iteration. Given any iteration, which is a subdivision of the original triangle by triangles, choose an interior point of any of these "atomic" triangles, and connect it to the three vertices of the corresponding atomic triangle. This increases the number of atomic triangles by 2 at each step. If G is an Apollonian graph on $n \geq 3$ nodes, one shows that $\alpha_G = \min(2, n-2)$. Notice, for $n \geq 4$ this is independent of n or the specific triangulation.

It is natural to ask what is known for nonchordal graphs. We mention one such result, also shown in the aforementioned 2016 paper [106].

Theorem 8.12   *Let C_n denote the cycle graph on n vertices (which is nonchordal for $n \geq 4$). Then $\mathcal{H}_{C_n} = [1, \infty)$ for all $n \geq 4$.*

Remarkably, this is the same combinatorial recipe as for chordal graphs (in Theorem 8.11)!

We end with some questions, which can be avenues for further research into this nascent topic.

Question 8.13

(i) Compute the critical exponent (and set of powers preserving positivity) for graphs other than the ones discussed above. In particular, compute α_G for all $G = (V, E)$ with $|V| \leq 5$.

(ii) For all graphs G with known critical exponent, the critical exponent turns out to be $r - 2$, where r is the largest integer, such that G contains either K_r or $K_r^{(1)}$. Does the same result hold for all graphs?

(iii) In fact, more is true in all known cases: $\mathcal{H}_G = \mathbb{Z}^{\geq 0} \cup [\alpha_G, \infty)$. Is this true for all graphs?

(iv) Taking a step back: can one show that the critical exponent of a graph is an integer (perhaps without computing it explicitly)?

(v) Does the critical exponent have connections to – or can it be expressed in terms of – other, purely combinatorial graph invariants?

(vi) More generally, is it possible to classify the entrywise functions that preserve positivity on $\mathbb{P}_G$, for G a noncomplete, nontree graph? Perhaps the simplest example is a cycle $G = C_n$.

9

Loewner Convexity and Single Matrix Encoders of Preservers

In Chapter 8, we classified the entrywise powers that are Loewner monotone or superadditive on all matrices in $\mathbb{P}_n([0, \infty))$. In this chapter, we similarly classify the Loewner convex powers on $\mathbb{P}_n$ (a notion that is not yet defined). Before doing so, we show that there exist *individual* matrices that turn out to encode the sets of entrywise powers preserving Loewner positivity and monotonicity.

9.1 Individual Matrices Encoding Loewner Positive Powers

We begin with the Loewner positive powers and recall Jain's Theorem 5.7. This was strengthened by Jain in her 2020 paper in [132]:

Theorem 9.1 (Jain)   *Suppose $n \geq 2$ is an integer, and $x_1, x_2, \ldots, x_n$ are pairwise distinct real numbers, such that $1 + x_j x_k > 0$ for all j, k. Let $C := (1 + x_j x_k)_{j,k=1}^n$. Then $C^{\circ \alpha}$ is positive semidefinite if and only if $\alpha \in \mathbb{Z}^{\geq 0} \cup [n-2, \infty)$.*

The remainder of this section is devoted to proving Theorem 9.1, beginning with the following notation:

Definition 9.2   Given a real tuple $\mathbf{x} = (x_1, \ldots, x_n)$, define $A_{\mathbf{x}} := -\infty$ if all $x_j \leq 0$, and $-1/\max_j x_j$ otherwise. Similarly, define $B_{\mathbf{x}} := \infty$ if all $x_j \geq 0$, and $-1/\min_j x_j$ otherwise.

Here are a few properties of $A_{\mathbf{x}}, B_{\mathbf{x}}$; the details are straightforward verifications, which are left to the reader.

Lemma 9.3

(1) *Suppose $x \in \mathbb{R}$. Then $1 + yx > 0$ for a real scalar y, if and only if $\mathrm{sgn}(x)y \in (-1/|x|, \infty)$, where we set $1/|x| := \infty$ if $x = 0$.*
(2) *Given real scalars $x_1, \ldots, x_n, y$, we have $1 + yx_j > 0$ for all j if and only if $y \in (A_{\mathbf{x}}, B_{\mathbf{x}})$.*
(3) *$A_{-\mathbf{x}} = -B_{\mathbf{x}}$ and $A_{\mathbf{x}} < 0 < B_{\mathbf{x}}$ for all $\mathbf{x} \in \mathbb{R}^n$.*

Proof sketch   The first part follows by using $x = \mathrm{sgn}(x)|x|$; the second follows by intersecting the solution-intervals for each x_j. The final assertion is a consequence of the first two. $\square$

We now show an intermediate result that resembles Descartes' rule of signs (Lemma 4.2), except that it holds for powers of $1 + ux$ rather than $\exp(ux)$:

Proposition 9.4   *Fix a real number r, an integer $n \geq 1$, and two real tuples $\mathbf{c} = (c_1, \ldots, c_n) \neq 0$ and $\mathbf{x} = (x_1, \ldots, x_n)$ with pairwise distinct x_j. Then the function*

$$\varphi_{\mathbf{x}, \mathbf{c}, r} : (A_{\mathbf{x}}, B_{\mathbf{x}}) \to \mathbb{R}, \qquad u \mapsto \sum_{j=1}^{n} c_j (1 + ux_j)^r,$$

either is identically zero or has at most $n - 1$ zeros, counting multiplicities.

Proof   The proof of this Descartes-type result once again employs the trick by Poulain and Laguerre – namely, to use Rolle's theorem and induct. If $r = 0$, then the result is straightforward, so assume henceforth that $r \neq 0$. Let $s = S^-(\mathbf{c})$ denote the number of sign changes in the nonzero tuple $\mathbf{c} = (c_1, \ldots, c_n)$. Now claim more strongly that the number of zeros is at most s. The proof is by induction on $n \geq 1$ and then on $s \in [0, n-1]$. The base case of $n = 1$ is clear; and, for any n, the base case of $s = 0$ is also immediate. Thus, suppose $n \geq 2$ and $s \geq 1$, and suppose the result holds for all tuples $\mathbf{c}$ of length $< n$, as well as for all tuples $\mathbf{c} \in \mathbb{R}^n \setminus \{0\}$ with at most $s - 1$ sign changes. Because of this, we may also assume that all c_j are nonzero and $S^-(\mathbf{c}) = s$.

We begin by relabeling the x_j if required, to lie in increasing order

$$x_1 < \cdots < x_n.$$

Now, suppose there does not exist $0 < k \leq n$, such that $c_{k-1}c_k < 0$ (there is a sign change here in the tuple $\mathbf{c}$, which is also relabeled corresponding to the x_j if required) and $x_k > 0$. Then $x_k \leq 0$, so that $x_{k-1} < 0$. Now work with the tuples $-\mathbf{x}$ and $\mathbf{c}' := (c_n, \ldots, c_1)$, i.e.,

$$-x_n < -x_{n-1} < \cdots < -x_1, \qquad \varphi_{-\mathbf{x}, \mathbf{c}', r}(v) := \sum_{j=n}^{1} c_j (1 - ux_j)^r.$$

Here $v = -u \in (-B_{\mathbf{x}}, -A_{\mathbf{x}}) = (A_{-\mathbf{x}}, B_{-\mathbf{x}})$ by Lemma 9.3, so the result for $\varphi_{-\mathbf{x}, \mathbf{c}', r}(v)$ would prove that for $\varphi_{\mathbf{x}, \mathbf{c}, r}(u)$. Using this workaround if needed, it follows that there exists $1 \le k \le n$, such that $c_{k-1}c_k < 0$ and $x_k > 0$. In particular, there exists $v > 0$, such that

$$1 - vx_n < \cdots < 1 - vx_k < 0 < 1 - vx_{k-1} < \cdots < 1 - vx_1.$$

Define

$$\psi : (A_{\mathbf{x}}, B_{\mathbf{x}}) \to \mathbb{R}, \qquad \psi(u) := \sum_{j=1}^{n} c_j (1 - vx_j)(1 + ux_j)^{r-1}.$$

By choice of v, the sequence $(c_1(1 - vx_1), \ldots, c_n(1 - vx_n))$ has precisely $s - 1$ sign changes, so ψ has at most $s - 1$ zeros. Now, for $u \in (A_{\mathbf{x}}, B_{\mathbf{x}})$, we compute

$$\psi(u) = \sum_{j=1}^{n} c_j (1 + ux_j - (u + v)x_j)(1 + ux_j)^{r-1}$$

$$= \varphi_{\mathbf{x}, \mathbf{c}, r}(u) - (u + v) \sum_{j=1}^{n} c_j x_j (1 + ux_j)^{r-1} = -\frac{(u + v)^{r+1}}{r} h'(u),$$

where $h(u) := (u + v)^{-r} \varphi_{\mathbf{x}, \mathbf{c}, r}(u)$ and $r \neq 0$. Since $x_k > 0$, we obtain from above

$$u \in (A_{\mathbf{x}}, B_{\mathbf{x}}) \quad \Longrightarrow \quad u + v > A_{\mathbf{x}} + v = v - x_n^{-1} > v - x_k^{-1} > 0.$$

Thus, $u \mapsto u + v$ is positive on $(A_{\mathbf{x}}, B_{\mathbf{x}})$, hence $h: (A_{\mathbf{x}}, B_{\mathbf{x}}) \to \mathbb{R}$ is well defined. From above, ψ has at most $s - 1$ zeros on $(A_{\mathbf{x}}, B_{\mathbf{x}})$, hence so does h'. But then by Rolle's theorem, h has at most s zeros on $(A_{\mathbf{x}}, B_{\mathbf{x}})$, and hence, so does $\varphi_{\mathbf{x}, \mathbf{c}, r}$. $\square$

A second intermediate result involves a homotopy argument that will be crucial in proving Theorem 9.1:

Proposition 9.5 *Let $n \ge 2$ be an integer and fix real scalars*

$$x_1 < \cdots < x_n, \qquad 0 < y_1 < \cdots < y_n,$$

such that $1 + x_j x_k > 0$ for all j,k. Then there exists $\epsilon_0 > 0$, such that for all $0 < \epsilon \le \epsilon_0$, the linear homotopies (between x_j and ϵy_j)

$$x_j^{(\epsilon)}(t) := x_j + t(\epsilon y_j - x_j), \qquad t \in [0, 1]$$

all satisfy

$$1 + x_j^{(\epsilon)}(t)x_k^{(\epsilon)}(t) > 0, \qquad \forall j,k = 1,\ldots,n, \ t \in [0,1].$$

Notice that this is not immediate, nor even true for arbitrary ϵ. For instance, suppose $n = 2$, $\epsilon = 1$, and $(y_1, y_2) = (1,2)$. If, say, $(x_1, x_2) = (-199,0)$, then an easy "completion of squares" shows that the assertion is false at most times in the homotopy: $1 + x_1^{(1)}(t)x_2^{(1)}(t) \leq 0$ whenever t is in the range,

$$t \in \left[\frac{398}{800} - \frac{1}{20}\sqrt{\frac{398^2}{40^2} - 1}, \ \frac{398}{800} + \frac{1}{20}\sqrt{\frac{398^2}{40^2} - 1} \right] \supset [0.0026, 0.9924].$$

Similarly, if, say, $(x_1, x_2) = (-8.5, 0.1)$, then

$$1 + x_1^{(1)}(t)x_2^{(1)}(t) \leq 0, \quad \forall t \in \left[\frac{8 - \sqrt{61}}{19}, \frac{8 + \sqrt{61}}{19} \right] \supset [0.01, 0.8321].$$

Thus, the ϵ in the statement is crucial for the result to hold.

Proof of Proposition 9.5   We begin with three observations; in all of them, $x_j(t) = x_j^{(\epsilon)}(t)$ for some fixed $\epsilon > 0$, and all j, t. First, we have $x_1(t) < \cdots < x_n(t)$ for all $t \in [0,1]$.

Second, if $x_1 \geq 0$, then it is clear that $x_j(t) \geq 0$ for all $j \in [1,n]$ and all $t \in [0,1]$, and the result is immediate. Thus, we suppose henceforth that $x_1 < 0$.

Third, if there exist integers $j < k$ and a time $t \in [0,1]$, such that $1 + x_j(t)x_k(t) \leq 0$, then $x_j(t) < 0 < x_k(t)$, and hence $x_1(t) < 0 < x_n(t)$. One then verifies easily that $1 + x_1(t)x_n(t) \leq 1 + x_j(t)x_k(t) \leq 0$.

Thus, for every choice of x_j, y_j as above, with $x_1 < 0$, it suffices to produce $\epsilon_0 > 0$, such that $1 + x_1^{(\epsilon)}(t)x_n^{(\epsilon)}(t) > 0$ for all $t \in (0,1)$ and all $0 < \epsilon \leq \epsilon_0$. There are two cases, depending on the sign of x_n:

Case 1 $x_n \geq 0$. Then $x_n < 1/|x_1|$. We claim that $\epsilon_0 := 1/(|x_1|y_n)$ works; to see this, compute using the known inequalities on x_j, y_j

$$1 + x_1^{(\epsilon)}(t)x_n^{(\epsilon)}(t) = 1 + (t\epsilon y_1 + (1-t)x_1)(t\epsilon y_n + (1-t)x_n)$$

$$> 1 + (1-t)x_1(t\epsilon y_n + (1-t)x_n)$$

$$> 1 + (1-t)x_1(t\epsilon y_n + (1-t)/|x_1|),$$

where the (final) two inequalities are strict because $t \in (0,1)$. Continuing, this last expression equals

$$= 1 - (1-t)^2 + t(1-t)\epsilon y_n x_1 \geq t(2 - t - (1-t)\epsilon_0 y_n |x_1|) = t > 0.$$

Case 2 $x_n < 0$. For ϵ close to 0, define

$$f(\epsilon) := 1 - \frac{\epsilon^2 (x_n y_1 - x_1 y_n)^2}{4(\epsilon y_1 - x_1)(\epsilon y_n - x_n)}.$$

This function is continuous in ϵ and $f(0) > 0$. Hence, there exists $\epsilon_0 > 0$, such that $f(\epsilon) > 0$ for all $0 \leq \epsilon \leq \epsilon_0$.

We show that ϵ_0 satisfies the desired properties. Let $0 < \epsilon \leq \epsilon_0$, and set

$$t_j^{(\epsilon)} := -x_j / (\epsilon y_j - x_j), \qquad 1 \leq j \leq n.$$

Note that $x_j^{(\epsilon)}(t)$ is negative, zero, or positive when $t < t_j^{(\epsilon)}, t = t_j^{(\epsilon)}$, or $t > t_j^{(\epsilon)}$, respectively. Also note by the observations above, and since $t_j^{(\epsilon)}$ is the time at which $x_j^{(\epsilon)}(\cdot)$ vanishes, that

$$0 < t_n^{(\epsilon)} < t_{n-1}^{(\epsilon)} < \cdots < t_1^{(\epsilon)} < 1.$$

Now, if $0 \leq t \leq t_n^{(\epsilon)}$ or $t_1^{(\epsilon)} \leq t \leq 1$, then $x_1^{(\epsilon)}(t), x_n^{(\epsilon)}(t)$ are both nonpositive or nonnegative, respectively. Hence, $1 + x_1^{(\epsilon)}(t) x_n^{(\epsilon)}(t) \geq 1$, as desired.

Next, suppose $t \in (t_n^{(\epsilon)}, t_1^{(\epsilon)})$. Observe that

$$x_j^{(\epsilon)}(t) = t\epsilon y_j + (1-t)x_j = (t - t_j^{(\epsilon)})(\epsilon y_j - x_j), \qquad \forall j \in [1,n], \ t \in [0,1].$$

Now, using the AM–GM inequality, the proof is complete

$$1 + x_1^{(\epsilon)}(t) x_n^{(\epsilon)}(t) = 1 + (t - t_1^{(\epsilon)})(t - t_n^{(\epsilon)})(\epsilon y_1 - x_1)(\epsilon y_n - x_n)$$

$$\geq 1 - \frac{1}{4}(t_1^{(\epsilon)} - t_n^{(\epsilon)})^2 (\epsilon y_1 - x_1)(\epsilon y_n - x_n) = f(\epsilon) > 0.$$

$\square$

With all of the above results at hand, we can proceed:

Proof of Theorem 9.1   For ease of exposition, we write the proof in steps.

Step 1   The first observation is slightly more general than applies here. *Suppose $y_1, \ldots, y_n$ are distinct real numbers, such that $1 + y_k x_j > 0$ for all $1 \leq j, k \leq n$. Let $S := (1 + y_k x_j)$, and let r be real. If $r \in \{0, 1, \ldots, n-2\}$, then $S^{\circ r}$ has rank $r + 1$, else $S^{\circ r}$ is nonsingular.*

Indeed, for $r \in \{0, \ldots, n-2\}$, we have

$$S^{\circ r} = (W_\mathbf{y}^{(r)})^T D W_\mathbf{x}^{(r)}, \qquad \text{where } W_\mathbf{x}^{(r)} := \begin{pmatrix} 1 & 1 & \cdots & 1 \\ x_1 & x_2 & \cdots & x_n \\ \vdots & \vdots & \ddots & \vdots \\ x_1^r & x_2^r & \cdots & x_n^r \end{pmatrix},$$

and where $D_{(r+1)\times(r+1)}$ is a diagonal matrix with entries $\binom{r}{0}, \binom{r}{1}, \ldots, \binom{r}{r}$. Now D is nonsingular, and $W_{\mathbf{x}}^{(r)}, W_{\mathbf{y}}^{(r)}$ are submatrices of Vandermonde matrices and hence of full rank – so $S^{\circ r}$ has rank $r + 1$.

Now suppose $r \neq 0, 1, \ldots, n - 2$ and $S^{\circ r} \mathbf{c}^T = \mathbf{0}$ for some nonzero tuple $\mathbf{c} = (c_1, \ldots, c_n)$. Rewrite $S^{\circ r} \mathbf{c}^T = \mathbf{0}$ to obtain

$$\varphi_{\mathbf{x}, \mathbf{c}, r}(y_k) = \sum_{j=1}^{n} c_j (1 + y_k x_j)^r = 0, \qquad k = 1, \ldots, n.$$

Since $1 + y_k x_j > 0$ for all j, k, we have $y_k \in (A_{\mathbf{x}}, B_{\mathbf{x}})$ by Lemma 9.3(2). So $\varphi_{\mathbf{x}, \mathbf{c}, r} \equiv 0$ on $(A_{\mathbf{x}}, B_{\mathbf{x}})$, by Proposition 9.4, and hence $\varphi_{\mathbf{x}, \mathbf{c}, r}^{(l)}(0) = 0$ for $l = 0$, $1, \ldots, n - 1$ by Lemma 9.3(3). Reformulating this,

$$\sum_{j=1}^{n} (c_j r(r - 1) \cdots (r - l + 1)) x_j^l = 0, \qquad \forall l = 0, 1, \ldots, n - 1,$$

i.e., $W_{\mathbf{x}}^{(n-1)} D \mathbf{c}^T = \mathbf{0}$, where D is a diagonal matrix with diagonal entries

$$1, \quad r, \quad r(r - 1), \quad \ldots, \quad r(r - 1) \cdots (r - n + 2).$$

Now $W_{\mathbf{x}}^{(n-1)}$ is a nonsingular (Vandermonde) matrix, as is D by choice of r. Thus, the tuple $\mathbf{c}$ is zero, i.e., $S^{\circ r}$ is nonsingular.

Step 2   We now turn to the proof of the theorem. First, if $\alpha \in \mathbb{Z}^{\geq 0} \cup [n - 2, \infty)$, then $C^{\circ \alpha}$ is positive semidefinite by Theorem 5.3. Next, if $\alpha < 0$, then the leading 2×2 principal minor of $C(\mathbf{x})^{\circ \alpha}$ is easily seen to be negative. Finally, suppose $\alpha \in (0, n - 2) \setminus \mathbb{Z}$. Given a real vector $\mathbf{y} \in \mathbb{R}^n$, define $C(\mathbf{y}) := \mathbf{1}_{n \times n} + \mathbf{y}\mathbf{y}^T$. Now apply the previous step, fixing $r = \alpha$ and all $y_j = x_j$. Thus, $\det C(\mathbf{x})^{\circ \alpha} \neq 0$ for every $\mathbf{x}$ with all $1 + x_j x_k > 0$.

Now recall the proof of Theorem 5.3, and the subsequent remarks. Thus, if $x_0 < \min_j x_j$, then there exists $\epsilon_1 > 0$, such that $C(\sqrt{\epsilon}(\mathbf{x} - x_0 \mathbf{1}))^{\circ \alpha}$ has a negative eigenvalue for all $0 < \epsilon < \epsilon_1$. Now consider the linear homotopy

$$\mathbf{x}(t) := (1 - t + t \sqrt{\epsilon_2})\mathbf{x} - t \sqrt{\epsilon_2} x_0 \mathbf{1}, \qquad t \in [0, 1],$$

which goes from $\mathbf{x}$ to $\sqrt{\epsilon_2}(\mathbf{x} - x_0 \mathbf{1})$ as t goes from 0 to 1. Here we choose $\epsilon_2 \in (0, \epsilon_1)$, such that $\epsilon_0 := \sqrt{\epsilon_2}$ satisfies the conclusions of Proposition 9.5 for x_j as above and $y_j := x_j - x_0$ (suitably relabeled to be in increasing order if desired).

Again applying the previous step (for the same fixed $r = \alpha$), $\det C(\mathbf{x}(t))^{\circ \alpha} \neq 0$ for all $t \in [0, 1]$. We also know that $C(\mathbf{x}(1))^{\circ \alpha}$ has a negative eigenvalue, hence $\lambda_{\min}(C(\mathbf{x}(1))) < 0$. Now claim by the "continuity of eigenvalues" that

$\lambda_{\min}(C(\mathbf{x}(t))) < 0$ for all $t \in [0,1]$, and in particular at $t = 0$. This is shown in the next step and completes the proof.

Step 3   The claim in the preceding paragraph follows from this more general fact: *Suppose* $C\colon [0,1] \to \mathbb{P}_n(\mathbb{C})$ *is a continuous Hermitian matrix-valued function, such that each $C(t)$ is nonsingular. If $C(1)$ has a negative eigenvalue, then so does $C(t)$ for all $t \in [0,1]$.*

To show this statement, let $X := \{t \in [0,1]\colon \lambda_{\min}(C(t)) \geq 0\}$. Since the cone of positive semidefinite matrices is closed, it follows that X is closed in $[0,1]$. Now the claim follows from the subclaim that $X^c := [0,1] \setminus X$ is also closed: since $[0,1]$ is connected and $1 \in X^c$, it follows that $X^c = [0,1]$ and so $0 \in X^c$ as desired.

To show the subclaim, let $\|C\| := \left(\sum_{j,k=1}^n |c_{jk}|^2\right)^{1/2}$ for a matrix $C = (c_{jk})$. It follows via the Cauchy–Schwartz inequality that all eigenvalues of C lie in $[-\|C\|, \|C\|]$. Now, given a sequence $t_n \in X^c$ that converges to $t_0 \in [0,1]$, all entries of $\{C(t_n) : n \geq 1\}$ lie in a compact set, hence so do the corresponding minimum eigenvalues $\lambda_{\min}(C(t_n))$. Pick a subsequence n_k, such that the sequence $\lambda_{\min}(C(t_{n_k}))$ is convergent, with limit λ_0, say. Now $\lambda_0 \leq 0$. Also pick a unit-length eigenvector v_n of $C(t_n)$ corresponding to the eigenvalue $\lambda_{\min}(C(t_n))$; as the unit complex sphere is compact, there is a further subsubsequence n_{k_l}, such that $v_{n_{k_l}} \to v_0$ as $l \to \infty$, with v_0 also of unit norm.

With these choices at hand, write the equation $C(t_{n_{k_l}})v_{n_{k_l}} = \lambda_{\min}(C(t_{n_{k_l}}))$ $v_{n_{k_l}}$ and let $l \to \infty$. Then $C(t_0)v_0 = \lambda_0 v_0$, with $\lambda_0 \leq 0$. It follows from the hypotheses that $\lambda_{\min}(C(t_0)) < 0$, and the proof is complete.   □

9.2 Matrices Encoding Loewner Monotone Powers

We now turn to Loewner monotonicity (recall Theorem 8.7). The next result – again by Jain in 2020 [132] – shows that, akin to Theorem 9.1, there exist individual matrices that encode the Loewner monotone powers:

Corollary 9.6   *Suppose $n \geq 1$ and $x_1, \ldots, x_n$ are distinct nonzero real numbers, such that $1 + x_j x_k > 0$ for all j, k. Let $\mathbf{x} := (x_1, \ldots, x_n)^T$ and $\alpha \in \mathbb{R}$. Then $(\mathbf{1}_{n \times n} + \mathbf{x}\mathbf{x}^T)^{\circ \alpha} \geq \mathbf{1}_{n \times n}$ if and only if $\alpha \in \mathbb{Z}^{\geq 0} \cup [n - 1, \infty)$, if and only if x^α is Loewner monotone on $\mathbb{P}_n((0, \infty))$.*

Notice that here we cannot take $x_j = 0$ for any j; if, for instance, $x_n = 0$ and we call the matrix X, then the monotonicity of $X^{\circ \alpha}$ over $\mathbf{1}_{n \times n}$ is actually equivalent to the positivity of $X^{\circ \alpha}$, and so the result fails to hold.

Proof   If $\alpha \in \mathbb{Z}^{\geq 0} \cup [n - 1, \infty)$, then Theorem 8.7 implies x^α is Loewner monotone on $\mathbb{P}_n((0, \infty))$, hence on $X := \mathbf{1}_{n \times n} + \mathbf{xx}^T \geq \mathbf{1}_{n \times n}$. Conversely, suppose x^α is Loewner monotone on $X \geq \mathbf{1}_{n \times n}$. Let $\mathbf{x}' := (\mathbf{x}^T, 0)^T \in \mathbb{R}^{n+1}$, then

$$\widetilde{X} := \mathbf{1}_{(n+1) \times (n+1)} + \mathbf{x}'(\mathbf{x}')^T = \begin{pmatrix} X & \mathbf{1} \\ \mathbf{1}^T & 1 \end{pmatrix}$$

satisfies the hypotheses of Theorem 9.1. Now, by the theory of Schur complements (Theorem 1.28), $X^{\circ\alpha} \geq \mathbf{1}_{n \times n}$ if and only if $\widetilde{X}^{\circ\alpha} \in \mathbb{P}_{n+1}$. But this is if and only if $\alpha \in \mathbb{Z}^{\geq 0} \cup [n - 1, \infty)$, by Theorem 9.1.   □

9.3 Loewner Convex Powers and Individual Matrices Encoding Them

Finally, we turn to the entrywise powers preserving Loewner convexity.

Definition 9.7   Let $I \subset \mathbb{R}$ and $n \in \mathbb{N}$. A function $f : I \to \mathbb{R}$ is said to be *Loewner convex* on a subset $V \subset \mathbb{P}_n(I)$ if $f[\lambda A + (1 - \lambda)B] \leq \lambda f[A] + (1 - \lambda)f[B]$, whenever $A \geq B \geq 0_{n \times n}$ lie in V, and $\lambda \in [0, 1]$.

The final theorem in this chapter classifies the Loewner convex powers in the spirit of Theorems 5.3, 8.7, and 8.8. It shows in particular that there is a critical exponent for convexity as well. It also shows the encoding of these powers by individual matrices, in the spirit of Corollary 9.6:

Theorem 9.8 (Loewner convex entrywise powers)   *Fix an integer $n \geq 1$ and a scalar $\alpha \in \mathbb{R}$. The following are equivalent:*

(1) *The entrywise power x^α is Loewner convex on $\mathbb{P}_n([0, \infty))$.*
(2) *Fix distinct nonzero real numbers $x_1, \ldots, x_n$, such that $1 + x_j x_k > 0$ for all j, k. Then x^α is Loewner convex on $A := (1 + x_j x_k)_{j,k=1}^n \geq B = \mathbf{1}_{n \times n} \geq 0$.*
(3) $\alpha \in \mathbb{Z}^{\geq 0} \cup [n, \infty)$.

In particular, the critical exponent for Loewner convexity on $\mathbb{P}_n$ is n.

Thus, there are rank-2 Hankel *TN* matrices (with $x_j = x_0^j$ for $x_0 \in (0, \infty) \setminus \{1\}$), which encode the Loewner convex powers.

To prove this result, we require a preliminary result connecting Loewner convex functions with Loewner monotone ones. We also prove a parallel result connecting monotone maps to positive ones.

Proposition 9.9   *Suppose $n \geq 1$ and $A \geq B \geq 0_{n \times n}$ are positive semidefinite matrices with real entries, such that $A - B = uu^T$, with u having all nonzero entries. Fix any open interval I containing the entries of A, B, and suppose $f : I \to \mathbb{R}$ is differentiable.*

(1) *Then both notions of the "interval" $[B, A]$ agree, i.e.,*

$$\{C : B \leq C \leq A\} = \{\lambda A + (1 - \lambda)B : \lambda \in [0, 1]\}.$$

(2) *If $f[-]$ is Loewner monotone on the interval $[B, A]$, then $f'[-]$ is Loewner positive on (B, A). The converse holds for arbitrary matrices $0 \leq B \leq A$.*

(3) *If $f[-]$ is Loewner convex on the interval $[B, A]$, then $f'[-]$ is Loewner monotone on (B, A). The converse holds for arbitrary matrices $0 \leq B \leq A$.*

Proof

(1) That the left-hand side contains the right is straightforward. Conversely, if $B \leq C \leq A$, then $0 \leq C - B \leq A - B$, which has rank 1. Write $A - B = uu^T$; now, if $u^T v = 0$, then $\| \sqrt{C - B} \cdot v \|^2 = v^T (C - B)v = 0$, so $(C - B)v = 0$. This inclusion of kernels shows that $\ker(C - B)$ has a codimension of at most one. If $C \neq B$, then $\ker(C - B) = \ker u^T$ and $C - B$ has column space spanned by u, by the orthogonality of eigenspaces of Hermitian matrices for different eigenvalues. Thus, $C - B = \lambda uu^T$ for some $\lambda \in (0, 1]$. But then,

$$C = B + \lambda(A - B) = \lambda A + (1 - \lambda)B,$$

as desired.

(2) Suppose f' is Loewner positive on (B, A) (for any $0 \leq B \leq A$). We show that f is monotone on this interval by using the integration trick (5.2) (see also Theorem 5.9). Indeed,

$$f[A] - f[B] = \int_0^1 (A - B) \circ f'[\lambda A + (1 - \lambda)B] \, d\lambda.$$

By assumption and the Schur product theorem, the integrand is positive semidefinite, and hence so is the left-hand side, as desired. The same argument applies to show that $f[A_\lambda] \geq f[A_\mu]$, where $A_\lambda := \lambda A + (1 - \lambda)B$ and $0 \leq \mu \leq \lambda \leq 1$.

Conversely, suppose $f[A_\lambda] \geq f[A_\mu]$ for all $0 \leq \mu \leq \lambda \leq 1$. Now, given $\lambda \in (0, 1)$, let $0 < h \leq 1 - \lambda$, then $f[A_{\lambda+h}] \geq f[A_\lambda]$, so

$$0 \leq \lim_{h \to 0^+} \frac{1}{h} \left(f[A_{\lambda+h}] - f[A_\lambda] \right)$$

$$= \lim_{h \to 0^+} \frac{1}{h} \left(f[\lambda A + (1-\lambda)B + h(A-B)] - f[\lambda A + (1-\lambda)B] \right)$$

$$= f'[A_\lambda] \circ (A - B).$$

By the assumptions, $(A-B)^{\circ(-1)}$ is also a rank-1 positive semidefinite matrix with all nonzero entries, so taking the Schur product, we have $f'[A_\lambda] \geq 0$ for all $\lambda \in (0,1)$, as desired.

(3) Suppose f' is Loewner monotone on (B, A) (for any $0 \leq B \leq A$). As above, we use the integration trick to show that f is convex, beginning with

$$f[(A+B)/2] - f[B] = \frac{1}{2} \int_0^1 (A-B) \circ f' \left[\lambda \frac{A+B}{2} + (1-\lambda)B \right] d\lambda,$$

$$\frac{f[A] + f[B]}{2} - f[B] = \frac{f[A] - f[B]}{2}$$

$$= \frac{1}{2} \int_0^1 (A-B) \circ f'[\lambda A + (1-\lambda)B] \, d\lambda. \tag{9.1}$$

Now, by the hypotheses on f' and the Schur product theorem, it follows that

$$(A-B) \circ f'[\lambda A + (1-\lambda)B] \geq (A-B) \circ f' \left[\lambda \frac{A+B}{2} + (1-\lambda)B \right].$$

This, combined with (9.1), yields $f[(A+B)/2] \leq (f[A] + f[B])/2$. One now proves by induction – first on N and then on j – that

$$f \left[\frac{j}{2^N} A + \left(1 - \frac{j}{2^N} \right) B \right] \leq \frac{j}{2^N} f[A] + \left(1 - \frac{j}{2^N} \right) f[B],$$

$$\forall N \geq 1, \ 1 \leq j \leq 2^N.$$

Now, given any $\lambda \in [0,1]$, approximate λ by a sequence of dyadic rationals $j/2^N$ as above, and use the continuity of $f[-]$ and the preceding inequality to conclude that $f[-]$ is Loewner convex on $\{B, A\}$. The same argument can be adapted to show that $f[-]$ is Loewner convex on $\{A_\lambda, A_\mu\}$, as in the preceding part.

Conversely, since we have $f[\lambda A + (1-\lambda)B] \leq \lambda f[A] + (1-\lambda)f[B]$ for $0 \leq \lambda \leq 1$, it follows for $\lambda \in (0,1)$ that

$$\frac{f[B + \lambda(A - B)] - f[B]}{\lambda} \leq f[A] - f[B],$$

$$\frac{f[A + (1 - \lambda)(B - A)] - f[A]}{1 - \lambda} \leq f[B] - f[A].$$

Letting $\lambda \to 0^+$ and $\lambda \to 1^-$, respectively, yields

$$(A - B) \circ f'[B] \leq f[A] - f[B], \qquad (B - A) \circ f'[A] \leq f[B] - f[A].$$

Adding these yields, $(A - B) \circ (f'[A] - f'[B]) \geq 0$. Finally, $A - B$ has all nonzero entries, so has a rank-1 Schur-inverse; taking the Schur product with this yields $f'[A] \geq f'[B]$. As above, the same argument can be adapted to show that $f'[-]$ is Loewner monotone on $\{A_\lambda, A_\mu\}$. □

Finally, we have:

Proof of Theorem 9.8   Clearly, (1) $\implies$ (2). Now setting $f(x) := x^\alpha$, both (2) $\implies$ (3) and (3) $\implies$ (1) follow via Proposition 9.9(3) and Corollary 9.6. □

The above results on individual (pairs of) matrices encoding the entrywise powers preserving Loewner positivity, monotonicity, and convexity naturally lead to the following question.

Question 9.10   Given $n \geq 1$, do there exist matrices $A, B \in \mathbb{P}_n((0, \infty))$, such that $(A + B)^{\circ \alpha} \geq A^{\circ \alpha} + B^{\circ \alpha}$ if and only if $\alpha \in \mathbb{Z}^{\geq 0} \cup [n, \infty)$? In other words, for each $n \geq 1$, is the set of Loewner superadditive entrywise powers on $\mathbb{P}_n([0, \infty))$ (see Theorem 8.8) also encoded by a single pair of matrices?

We provide a partial solution here, for the matrices studied above. Suppose $u = (u_1, \ldots, u_n)^T \in (0, \infty)^n$ has pairwise distinct coordinates. Let $A := \mathbf{1}_{n \times n}, B := uu^T$. By computations similar to the proof of Theorem 8.4, it follows that

$$(A + B)^{\circ \alpha} \geq A^{\circ \alpha} + B^{\circ \alpha} \quad \Longleftrightarrow \quad \begin{pmatrix} 1 + uu^T & \mathbf{1}_{n \times 1} & u \\ \mathbf{1}_{1 \times n} & 1 & 0 \\ u^T & 0 & 1 \end{pmatrix}^{\circ \alpha} \in \mathbb{P}_{n+2}.$$

$$(9.2)$$

When $\alpha = 1$, denote the matrix on the right in (9.2) by $M(u)$; this is easily seen to have rank 2. Now considering any diagonal entry of the inequality on the left in (9.2), we obtain $\alpha \geq 1$. By Theorem 8.8 and Remark 8.9, it suffices to now assume $\alpha \in (1, n) \setminus \mathbb{Z}$. But if $\alpha \in (1, n - 1)$, then Theorem 9.1 yields the desired result, by considering the leading principal $(n+1) \times (n+1)$ submatrices in the preceding inequality on the right in (9.2).

Thus, it remains to show that for $\alpha \in (n-1, n)$, the matrix $M(u)^{\circ\alpha} \notin \mathbb{P}_n$ for all u with pairwise distinct, positive coordinates. In fact, we claim that it suffices to show for $\alpha \in (n-1, n)$ that $M(u)^{\circ\alpha}$ is nonsingular for all u as above. Indeed, this would imply by a different homotopy argument that $M(\sqrt{\epsilon}u)^{\circ\alpha}$ is nonsingular for all $\epsilon > 0$; but for small enough $\epsilon > 0$ the proof of Theorem 8.8 shows that $M(\sqrt{\epsilon}u)^{\circ\alpha}$ has a negative eigenvalue, so the same holds for all $\epsilon > 0$ by the continuity of eigenvalues (or see Step 3 in the proof of Theorem 9.1).

In light of this discussion, we end this chapter with a question closely related to the preceding question above.

Question 9.11 Suppose $n \geq 2$ and $\alpha \in (n-1, n)$. Is $M(u)^{\circ\alpha} \in \mathbb{P}_{n+2}$, where $u \in (0, \infty)^n$ has pairwise distinct coordinates and $M(u)$ is as in (9.2)?

10

Exercises

Question 10.1   Prove that every real symmetric matrix is a linear combination of two positive semidefinite matrices.

Question 10.2   Given real-valued random variables $X_1, \ldots, X_n$ for some $n \geq 1$, with each X_j having finite first and second moments, prove that the covariance matrix $(\mathbb{E}(X_j X_k) - \mathbb{E}(X_j)\mathbb{E}(X_k))_{j,k=1}^n$ is positive semidefinite. (In particular, the same is true for correlation matrices, i.e., when each X_j has variance one.)

Question 10.3   Prove Example 1.25 – i.e., graph Laplacian matrices are positive semidefinite. In fact, show this more generally for weighted graphs.

Question 10.4 (Gram matrices)   Suppose $m, n \geq 1$ are integers and $A = (\langle x_j, x_k \rangle)_{j,k=1}^m$ is the Gram matrix of a finite set of vectors $x_1, \ldots, x_m \in \mathbb{R}^n$.

(i) Show that A is positive semidefinite, with rank at most n (irrespective of m).

(ii) Find a set of vectors $x_1, \ldots, x_n$ whose Gram matrix A has rank n.

Question 10.5   Fix a triple (n, a, b), with $n > 0$ an integer and $a, b \in \mathbb{R}$. Compute the spectrum of the $n \times n$ matrix with all diagonal entries a and the remaining entries b. In particular, classify the triples for which this matrix is positive semidefinite.

Question 10.6   Here is a family of matrices arising from physics. Consider $2n$ slots, n of which are occupied by particles. Thus, there are $\binom{2n}{n}$ possible configurations, which we denote by 0-1 strings of length $2n$, with n zeros and n ones.

Define the *edit distance* (or *Hamming distance*) $d(\mathbf{s}, \mathbf{s}')$ between two such configurations to be the number of position with differing coordinates. Finally, define the square matrix A of dimension $\binom{2n}{n}$ to have $(\mathbf{s}, \mathbf{s}')$th entry $1 - \frac{d(\mathbf{s}, \mathbf{s}')}{2n}$. Prove that the matrix A is positive semidefinite.

Question 10.7   Here is an example of positive semidefiniteness arising from a combinatorial problem. Given integers $n, r > 0$ and an alphabet with r letters $\{1, \ldots, r\}$, consider the set $S_{(n)}$ of all words/strings of length n. Also fix nonnegative *weights* $w_1, \ldots, w_r$. Now given two words $\mathbf{s} = (s_1, \ldots, s_n)$ and $\mathbf{s}' = (s_1', \ldots, s_n')$ of length n, define their *interaction* to be

$$f(\mathbf{s}, \mathbf{s}') := \sum_{j=1}^{r} w_j \cdot \#\{1 \leq k \leq n : s_k = s_k' = j\}.$$

(i) Prove that the $S_{(n)} \times S_{(n)}$ matrix with $(\mathbf{s}, \mathbf{s}')$-entry $f(\mathbf{s}, \mathbf{s}')$ is positive semidefinite. (Hint: Consider the refinement in which we use weights $w_{jk} \in [0, \infty)$ whenever both strings have letter j in position k.)
(ii) Prove that the positivity of the matrices A arising from physics (above) is in fact a special case of this combinatorial problem.

Question 10.8   Suppose $d \geq 0$ and we consider a star graph G on $d + 1$ nodes, where node 1 is adjacent to all other nodes. Let

$$A = \begin{pmatrix} p_1 & \alpha_2 & \cdots & \alpha_{d+1} \\ \alpha_2 & p_2 & & 0 \\ \vdots & & \ddots & \\ \alpha_{d+1} & 0 & & p_{d+1} \end{pmatrix} \in \mathbb{R}^{(d+1) \times (d+1)}$$

be a real symmetric matrix, with pattern of zeros according to G.

(i) Show that $\det A = \prod_{j=1}^{d+1} p_j - \sum_{j>1} \alpha_j^2 \prod_{k=2, \, k \neq j}^{d+1} p_k$. (Hint: First do this for all $p_j \neq 0$, then extend by continuity to all p_j since the determinant is a polynomial function in the p_j, hence continuous.)
(ii) Show that $A \in \mathbb{P}_G(\mathbb{R})$ if and only if all $p_j \geq 0$ and $\det(A) \geq 0$.

Question 10.9 (Minimum matrices)

(i) Suppose $x_1, \ldots, x_n$ are nonnegative real numbers. Show that the matrix with (j, k)-entry $\min(x_j, x_k)$ is positive semidefinite. (Work this out in two ways: (a) write the matrix as a sum of rank-1 constant-entry-padded matrices; or (b) take Schur complements and use the induction hypothesis.)

(ii) Show that if $0 < x_1 < x_2 < \cdots < x_n$, then the matrix in the preceding part is positive definite, with determinant $x_1 \prod_{j \geq 1} (x_{j+1} - x_j)$.

(iii) Show next that if $m_1, \ldots, m_n$ are nonnegative integers, and $p \geq 2$ is a prime integer, then the matrix with entries $p^{\min(m_j, m_k)}$ is positive semidefinite.

(iv) Finally, if $l_1, \ldots, l_n \geq 1$ are positive integers, then prove that their gcd matrix – i.e., the matrix with (j, k) entry $\gcd(l_j, l_k)$ – is positive semidefinite.

Question 10.10 (Toeplitz cosine matrices)   Recall from Lemma 1.15 and the subsequent remarks that the "Rudin" matrix $(\alpha + \beta \cos((j - k)\theta))_{j,k=1}^n$ has rank at most 3 (where $\alpha, \beta, \theta \in \mathbb{R}$), and it is positive semidefinite if $\alpha, \beta \geq 0$. We now explore this situation in somewhat greater generality.

Given a measure μ on the circle S^1, define the *Fourier–Stieltjes matrix* F_μ to be the (bi-infinite) Toeplitz matrix with (j, k) entry $\int_{S^1} e^{i(k-j)\theta} \, d\mu$ for $j, k \in \mathbb{Z}$. (See Remark 2.7.)

(i) Find an atomic measure μ on (at most) three points in S^1, such that F_μ is precisely the above Rudin matrix $(\alpha + \beta \cos((j - k)\theta))_{j,k \in \mathbb{Z}}$.

(ii) Suppose μ is an atomic measure supported on S^1, with mass $\alpha_j \in \mathbb{C}$ at $e^{i\theta_j}$ for $j = 1, \ldots, r$. Show that the Fourier–Stieltjes matrix F_μ has rank at most r.

(iii) Suppose $\mu \geq 0$ is a Borel probability measure on S^1 that is "symmetric" about the X-axis; in other words, $\mu(A) = \mu(\overline{A})$ for all Borel measurable subsets $A \subset S^1$. (Here, $\overline{A} := \{\overline{z} : z \in A\}$.) Show that the matrix F_μ is real, symmetric, and in fact positive semidefinite. (Hint: First prove that $\int_{S^1} f(e^{i\theta})d\mu = \int_{S^1} f(e^{-i\theta})d\mu$ for arbitrary (bounded) Borel measurable functions $f : S^1 \to \mathbb{C}$.)

Question 10.11   If $C \in \mathbb{P}_n$ and $u \in \mathbb{R}^n$ are such that $uu^T - C$ is positive semidefinite, show that $C = cuu^T$ for some $c \in [0, 1]$. In particular if $C = vv^T$, then $v = cu$ for some $c \in [-1, 1]$.

Question 10.12   Show that $n \, \mathrm{Id}_{n \times n} - C$ is positive semidefinite for all correlation matrices $C \in \mathbb{P}_n$. Show that the coefficient n is sharp.

Question 10.13   Suppose A is a real symmetric matrix, and f is any map from the spectrum of A to $\mathbb{R}$. If $A = UDU^T$ is any spectral decomposition (i.e., U is orthogonal and D is diagonal), then show that $f(A) := Uf(D)U^T$ equals $g(A) = Ug(D)U^T$ for some (real) polynomial g. In particular, $f(A)$ is independent of the choice of U, D. Here $f(D)$ denotes the diagonal matrix with diagonal entries $f(d_{jj})$.

Question 10.14 (Geometric mean)   The *geometric mean* of two positive definite $n \times n$ matrices is given by

$$A\#B := A^{1/2}(A^{-1/2}BA^{-1/2})^{1/2}A^{1/2}.$$

(i) Verify that $A\#B$ is the unique positive definite solution X to the Riccati equation $XA^{-1}X = B$. (Hint: First do this for $A = \mathrm{Id}$.)
(ii) Consequently, show that $A\#B = B\#A$, and

$$(C^{-1}AC)\#(C^{-1}BC) = C^{-1}(A\#B)C$$

for positive definite A, B and unitary C.
(iii) When A, B commute, show that $A\#B = (AB)^{1/2}$.
(iv) Let the *Riemannian distance* between two positive definite $n \times n$ matrices be given by $d(A, B) := \|(\log(\lambda_1), \ldots, \log(\lambda_n))\|$, where $\| \cdot \|$ is the 2-norm and λ_j are the eigenvalues of AB^{-1} (why are all of these positive real numbers?).
   (a) Show that $d(\cdot, \cdot)$ is a metric on positive definite matrices.
   (b) Show that $A\#B$ is the unique matrix, such that $d(A, X) = d(A, B)/2 = d(X, B)$.

Question 10.15   Compute the Moore–Penrose inverse of the following matrices:

(i) The column vector $u \in \mathbb{R}^n$ for some integer $n \geq 1$.
(ii) The matrix uv^T, where $u \in \mathbb{R}^n$ and $v \in \mathbb{R}^m$ for some integers $m, n \geq 1$.
(iii) $\sum_{j=1}^{k} c_j u_j u_j^T$, where $u_1, \ldots, u_k \in \mathbb{R}^n$ are orthonormal, and $c_j \in \mathbb{R}$ are nonzero scalars.
(iv) More generally, if the columns of A are linearly independent, show that $A^\dagger = (A^*A)^{-1}A^*$.

Question 10.16 (The decompression trick, see Belton et al. [23])

(i) Fix integers $m, n \geq 1$ and real scalars a, b, c. Show that $A := \begin{pmatrix} a & b \\ b & c \end{pmatrix}$ is positive semidefinite, if and only if the "decompressed" block matrix

$$\Sigma_{(m,n)}^{\uparrow}(A) := \begin{pmatrix} a\mathbf{1}_{n \times n} & b\mathbf{1}_{n \times m} \\ b\mathbf{1}_{m \times n} & c\mathbf{1}_{m \times m} \end{pmatrix}$$

is positive semidefinite.
(ii) The goal is now to extend and prove the previous statement to arbitrary real symmetric matrices, with a precise determination of the eigenvalues of the decompressed matrix. Thus, first given integers $m_1, \ldots, m_n \geq 1$, define $\mathbf{m} := (m_1, \ldots, m_n)$, and the decompression operator

$$\Sigma_{\mathbf{m}}^{\uparrow} : \mathbb{R}^{n \times n} \to \mathbb{R}^{N \times N}, \quad \text{with} \quad N := \sum_{j=1}^{n} m_j,$$

by setting $\Sigma_{\mathbf{m}}^{\uparrow}(A)$ to be the block matrix $(a_{jk}\mathbf{1}_{m_j \times m_k})_{j,k=1}^{n}$.

First prove that $\Sigma_{\mathbf{m}}^{\uparrow}(A) = \mathcal{W}_{\mathbf{m}} A \mathcal{W}_{\mathbf{m}}^{T}$, where $\mathcal{W}_{\mathbf{m}} \in \mathbb{R}^{N \times n}$ is the matrix with (j,k) entry 1 if $m_1 + \cdots + m_{k-1} < j \le m_1 + \cdots + m_k$, and 0 otherwise.

(iii) Now suppose $A \in \mathbb{R}^{n \times n}$ is symmetric. Using Equation (2.1) and the results from part (ii), prove that the eigenvalues of the decompressed matrix $\Sigma_{\mathbf{m}}^{\uparrow}(A)$ are 0 with multiplicity $N - n$, and the n eigenvalues of $\sqrt{D} A \sqrt{D}$, where D is the diagonal matrix with entries $m_1, \ldots, m_n$.

Question 10.17 (Schur product theorem for positive definite matrices)   This result says that if $A, B \in \mathbb{P}_n$ are positive *definite*, then so is their Schur product $A \circ B$. Akin to Theorem 2.10, this question outlines four proofs.

(i) Prove this using the first proof of Theorem 2.10, i.e., Kronecker products.

(ii) Suppose $\{w_k : 1 \le k \le n\}$ is a basis of $\mathbb{R}^n$, and $u, v \in \mathbb{R}^n$. Prove that if u is orthogonal to all "Schur product vectors" $v \circ w_k$, then $u \perp v$.

(iii) Use the preceding part to prove the Schur product theorem for positive definite matrices, using the second proof of Theorem 2.10, i.e., spectral decompositions.

(iv) Given integers $m, n \ge 1$ and a nonzero real matrix $N_{m \times n}$, prove that $N^T N$ has a strictly positive trace. (In other words, the trace, sending (A, B) to $\mathrm{tr}(A^T B)$, is an inner product on $\mathbb{R}^{m \times n}$.)

(v) Use part (iv) to prove the Schur product theorem for positive definite matrices, using the third proof of Theorem 2.10, i.e., Equation (2.4).

(vi) Prove the result in a fourth different way using Theorem 2.10 (for positive semidefinite matrices) and the fact that perturbations (by multiples of the identity) of positive definite matrices are positive semidefinite.

Question 10.18 (Khare [142])   This question extends the nonzero lower bound for the Schur product in Theorem 2.12, to matrices $A_{n \times a}, B_{n \times b}$ of unequal sizes (where $n, a, b \ge 1$). Show that

$$AA^* \circ BB^* \ge \frac{1}{\min(\mathrm{rk}\, AA^*, \, \mathrm{rk}\, BB^*)} d_{A_0 B_0^T} d_{A_0 B_0^T}^*,$$

where A_0 appends $p + \max(a,b) - a$ zero-columns to the right of A, and B_0 appends $p + \max(a,b) - b$ zero-columns to the right of B, for some $p \ge 0$.

Question 10.19   Suppose $x_1, \ldots, x_n$ are pairwise distinct positive real numbers (so they are not necessarily strictly increasing or decreasing) and $\alpha_1, \ldots, \alpha_n$ are a strictly increasing set of real exponents. If $V = \left(x_j^{\alpha_k} \right)$ denotes the corresponding generalized Vandermonde matrix, show that $V^T V$ is totally positive (*TP*).

Question 10.20   Fix an integer $n \geq 2$.

(i) Show that if $\pi \in S_n$ is a permutation not equal to 1 or
$w_\circ := (n, n-1, \ldots, 2, 1)$, then there exist indices $1 \leq i < j < k \leq n$,
such that $\pi_i < \pi_j > \pi_k$ or $\pi_i > \pi_j < \pi_k$.
(ii) Let $2 \leq p \leq n$ be an integer. Classify the permutations $\pi \in S_n$, such that
applying π to both the row-set and the column-set of an $n \times n$ TP_p
matrix yields a TP_p matrix.

Question 10.21   Fix integers $1 \leq p \leq n$. Show using Theorem 3.6 that the set of TP_p Hankel $n \times n$ matrices is a convex cone, closed under entrywise products. Is this cone closed?

Question 10.22   This question discusses various density phenomena, for real square $n \times n$ matrices, under the usual Euclidean norm on $\mathbb{R}^{n^2}$.

(i) Show that for any graph G on n vertices, the positive definite matrices
in $\mathbb{P}_G$ are dense in $\mathbb{P}_G$. (Show also that $\mathbb{P}_G$ is a closed convex cone,
which is also closed under entrywise products.)
(ii) Show that the real symmetric $n \times n$ Toeplitz nonsingular matrices are
dense in the real symmetric $n \times n$ Toeplitz matrices.
(iii) Show that given integers $1 \leq p \leq n$, the set of Hankel $n \times n$ TP_p
matrices are dense in the Hankel $n \times n$ TN_p matrices.

Question 10.23   Show Corollaries 3.7 and 3.8.

Question 10.24   Here are some (negative) examples related to Theorem 3.6 and the cone of totally nonnegative (*TN*) Hankel matrices.

(i) Suppose A, B are symmetric and *TN* matrices. Then $A + B$ need not
even have nonnegative 2×2 minors. (So, the Hankel assumption is
crucial.) In fact, find two 3×3 matrices, with all entries in $\{0, 1\}$, for
which this holds.
(ii) Improve Sylvester's criterion for positive definite matrices: A is
positive definite if and only if all leading principal minors are positive.
In particular, the Fekete–Schoenberg Lemma 3.12 has an analogue for
positive definite matrices.
(iii) Check that the previous part does not hold if we change the relevant
words to "positive semidefinite" and "nonnegative." In fact, come up

with a 3×3 example of a symmetric matrix with all entries in $\{0, -1\}$,
such that all *contiguous* minors are nonnegative, but the matrix is not
positive semidefinite. Thus, the Fekete–Schoenberg Lemma 3.12 does
not have an analogue for positive semidefinite matrices.

(iv) Similarly, the following "nonnegative" analogue of Lemma 3.12 fails to
hold: if all 1×1 and 2×2 contiguous minors are nonnegative, then so
are all 2×2 minors. Indeed, consider the matrix $\begin{pmatrix} 0 & 0 & 1 \\ 1 & 0 & 0 \end{pmatrix}$.

(v) Check, however, that the previous assertion does indeed hold if we insist
on the entries being positive rather than nonnegative.

Question 10.25   Given a real tuple $\mathbf{x} = (x_1, \ldots, x_n)$, recall that its *Vandermonde determinant* is $V(\mathbf{x}) := \prod_{1 \leq j < k \leq n}(x_k - x_j)$. Now prove that for any
nonempty subsets $X, Y \subset \mathbb{R}$, a kernel $K : X \times Y \to \mathbb{R}$ is TN_p if and only if for
all $1 \leq n \leq \min(p, |X|, |Y|)$ and tuples $\mathbf{x}, \mathbf{y} \in \mathbb{R}^n$,

$$V(\mathbf{x})V(\mathbf{y}) \det(K(x_j, y_k))_{j,k=1}^n \geq 0.$$

Similarly, K is TP_p if the preceding inequality is always strict, for $\mathbf{x}, \mathbf{y}$ with
pairwise distinct coordinates.

Question 10.26 (The correlation trick)   Recall that a positive semidefinite
matrix is a correlation matrix if all diagonal entries are 1.

(i) Prove that for every positive definite matrix A, there exists a unique
positive definite diagonal matrix D and correlation matrix C, such that
$A = DCD$.

(ii) Fix a dimension $n \geq 1$. Does the procedure in part (i) recover all $n \times n$
correlation matrices? Prove or find a counterexample.

(iii) Prove that A and C have the same *pattern of zeros* and the same rank.
By the former, we mean that if $a_{jk} = 0$ for some j, k, then $c_{jk} = 0$ as
well.

(iv) Suppose $A \in \mathbb{P}_n$ has positive entries, and $\alpha \in [0, \infty)$. Prove that $A^{\circ \alpha}$ is
positive semidefinite if and only if $C^{\circ \alpha}$ is also. By part (iii), this reduces
the study of entrywise powers preserving positivity on (structured)
matrices of a fixed size to that on (structured) *correlation* matrices of a
fixed size. The advantage in studying the latter, restricted problem is that
the test set is a compact subset of $\mathbb{R}^{n^2}$.

(v) Suppose A is positive definite. Prove that if A is TP/TN of order p, then
so is C – and vice versa.

Question 10.27

(i) Consider a diagonal matrix $D_{n \times n}$. Let $J, K \subset [n]$ be of the same size. Then either $J = K$ or $D_{J \times K}$ contains a row and column of zeros.

(ii) Now show that if $m, n > 0$ and $1 \le p \le \min(m, n)$ are integers, and $D_{m \times m}, D'_{n \times n}$ are positive definite diagonal matrices, then $D A_{m \times n} D'$ is TP_p (or TN_p) if and only if A is TP_p (or TN_p).

(iii) More generally, suppose X, Y are totally ordered sets, and $f : X \to (0, \infty)$, $g : Y \to (0, \infty)$ are arbitrary functions. Show that a kernel $K : X \times Y \to \mathbb{R}$ is TP_p (or TN_p) if and only if $\widetilde{K} : X \times Y \to \mathbb{R}$ also, given by $\widetilde{K}(x, y) := f(x) K(x, y) g(y)$.

(iv) Show that if $X, Y \subset \mathbb{R}$ and $K : X \times Y \to \mathbb{R}$ is TP_p (or TN_p), then so is the kernel $K(-x, -y)$.

Question 10.28 (k-PMP matrices, see Belton et al. [19]) Similar to the notion of TP/TN matrices of a certain order, we now define the analogous notions for positive semidefiniteness. Given an integer $k \ge 0$, a real symmetric matrix is said to be k-PMP ("principal minor positive") if every principal $r \times r$ minor is nonnegative, for all $1 \le r \le k$. Thus, 0-PMP matrices are merely real and symmetric, 1-PMP matrices are real and symmetric with nonnegative diagonal entries, and n-PMP $n \times n$ matrices are positive semidefinite. (This is akin to TP_p matrices, which interpolate between matrices with positive entries for $p = 1$ and TP matrices for "maximal" p.)

(i) Show that for all integers $0 \le k \le n$, the set of k-PMP matrices in $\mathbb{R}^{n \times n}$ is a closed convex cone, which is also closed under entrywise products.

(ii) Let $A = \lambda \operatorname{Id}_{n \times n} - 1_{n \times n}$, for $\lambda \in \mathbb{R}$. Show that A is not k-PMP if and only if $\lambda < k$. (In particular, k-PMP is not the same as $(k + 1)$-PMP.)

(iii) Show that a matrix that is k-PMP but not $(k + 1)$-PMP (so it has size at least $k + 1$) has at least k positive eigenvalues and at least one negative eigenvalue. (In fact, this cannot be improved upon!) If needed, use the *Cauchy interlacing theorem*, which says that if a real symmetric matrix $A_{n \times n}$ has eigenvalues $\lambda_1 \le \cdots \le \lambda_n$, and a principal submatrix $B_{m \times m}$ of A has eigenvalues $\mu_1 \le \cdots \le \mu_m$, then $\lambda_j \le \mu_j \le \lambda_{n - m + j}$ for all $1 \le j \le m$.

Question 10.29 (Positive $\{0, 1\}$-matrices, see Horn [126]; also [19, 109]) Suppose G is a finite simple graph with node set $\{1, \ldots, n\}$, with $n \times n$ adjacency matrix A_G having (j, k) entry 0 for $j = k$ or $(j, k) \notin E(G)$, and 1 otherwise. Then the following are equivalent

(i) $\mathrm{Id}_{n \times n} + A_G$ is positive semidefinite.
(ii) $\mathrm{Id}_{n \times n} + A_G$ is 3-PMP.
(iii) G is a disconnected union of complete graphs.

Question 10.30 (Positive semidefinite $\{0, \pm 1\}$ matrices, see Hershkowitz et al. [117]; also [19])   Suppose $A \in \mathbb{R}^{n \times n}$ is a 3-PMP matrix with all entries in $\{-1, 0, 1\}$.

(i) Prove that there exists a permutation $\sigma \in S_n$, such that by permuting the rows and columns each by σ, the matrix A is block diagonal, with at most one diagonal block consisting of zeros, and all other diagonal blocks B consisting of matrices with entries ± 1. (Hint: Apply the previous problem to a 3-PMP matrix obtained from A.)
(ii) Now prove that every nonzero diagonal block $B_{m \times m}$ (as in part (i)) has rank 1, i.e., there exists a vector $u \in \mathbb{R}^m$ with entries ± 1, such that $B = uu^T$. (Hint: Let $m \geq 3$ and let u be any nonzero column of B, say the ith; then by considering the principal minor with the i, j, k columns, show that the (j, k)-entry of B equals $u_j u_k$.)

Question 10.31 (Jain [131])   Fix an integer $n \geq 2$. As discussed in Remark 5.6, the entrywise powers preserving positivity on $n \times n$ Hankel matrices of rank at most 2 are precisely those preserving positivity on all $n \times n$ matrices, namely, $\mathbb{Z}^{\geq 0} \cup [n - 2, \infty)$. Even more strongly, Theorem 9.1 specializes (via $x_j = u_0^{j-1}$ for any $u_0 \in (0, \infty) \setminus \{1\}$) to provide a family of individual Hankel encoders of this set of entrywise powers.

The goal of this question is to show that the same holds for Toeplitz matrices. Namely, there exists a family of rank-2 Toeplitz matrices $A(\theta) \in \mathbb{P}_n((0, 1])$, such that if $\alpha < 0$ or $\alpha \in (0, n) \setminus \mathbb{Z}$, then $A(\theta)^{\circ \alpha} \notin \mathbb{P}_n$.

(i) For $\theta \in (0, \pi/(2n - 2))$, define $D(\theta)$ to be the $n \times n$ diagonal matrix with (j, j) entry $\cos((j - 1)\theta)$, and

$$A'(\theta) := (1 + \tan((j - 1)\theta)\tan((k - 1)\theta))_{j,k=1}^n,$$
$$A(\theta) := (\cos((j - k)\theta))_{j,k=1}^n.$$

Verify that $D(\theta)$ is positive definite, and $A'(\theta), A(\theta) \in \mathbb{P}_n((0, \infty))$ and have rank 2. Also verify that $A(\theta)$ is a Toeplitz matrix and
$A(\theta) = D(\theta)A'(\theta)D(\theta)$.
(ii) Using Theorem 9.1, show that $A(\theta)^{\circ \alpha}$ is not positive semidefinite if $\alpha < 0$ or $\alpha \in (0, n) \setminus \mathbb{Z}$.

Question 10.32   Verify Remark 8.9.

Question 10.33   Given $\alpha \in \mathbb{R}$, verify that the power function $x \mapsto x^\alpha$ is superadditive on $(0, \infty)$ if and only if $\alpha \geq 1$.

Question 10.34   Suppose a (finite, simple) connected graph G has vertex set $V = V_1 \sqcup S \sqcup V_2$, with all paths from V_1 to V_2 passing through S (a "separating" set). Assume that the subgraph of G induced on S is complete. Let G_j be the induced subgraph of G on the vertex set $V_j \cup S$, for $j = 1, 2$.

(i) Show that every matrix $A \in \mathbb{P}_G(\mathbb{R})$ can be decomposed as a sum $A = A_1 + A_2$, with $A_j \in \mathbb{P}_{G_j}(\mathbb{R})$. The converse is clear.
(ii) Suppose $f : \mathbb{R} \to \mathbb{R}$ is such that $f[-]$ preserves positivity on $\mathbb{P}_{G_j}(\mathbb{R})$ and $f[-]$ is Loewner superadditive on $\mathbb{P}_S(\mathbb{R})$. Show that $f[-]$ preserves positivity on $\mathbb{P}_G(\mathbb{R})$.

Question 10.35 (Coalescences, see Guillot et al. [106])   Given graphs G_1, G_2, by their *coalescence* $G_1 \odot G_2$ we mean any graph obtained by gluing the two graphs along one vertex of each. Repeating this procedure inductively yields a coalescence of graphs $G_1, \ldots, G_k$ for any $k \geq 2$.

Now suppose $G_1, \ldots, G_k$ are connected graphs (with $k \geq 2$), and each of them has at least one edge. Let G denote any coalescence of $G_1, \ldots, G_k$, and suppose $f : I \to \mathbb{R}$, where I is either $[0, \infty)$ or $\mathbb{R}$. (In other words, work out the following in both cases.) If $f(0) = 0$, then show that $f[-]$ preserves positivity on $\mathbb{P}_G(I)$ if and only if:

(i) $f[-]$ preserves positivity on $\mathbb{P}_{G_j}(I)$ for all j; and
(ii) f is continuous and superadditive on $[0, \infty)$.

In particular, either f is always positive on $(0, \infty)$, or $f \equiv 0$ on $\mathbb{R}$.

Question 10.36 (General positivity preservers on $\mathbb{P}_G$, see Guillot et al. [106])   Suppose G is a graph on a vertex set V with $|V| = n \geq 1$. An ordering $(v_1, \ldots, v_n)$ of V is a *perfect elimination ordering* if for all $1 \leq j \leq n$, the neighbors of the vertex v_j in the subgraph G_j of G induced by $\{v_1, \ldots, v_j\}$ form a complete subgraph, or clique.

It is a fact that G is chordal if and only if it has a perfect elimination ordering. Now suppose $(v_1, \ldots, v_n)$ is such an ordering. Let $c = \omega(G)$ denote the size of the largest complete subgraph (or clique) in G, and define

$$d := \max_{j=1,\ldots,n} \deg_{G_j}(v_j).$$

Suppose $f : \mathbb{R} \to \mathbb{R}$ is such that $f[-]$ preserves positivity on rank-1 matrices in $\mathbb{P}_c(\mathbb{R})$, and $f[M + uu^T] \geq f[M] + f[uu^T]$ for $M \in \mathbb{P}_d(\mathbb{R})$ and $u \in \mathbb{R}^d$. Show that $f[-]$ preserves positivity on $\mathbb{P}_G(\mathbb{R})$. (Hint: Proceed by induction on j to show that $f[-]\colon \mathbb{P}_{G_j}(\mathbb{R}) \to \mathbb{P}_j$, using similar arguments to those in the proof of Theorem 7.4.)

Question 10.37   Suppose $f\colon [0,\infty) \to [0,\infty)$ is continuous at 0^+, positive outside 0, and superadditive. Suppose $A \in \mathbb{P}_n([0,\infty))$ for some integer $n \geq 1$.

 (i) If $f[A]$ is positive semidefinite, show that $f[A + \epsilon \,\mathrm{Id}_n]$ is positive definite for all $\epsilon > 0$.

(ii) If $f[A]$ is not positive semidefinite, show that $f[A + \epsilon \,\mathrm{Id}_n]$ is not positive (semi)definite for all $\epsilon > 0$ small enough.

(iii) Now classify the set of entrywise powers preserving the class of *positive definite* matrices in $\mathbb{P}_G([0,\infty))$, for a chordal graph G.

Question 10.38   Suppose G is a finite simple connected graph that is not complete and G' is obtained from G by attaching a pendant edge to a vertex. Suppose $I = [0,\infty)$ and $f\colon I \to I$ is multiplicatively midconvex and superadditive. Prove that $f[-]$ preserves $\mathbb{P}_G(I)$ if and only if it preserves $\mathbb{P}_{G'}(I)$. (Hint: Use similar arguments to those in the proof of Theorem 7.4.)

PART TWO

ENTRYWISE FUNCTIONS PRESERVING POSITIVITY IN ALL DIMENSIONS

11

History – Schoenberg, Rudin, Vasudeva, and Metric Geometry

In Part II, we take a step back and explore the foundational results on entrywise preservers of positive semidefiniteness – as well as the rich history that motivated these results.

11.1 History of the Problem

In the forthcoming chapters in Part II, we will answer the question:

"Which functions, when applied entrywise, preserve positivity (i.e., positive semidefiniteness)?"

(Henceforth, we use the word "positivity" to denote "positive semidefiniteness.") This question has been the focus of a concerted effort and significant research activity over the past century. It began with the Schur product theorem (1911, [216]) and the following consequence:

Lemma 11.1 (Pólya and Szegő, 1925, [185]) *Suppose a power series $f(x) = \sum_{k=0}^{\infty} c_k x^k$ is convergent on $I \subset \mathbb{R}$ and $c_k \geq 0$ for all $k \geq 0$. Then $f[-]: \mathbb{P}_n(I) \to \mathbb{P}_n(\mathbb{R})$ for all $n \geq 1$.*

Proof By induction and the Schur product theorem 2.10, $f(x) = x^k$ preserves positivity on $\mathbb{P}_n(\mathbb{R})$ for all integers $k \geq 0$ and $n \geq 1$, and hence sends $\mathbb{P}_n(I)$ to $\mathbb{P}_n(\mathbb{R})$. From this the lemma follows, using that $\mathbb{P}_n(\mathbb{R})$ is a closed convex cone. □

With Lemma 11.1 at hand, Pólya and Szegő asked if there exist any other functions that preserve positivity on $\mathbb{P}_n$ for all $n \geq 1$. A negative answer would essentially constitute the converse result to the Schur product theorem; and indeed, this was shown by Schur's student Schoenberg (who is well known for his substantial contribution to the theory of splines, among other areas), for continuous functions:

Theorem 11.2 (Schoenberg, 1942, [206])   *Suppose $I = [-1, 1]$ and $f : I \to \mathbb{R}$ is continuous. The following are equivalent:*

(1) *The entrywise map $f[-]$ preserves positivity on $\mathbb{P}_n(I)$ for all $n \geq 1$.*
(2) *The function f equals a convergent power series $\sum_{k=0}^{\infty} c_k x^k$ for all $x \in I$, with the Maclaurin coefficients $c_k \geq 0$ for all $k \geq 0$.*

Schoenberg's 1942 paper is well known in the analysis literature. In a sense, his Theorem 11.2 is the (harder) converse to the Schur product theorem, i.e., Lemma 11.1, which is the implication (2) $\implies$ (1). Some of these points were discussed in Section 7.2.

Schoenberg's theorem can also be stated for $I = (-1, 1)$. In this setting, the continuity hypothesis was subsequently removed from assertion (1) by Rudin, who, moreover, showed that in order to prove assertion (2) in Theorem 11.2, one does not need to work with the full test set $\bigcup_{n \geq 1} \mathbb{P}_n(I)$. Instead, it is possible to work only with low-rank Toeplitz matrices:

Theorem 11.3 (Rudin, 1959, [197])   *Suppose $I = (-1, 1)$ and $f : I \to \mathbb{R}$. Then the assertions in Schoenberg's theorem 11.2 are equivalent on I, and further equivalent to:*

(3) *$f[-]$ preserves positivity on the Toeplitz matrices in $\mathbb{P}_n(I)$ of rank ≤ 3, for all $n \geq 1$.*

Schoenberg's theorem also has a one-sided variant, over the semi-axis $I = (0, \infty)$:

Theorem 11.4 (Vasudeva, 1979 [230])   *Suppose $I = (0, \infty)$ and $f : I \to \mathbb{R}$. Then the two assertions of Schoenberg's theorem 11.2 are equivalent on I as well.*

Our goal in this part of the text is to prove stronger versions of the theorems of Schoenberg and Vasudeva. Specifically, we will (i) remove the continuity hypothesis, and (ii) work with severely reduced test sets in each dimension, consisting of only the *Hankel matrices of rank at most* 3. For instance, we will show Theorem 11.3, but with the word "Toeplitz" replaced by "Hankel." Similarly, we will show a strengthening of Theorem 11.4, using totally nonnegative (*TN*) Hankel matrices of rank at most 2. These results are stated and proved in this part of the text.

11.2 Digression: The Complex Case

In the aforementioned 1959 paper [197], Rudin made some observations about the complex case *a la* Pólya–Szegő, and presented a conjecture, which is now explained. First, observe that the Schur product theorem holds for complex Hermitian matrices as well, with the same proof via the spectral theorem:

"If A, B are $n \times n$ complex (Hermitian) positive semidefinite matrices, then so is $A \circ B$."

As a consequence, every monomial $z \mapsto z^k$ preserves positivity on $\mathbb{P}_n(\mathbb{C})$ for all integers $k \geq 0$ and $n \geq 1$. (Here $\mathbb{P}_n(\mathbb{C})$ comprises the complex Hermitian matrices $A_{n \times n}$, such that $u^* A u \geq 0$ for all $u \in \mathbb{C}^n$.) But more is true: the (entrywise) conjugation map also preserves positivity on $\mathbb{P}_n(\mathbb{C})$ for all $n \geq 1$. Now using the Schur product theorem, the functions

$$z \mapsto z^k (\overline{z})^m, \qquad k, m \geq 0$$

each preserve positivity on $\mathbb{P}_n(\mathbb{C})$, for all $n \geq 1$. Since $\mathbb{P}_n(\mathbb{C})$ is easily seen to be a closed convex cone also, Rudin observed that if a series

$$f(z) = \sum_{k,m \geq 0} c_{k,m} z^k (\overline{z})^m, \qquad \text{with } c_{k,m} \geq 0,$$

is convergent on the open disk $D(0, \rho) := \{z \in \mathbb{C} : |z| < \rho\}$, then $f[-]$ entrywise preserves positivity on $\mathbb{P}_n(D(0, \rho))$ for $n \geq 1$. Rudin conjectured that there are no other preservers. This was proved soon after:

Theorem 11.5 (Herz, 1963, [118])   *Suppose $I = D(0, 1) \subset \mathbb{C}$ and $f : I \to \mathbb{C}$. The following are equivalent:*

(1) *The entrywise map $f[-]$ preserves positivity on $\mathbb{P}_n(I)$ for all $n \geq 1$.*
(2) *f is of the form $f(z) = \sum_{k,m \geq 0} c_{k,m} z^k (\overline{z})^m$ on I, with $c_{k,m} \geq 0$ for all $k, m \geq 0$.*

For completeness, we also point out [181] for a recent, noncommutative variant of the Schur product and Schoenberg's theorem.

The real and complex cases of Schoenberg/Herz's theorems have been since proved using alternate tools. Christensen and Ressel showed Schoenberg's theorem 11.2 using Choquet's representation theorem, in 1978 [57]. In fact, they removed the continuity assumption of Schoenberg, while continuing to work over the domain $[-1, 1]$. In this case, the preservers turn out to be almost the same as in Schoenberg's theorem – i.e., limits of functions in the convex hull of the monomials – but now one needs to include the pointwise limits on $[-1, 1]$ of the odd powers and of the even powers:

Theorem 11.6 (Christensen and Ressel, 1978, [57])   *The following are equivalent:*

(1) *The entrywise map* $f[-]$ *preserves positivity on* $\mathbb{P}_n(I)$ *for all* $n \geq 1$.
(2) *The function* f *equals a power series plus two other terms:*

$$f(x) = \sum_{k=0}^{\infty} c_k x^k + c_{-1}[\mathbf{1}_{\{1\}}(x) - \mathbf{1}_{\{-1\}}(x)] + c_{-2}\mathbf{1}_{\{-1,1\}}(x), \qquad x \in I,$$

with the Maclaurin coefficients $c_k \geq 0$ *for all* $k \geq -2$ *and* $\sum_{k \geq -2} c_k < \infty$.

Christensen and Ressel also proved the complex analogue, for preservers of positivity on Gram matrices from unit complex spheres, in 1982:

Theorem 11.7 (Christensen and Ressel, 1982, [58])   *Suppose* $f : \overline{D}(0, 1) \to \mathbb{C}$ *is continuous on the closed unit disk, and* $\mathcal{H}$ *is an infinite-dimensional complex Hilbert space with unit sphere S. Then the following are equivalent:*

(1) f *is "positive definite" on S, in that for all* $n \geq 1$ *and points* $z_1, \ldots, z_n \in S$, *the matrix with* (j,k)-*entry* $f(\langle z_j, z_k \rangle)$ *is positive semidefinite.*
(2) $f(z)$ *has the unique series representation* $f(z) = \sum_{k,m \geq 0} c_{k,m} z^k (\overline{z})^m$, *with all* $c_{k,m} \geq 0$ *and* $\sum_{k,m \geq 0} c_{k,m} < \infty$.

This resembles Herz's theorem 11.5 (which proved Rudin's conjecture) similar to the relation between Schoenberg's theorem 11.2 and Rudin's theorem 11.3.

As a final remark, the Schoenberg/Rudin/Vasudeva/Herz results are reminiscent of an earlier, famous result, by Loewner in the parallel (and well studied) setting of the matrix functional calculus. Namely, given a complex Hermitian matrix A with eigenvalues in a real interval (a, b), a function $f : (a, b) \to \mathbb{R}$ acts on A as follows: let $A = UDU^*$ be a spectral decomposition of A; then $f(A) := Uf(D)U^*$, where $f(D)$ is the diagonal matrix with diagonal entries $f(d_{jj})$. Now, Loewner showed in 1934 [158] even before Schoenberg:

Theorem 11.8 (Loewner)   *Let* $-\infty \leq a < b \leq \infty$, *and* $f : (a, b) \to \mathbb{R}$. *The following are equivalent:*

(1) f *is matrix monotone: if* $A \leq B$ *are square matrices with eigenvalues in* (a, b), *then* $f(A) \leq f(B)$.

(2) f is C^1 on (a,b), and given $a < x_1 < \cdots < x_k < b$ for any $k \geq 1$, the Loewner matrix given by $L_f(x_j, x_k) := \frac{f(x_j) - f(x_k)}{x_j - x_k}$ if $j \neq k$, else $f'(x_j)$, is positive semidefinite.

(3) There exist real constants $p \geq 0$ and q, and a finite measure μ on $\mathbb{R} \setminus (a,b)$, such that

$$f(x) = q + px + \int_{\mathbb{R} \setminus (a,b)} \frac{1 + xy}{y - x} \, d\mu(y).$$

(4) There exists a function $\tilde{f}$ that is analytic on $(\mathbb{C} \setminus \mathbb{R}) \sqcup (a,b)$, such that (a) $f \equiv \tilde{f}|_{(a,b)}$ and (b) if $\Im z > 0$ then $\Im f(z) > 0$.

Notice similar to the preceding results, the emergence of analyticity from the dimension-free preservation of a matrix property. (In fact, one shows that Loewner monotone functions on $n \times n$ matrices are automatically C^{2n-3}.) This is also the case with a prior result of Rudin with Helson, Kahane, and Katznelson in 1959 [114], which directly motivated Rudin's 1959 paper discussed above. (The bibliographic notes at the end of this part provide a few more details.)

11.3 Origins: Menger, Fréchet, Schoenberg, and Metric Geometry

In this section and the next two (Sections 11.4 and 11.5), we study some of the historical origins of positive (semi)definite matrices. This class of matrices of course arises as Hessians of twice-differentiable functions at local minima; however, the branch of early twentieth-century mathematics that led to the development of positivity preservers is *metric geometry*. More precisely, the notion of a metric space – emerging from the works of Fréchet and Hausdorff – and isometric embeddings of such structures into Euclidean and Hilbert spaces, spheres, and hyperbolic and homogeneous spaces, were studied by Schoenberg, Bochner, and von Neumann among others; and it is this work that led to the study of matrix positivity and its preservation.

Definition 11.9   A *metric space* is a set X together with a metric $d \colon X \times X \to \mathbb{R}$, satisfying:

(1) *Positivity:* $d(x, y) \geq 0$ for all $x, y \in X$, with equality if and only if $x = y$.
(2) *Symmetry:* $d(x, y) = d(y, x)$ for all $x, y \in X$.
(3) *Triangle inequality:* $d(x, z) \leq d(x, y) + d(y, z)$ for all $x, y, z \in X$.

In this chapter, we will state and prove three results by Schoenberg, which explain his motivations in studying positivity and its preservers and serve to illustrate the (by now well explored) connection between metric geometry and matrix positivity. We begin with a sample result on metric embeddings, shown by Fréchet in 1910 [89]: *If (X, d) is a metric space with $|X| = n + 1$, then (X, d) isometrically embeds into* $(\mathbb{R}^n, \|\cdot\|_\infty)$.

Such results led to exploring which metric spaces isometrically embed into Euclidean spaces. Specifically, in Menger's 1931 paper [171] and Fréchet's 1935 paper [90] the authors explored the following question: *Given integers $n, r \geq 1$, characterize the tuples of $\binom{n+1}{2}$ positive real numbers that can denote the distances between the vertices of an $(n + 1)$-simplex in $\mathbb{R}^r$ but not in $\mathbb{R}^{r-1}$.* In other words, given a finite metric space X, what is the smallest r, if any, such that X isometrically embeds into $\mathbb{R}^r$?

In his 1935 paper [202] Schoenberg gave an alternate characterization of all such "admissible" tuples of distances. This characterization used ... matrix positivity!

Theorem 11.10 (Schoenberg, 1935, [202])   *Fix integers $n, r \geq 1$ and a finite set $X = \{x_0, \ldots, x_n\}$ together with a metric d on X. Then (X, d) isometrically embeds into some $\mathbb{R}^r$ (with the Euclidean distance/norm), if and only if the $n \times n$ matrix*

$$A := \left(d(x_0, x_j)^2 + d(x_0, x_k)^2 - d(x_j, x_k)^2 \right)_{j,k=1}^n \tag{11.1}$$

is positive semidefinite. Moreover, the smallest such r is precisely the rank of the matrix A.

This classical theorem is at the heart of multidimensional scaling; see, e.g., [64]. Additionally, the matrix A features later in this text when we study a result of Menger in Chapter 22; it is an alternate form of the *Cayley–Menger matrix* associated to the metric space X. In Chapter 22, we will also extend this theorem to embeddings of separable metric spaces.

Proof   If (X, d) isometrically embeds into $(\mathbb{R}^r, \|\cdot\| = \|\cdot\|_2)$, then

$$d(x_0, x_j)^2 + d(x_0, x_k)^2 - d(x_j, x_k)^2$$
$$= \|x_0 - x_j\|^2 + \|x_0 - x_k\|^2 - \|(x_0 - x_j) - (x_0 - x_k)\|^2 \tag{11.2}$$
$$= 2\langle x_0 - x_j, x_0 - x_k \rangle.$$

But then the matrix A in (11.1) is the Gram matrix of a set of vectors in $\mathbb{R}^r$, and hence, is positive semidefinite. In the rest of this chapter, we use Theorem 1.5 and Proposition 1.13 (and their proofs) without further reference. Thus, $A = B^T B$, where the columns of B are $x_0 - x_j \in \mathbb{R}^r$. But then A has rank at most the rank of B, hence at most r. Since (X, d) does not embed in $\mathbb{R}^{r-1}$, by the same argument A has rank precisely r.

Conversely, suppose the matrix A in (11.1) is positive semidefinite of rank r. First, consider the case when $r = n$, i.e., A is positive definite. By Theorem 1.5, $\frac{1}{2}A = B^T B$ for a square invertible matrix B. Thus, left-multiplication by B sends the r-simplex with vertices $\mathbf{0}, \mathbf{e}_1, \ldots, \mathbf{e}_r$ to an r-simplex, where $\mathbf{e}_j$ comprise the standard basis of $\mathbb{R}^r$.

Now claim that the assignment $x_0 \mapsto \mathbf{0}$, $x_j \mapsto B\mathbf{e}_j$ for $1 \leq j \leq r$, is an isometry: $X \to \mathbb{R}^r$ whose image, being the vertex set of an r-simplex, necessarily cannot embed inside $\mathbb{R}^{r-1}$. Indeed, computing these distances proves the claim, and hence the theorem for $r = n$:

$$d(B\mathbf{e}_j, \mathbf{0})^2 = \|B^T \mathbf{e}_j\|^2 = \frac{1}{2}\mathbf{e}_j^T A \mathbf{e}_j = \frac{a_{jj}}{2} = d(x_0, x_j)^2,$$

$$d(B\mathbf{e}_j, B\mathbf{e}_k)^2 = \|B\mathbf{e}_j - B\mathbf{e}_k\|^2 = \frac{a_{jj} + a_{kk}}{2} - a_{jk}$$

$$= d(x_0, x_j)^2 + d(x_0, x_k)^2$$

$$- \left(d(x_0, x_j)^2 + d(x_0, x_k)^2 - d(x_j, x_k)^2\right)$$

$$= d(x_j, x_k)^2.$$

Next, suppose $r < n$. Then $\frac{1}{2}A = B^T P B$ for some invertible matrix B, where P is the projection operator $\begin{pmatrix} \mathrm{Id}_{r \times r} & 0 \\ 0 & 0_{(n-r)\times(n-r)} \end{pmatrix}$. Let $\Delta := \{P B\mathbf{e}_j,$ $1 \leq j \leq n\} \sqcup \{\mathbf{0}\}$ denote the projection under P of the vertices of an $(n+1)$ simplex. Repeating the above proof shows that the map : $x_0 \mapsto \mathbf{0}$, $x_j \mapsto P B\mathbf{e}_j$ for $1 \leq j \leq n$, is an isometry from X onto Δ. By construction, Δ lies in the image of the projection P, hence in a copy of $\mathbb{R}^r$. But being the image under P of the vertex set of an n-simplex, Δ cannot lie in a copy of $\mathbb{R}^{r-1}$ (otherwise so would its span, which is all of $P(\mathbb{R}^n) \cong \mathbb{R}^r$). □

We end this part with an observation. A (real symmetric square) matrix $A'_{(n+1)\times(n+1)}$ is said to be *conditionally positive semidefinite* if $(\mathbf{u}')^T A' \mathbf{u}' \geq 0$ whenever $\sum_{j=0}^n u'_j = 0$. Such matrices are also studied in the literature (though not as much as positive semidefinite matrices). The following lemma reformulates Theorem 11.10 into the conditional positivity of a related matrix:

Lemma 11.11 *Let $X = \{x_0, \ldots, x_n\}$ be a finite set equipped with a metric d. Then the matrix $A_{n \times n}$ as in (11.1) is positive semidefinite if and only if the $(n+1) \times (n+1)$ matrix*

$$A' := (-d(x_j, x_k)^2)_{j,k=0}^n \tag{11.3}$$

is conditionally positive semidefinite.

In particular, Schoenberg's papers in the 1930s feature positive semidefinite matrices (Theorem 11.10) as well as conditionally positive semidefinite
matrices (Theorem 11.14). Certainly, the former class of matrices were a
popular and recurrent theme in the analysis literature, with contributions from
Carathéodory, Hausdorff, Hermite, Nevanlinna, Pick, Schur, and many others.

Proof of Lemma 11.11   Let $u_1, \ldots, u_n \in \mathbb{R}$ be arbitrary, and set $u_0 :=$
$-(u_1 + \cdots + u_n)$. Defining $\mathbf{u} := (u_1, \ldots, u_n)^T$ and $\mathbf{u}' := (u_0, \ldots, u_n)^T$,
we compute using that the diagonal entries of A' are zero

$$
(\mathbf{u}')^T A' \mathbf{u}' = \sum_{k=1}^{n} \left(\sum_{j=1}^{n} u_j \right) d(x_0, x_k)^2 u_k
$$

$$
+ \sum_{j=1}^{n} u_j d(x_j, x_0)^2 \left(\sum_{k=1}^{n} u_k \right) - \sum_{j,k=1}^{n} u_j d(x_j, x_k)^2 u_k
$$

$$
= \sum_{j,k=1}^{n} u_j u_k \left(d(x_0, x_k)^2 + d(x_j, x_0)^2 - d(x_j, x_k)^2 \right) = \mathbf{u}^T A \mathbf{u},
$$

for all $\mathbf{u} \in \mathbb{R}^n$. This proves the result.   □

11.4 Positivity Preservers: Schoenberg and Positive Definite Functions

We continue with our historical journey, this time into the origins of the entrywise calculus on positive matrices. As Theorem 11.10 and Lemma 11.11 show,
applying entrywise the function $-x^2$ to any distance matrix $(d(x_j, x_k))_{j,k=0}^{n}$
from Euclidean space yields a conditionally positive semidefinite matrix A'.

It is natural to want to remove the word "conditionally" from the above
result. Namely: which entrywise maps send distance matrices to positive
semidefinite matrices? These are precisely the positive definite functions:

Definition 11.12   Given a metric space (X, d), a function $f : [0, \infty) \to \mathbb{R}$
is *positive definite on X* if for any finite set of points $x_1, x_2, \ldots, x_n \in X$, the
matrix $f[(d(x_j, x_k))_{j,k=1}^{n}]$ is psd positive semidefinite.

By considering 2×2 distance matrices, note that positive definite functions
are not the same as positivity preservers; no distance matrix is positive
semidefinite unless all x_j are equal (in which case we get the zero matrix). On
a different note, given any metric space (X, d), the positive definite functions
on X form a closed convex cone, by Lemma 2.1.

In arriving at Theorem 11.2, Schoenberg was motivated by metric geometry – as we just studied – as well as the study of positive definite functions. The latter was also of interest to other mathematicians in that era: Bochner, Pólya, and von Neumann, to name a few. In fact, positive definite functions are what led to Schoenberg's Theorem 11.2 and the development of entrywise calculus. Note that Bochner – and previously Carathéodory, Herglotz, Mathias, and others – studied functions on groups G that were positive definite in the "more standard" sense – namely, where in the above definition $f : G \to \mathbb{C}$, and one substitutes $d(x_j, x_k)$ by $x_j^{-1} x_k$. The above definition seems due to Schoenberg, in his 1938 paper [205].

We now present – from this paper – another characterization by Schoenberg of metric embeddings into a Euclidean space $\mathbb{R}^r$, this time via positive definite functions. This requires a preliminary observation involving the positive definiteness of an even kernel:

Lemma 11.13   *Given $\sigma > 0$, the Gaussian kernel $T_{G_\sigma}(x, y) := \exp(-\sigma \|$ $x - y \|^2)$ – in other words, the function $\exp(-\sigma x^2)$ – is positive definite on $\mathbb{R}^r$ for all $r \geq 1$.*

Proof   Observe that the case of $\mathbb{R}^r$ for general r follows from the $r = 1$ case, via the Schur product theorem. In turn, the $r = 1$ case is a consequence of Pólya's lemma 4.9.   □

The following result of Schoenberg relates metric space embeddings with this positive definiteness of the Gaussian kernel:

Theorem 11.14 (Schoenberg, 1938 [205])   *A finite metric space (X, d) with $X = \{x_0, \dots, x_n\}$ embeds isometrically into $\mathbb{R}^r$ for some $r > 0$ (which turns out to be at most n), if and only if the $(n + 1) \times (n + 1)$ matrix with (j, k) entry*

$$\exp\left(-\sigma^2 d(x_j, x_k)^2\right), \qquad 0 \leq j, k \leq n$$

is positive semidefinite, along any sequence of nonzero scalars σ_m decreasing to 0^+ (equivalently, for all $\sigma \in \mathbb{R}$).

For another application of this result and those in Section 11.3, see Section 22.3.

Proof   Clearly, if (X, d) embeds isometrically into $\mathbb{R}^r$, then identifying the x_j with their images in $\mathbb{R}^r$, it follows by Lemma 11.13 that the matrix with (j, k) entry

$$\exp(-\sigma^2 \|x_j - x_k\|^2) = T_{G_1}(\sigma x_j, \sigma x_k)$$

is positive semidefinite for all $\sigma \in \mathbb{R}$.

Conversely, let $\sigma_m \downarrow 0^+$. From the positivity of the exponentiated distance matrices for σ_m, it follows for any vector $u := (u_0, \ldots, u_n)^T \in \mathbb{R}^{n+1}$ that

$$\sum_{j,k=0}^{n} u_j u_k \exp\left(-\sigma_m^2 d(x_j, x_k)^2\right) \geq 0.$$

Expanding into Taylor series and interchanging the infinite sum with the two finite sums,

$$\sum_{l=0}^{\infty} \frac{(-\sigma_m^2)^l}{l!} \sum_{j,k=0}^{n} u_j u_k d(x_j, x_k)^{2l} \geq 0, \qquad \forall m \geq 1.$$

Suppose we restrict to the vectors $\mathbf{u}'$ satisfying: $\sum_{j=0}^{n} u_j = 0$. Then the $l = 0$ term vanishes. Now dividing throughout by σ_m^2 and taking $m \to \infty$, the "leading term" in σ_m must be nonnegative. It follows that if $A' := \left(-d(x_j, x_k)^2\right)_{j,k=0}^{n}$, then $(\mathbf{u}')^T A' \mathbf{u}' \geq 0$ whenever $\sum_j u_j = 0$. By Lemma 11.11 and Theorem 11.10, (X, d) embeds isometrically into $\mathbb{R}^r$, where $r \leq n$ denotes the rank of the matrix $A_{n \times n}$ as in (11.1). □

11.5 From Spheres to Correlation Matrices to Positivity Preservers

The previous result, Theorem 11.14, says that Euclidean spaces $\mathbb{R}^r$ – or their direct limit / union $\mathbb{R}^\infty$ (which should more accurately be denoted $\mathbb{R}^{\mathbb{N}}$), or even its completion ℓ^2 of square-summable real sequences (which Schoenberg and others called *Hilbert space*) – can be characterized by the property that the maps

$$\exp(-\sigma^2 x^2), \qquad \sigma \in (0, \rho) \tag{11.4}$$

are all positive definite on each (finite) metric subspace. As we saw, such a characterization holds for each $\rho > 0$.[1]

Given this characterization, it is natural to seek out similar characterizations of distinguished submanifolds M in $\mathbb{R}^r$ or $\mathbb{R}^\infty$ or ℓ^2. In fact, in the aforementioned 1935 paper [202], Schoenberg showed the first such classification result, for $M = S^{r-1}$ a unit sphere – as well as for the *Hilbert sphere* S^∞. Note here that the unit sphere $S^{r-1} := \{x \in \mathbb{R}^r : \|x\|^2 = 1\}$, while the Hilbert sphere

---

[1] A related result on positive definite functions on – or Hilbert space embeddings of – a *topological* space X is by Kolmogorov around 1940 [151]. He showed that a continuous function $K: X \times X \to \mathbb{C}$ is positive definite if and only if there exists a Hilbert space $\mathcal{H}$ and a norm-continuous map $\varphi: X \to \mathcal{H}$, such that $K(x_1, x_2) = \langle \varphi(x_1), \varphi(x_2) \rangle$ for all $x_1, x_2 \in X$.

$S^\infty \subset \ell^2$ is the subset of all square-summable sequences with unit ℓ^2-norm. (This is the *closure* of the set of all real sequences with finitely many nonzero coordinates and unit ℓ^2-norm – which is the unit sphere $\bigcup_{r \geq 1} S^{r-1}$ in $\bigcup_{r \geq 1} \mathbb{R}^r$.)

One defines a rotationally invariant metric on S^∞ (hence on each S^{r-1}) as follows. The distance between x and $-x$ is π, and given points $x \neq \pm y$ in S^∞, there exists a unique plane passing through x, y, and the origin. This plane intersects the sphere S^∞ in a unit circle S^1 given by

$$\left\{ \alpha x + \beta y : \alpha, \beta \in \mathbb{R}, \ 1 = \|\alpha x + \beta y\|^2 = \alpha^2 + \beta^2 + 2\alpha\beta\langle x, y \rangle \right\} \subset S^\infty,$$

and we let $d(x, y)$ denote the angle – i.e., arclength – between x and y

$$d(x, y) := \sphericalangle(x, y) = \arccos(\langle x, y \rangle) \in [0, \pi].$$

Now we come to Schoenberg's characterization for metric embeddings into Euclidean spheres. He showed that in contrast to the family (11.4) of positive definite functions for Euclidean spaces, for spheres it suffices to consider *a single function*! This function is the cosine:

Proposition 11.15 (Schoenberg, 1935, [202])   *Let (X, d) be a finite metric space with $X = \{x_1, \ldots, x_n\}$. Fix an integer $r \geq 2$. Then X isometrically embeds into S^{r-1} but not S^{r-2}, if and only if $d(x_j, x_k) \leq \pi$ for all $1 \leq j, k \leq n$ and the matrix $(\cos d(x_j, x_k))_{j,k=1}^n$ is positive semidefinite of rank r.*

In particular, X embeds isometrically into the Hilbert sphere S^∞ – with the spherical metric – if and only if (a) $\mathrm{diam}(X) \leq \pi$ and (b) $\cos(\cdot)$ is positive definite on X.

Thus, matrix positivity is also intimately connected with spherical embeddings, which may not be surprising given Theorem 11.10.

Proof   If there exists an isometric embedding $\varphi \colon X \hookrightarrow S^{r-1}$ as claimed, we have as above

$$\cos d(x_j, x_k) = \cos \sphericalangle(\varphi(x_j), \varphi(x_k)) = \langle \varphi(x_j), \varphi(x_k) \rangle,$$

which yields a Gram matrix of rank at most r, hence exactly r (since X does not embed isometrically into S^{r-2}). Moreover, the spherical distance between x_j, x_k (for $j, k > 0$) is at most π, as desired.

Conversely, since $A := (\cos d(x_j, x_k))_{j,k=1}^n$ is positive, it is a Gram matrix (of rank r), and hence $A = B^T B$ for some $B_{r \times n}$ of rank r by Theorem 1.5. Let $y_j \in \mathbb{R}^r$ denote the columns of B; then clearly $y_j \in S^{r-1}\ \forall j$; moreover,

$$\cos \sphericalangle(y_j, y_k) = \langle y_j, y_k \rangle = a_{jk} = \cos d(x_j, x_k), \qquad \forall 1 \leq j, k \leq n.$$

Since $d(x_j, x_k)$ lies in $[0, \pi]$ by assumption, as does $\sphericalangle(y_j, y_k)$, we obtain an isometry $\varphi \colon X \to S^{r-1}$, sending $x_j \mapsto y_j$ for all $j > 0$. Finally, $\mathrm{im}(\varphi)$ is not contained in S^{r-2}, for otherwise A would have rank at most $r - 1$. This shows the result for S^{r-1}; the case of S^∞ is similar.   □

The proof of Proposition 11.15 shows that $\cos(\cdot)$ is a positive definite function on unit spheres of all dimensions.

Note that Proposition 11.15 and the preceding two theorems by Schoenberg in the 1930s:

(1) characterize metric space embeddings into Euclidean spaces via matrix positivity;
(2) characterize metric space embeddings into Euclidean spaces via the positive definite functions $\exp(-\sigma^2 (\cdot)^2)$ on $\mathbb{R}^r$ or $\mathbb{R}^\infty$ (so this involves positive matrices); and
(3) characterize metric space embeddings into Euclidean spheres S^{r-1} or S^∞ (with the spherical metric) via the positive definite function $\cos(\cdot)$ on S^∞.

Around the same time (in the 1930s), S. Bochner [45, 46] had classified all of the positive definite functions on $\mathbb{R}$. This result was extended in 1940 simultaneously by Weil, Povzner, and Raikov to classify the positive definite functions on any locally compact abelian group. Amid this backdrop, in his 1942 paper [206] Schoenberg was interested in understanding the positive definite functions of the form $f \circ \cos \colon [-1, 1] \to \mathbb{R}$ on a unit sphere $S^{r-1} \subset \mathbb{R}^r$, where $r \geq 2$.

To present Schoenberg's result, first consider the $r = 2$ case. As mentioned above, distance (i.e., angle) matrices are not positive semidefinite; but if one applies the cosine function entrywise, then we obtain the matrix with (j, k) entry $\cos(\theta_j - \theta_k)$, and this is positive semidefinite by Lemma 1.15. But now $f[-]$ preserves positivity on a set of Toeplitz matrices (among others), by Lemma 1.15 and the subsequent discussion. For general dimension $r \geq 2$, we have $\cos(d(x_j, x_k)) = \langle x_j, x_k \rangle$ (see also the proof of Proposition 11.15), so $\cos[(d(x_j, x_k))_{j,k}]$ always yields Gram matrices. Hence, $f[-]$ would once again preserve positivity on a set of positive matrices. It was this class of functions that Schoenberg characterized:

Theorem 11.16 (Schoenberg, 1942, [206])   *Suppose* $f \colon [-1, 1] \to \mathbb{R}$ *is continuous, and* $r \geq 2$ *is an integer. Then the following are equivalent:*

(1) $(f \circ \cos)$ *is positive definite on* S^{r-1}.

(2) *The function* $f(x) = \sum_{k=0}^{\infty} c_k C_k^{\left(\frac{r-2}{2}\right)}(x)$, *where* $c_k \geq 0, \forall k$, *and* $\left\{ C_k^{\left(\frac{r-2}{2}\right)}(x) : k \geq 0 \right\}$ *comprise the first Chebyshev or Gegenbauer family of orthogonal polynomials.*

Remark 11.17 Theorem 11.16 has an interesting reformulation in terms of entrywise positivity preservers on correlation matrices. Recall that on the unit sphere S^{r-1}, applying $\cos[-]$ entrywise to a distance matrix of points x_j yields precisely the Gram matrix with entries $\langle x_j, x_k \rangle$, which is positive of rank at most r. Moreover, as the vectors x_j lie on the unit sphere, the diagonal entries are all 1 and hence we obtain a correlation matrix. Putting these facts together, $f \circ \cos$ is positive definite on S^{r-1} if and only if $f[-]$ preserves positivity on all correlation matrices of arbitrary size but rank at most r. Thus, Schoenberg's works in 1935 [202] and 1942 [206] already contained connections to entrywise preservers of correlation matrices, which brings us around to the modern-day motivations that arise from precisely this question (now arising in high-dimensional covariance estimation, and discussed in Section 7.1).

Remark 11.18 Schoenberg's work has been followed by numerous papers attempting to understand positive definite functions on locally compact groups, spheres, two-point homogeneous metric spaces, and products of these. See, for example, [13, 14, 15, 31, 32, 47, 56, 79, 103, 167, 168, 240, 241, 245] for a selection of works. The connection to spheres has also led to work in statistics on spatio-temporal covariance functions on spheres, modeling the earth as a sphere [99, 187, 237]. (Note that a metric space X is n-point homogeneous [235] if for all $1 \leq p \leq n$ and subsets $X_1, X_2 \subset X$ of size p, every isometry: $X_1 \to X_2$ extends to a self-isometry of X. This was first studied by Birkhoff [39] and differs from the more widespread usage for spaces G/H. We will study this further in Chapter 22.)

Remark 11.19 Schoenberg's (and subsequent) work on finite- and infinite-dimensional spheres has many other applications. One area of recent activity involves sphere packing, spherical codes, and configurations of points on spheres that maximize the minimal distance or some potential function. See, for example, the work of Cohn with coauthors in [59, 60, 61, 62]; and Musin in 2008 [179].

Returning to the above discussion on Theorem 11.16, if instead we let $r = \infty$, then the corresponding result would classify positivity preservers on all correlation matrices (without rank constraints) by Remark 11.17. And indeed, Schoenberg achieves this goal in the same paper:

Theorem 11.20 (Schoenberg, 1942, [205])   *Suppose* $f : [-1, 1] \to \mathbb{R}$ *is continuous. Then* $f \circ \cos$ *is positive definite on* S^∞ *if and only if there exist scalars* $c_k \geq 0$, *such that*

$$f(\cos \theta) = \sum_{k \geq 0} c_k \cos^k \theta, \qquad \theta \in [0, \pi].$$

Notice here that $\cos^k \theta$ is positive definite on S^∞ for all integers $k \geq 0$, by Proposition 11.15 and the Schur product theorem. Hence, so is $\sum_{k \geq 0} c_k \cos^k \theta$ if all $c_k \geq 0$.

Freed from the sphere context, the preceding theorem says that a continuous function $f : [-1, 1] \to \mathbb{R}$ preserves positivity when applied entrywise to all correlation matrices, if and only if $f(x) = \sum_{k \geq 0} c_k x^k$ on $[-1, 1]$ with all $c_k \geq 0$. This finally explains how and why Schoenberg arrived at his celebrated converse to the Schur product theorem – namely, Theorem 11.2 on entrywise positivity preservers.

11.6 Digression on Ultraspherical Polynomials

Before proceeding further, we describe the orthogonal polynomials $C_k^{(\alpha)}(x)$ for $k \geq 0$, where $\alpha = \alpha(r) = (r - 2)/2$. Given $r \geq 2$, note that $\alpha(r)$ ranges over the nonnegative half-integers. Though not used below, here are several different (equivalent) definitions of the polynomials $C_k^{(\alpha)}$ for general real $\alpha \geq 0$.

First, if $\alpha = 0$, then $C_k^{(0)}(x) := T_k(x)$, the Chebyshev polynomials of the first kind

$$T_0(x) = 1, \quad T_1(x) = x, \quad T_2(x) = 2x^2 - 1, \quad \ldots, \quad T_k(\cos(\theta)) = \cos(k\theta)$$

for all $k \geq 0$. A second way to compute the polynomials $C_k^{(0)}(x)$ is through their *generating function*

$$\frac{1 - xt}{1 - 2xt + t^2} = \sum_{k=0}^{\infty} C_k^{(0)}(x) t^k.$$

For higher α: setting $\alpha = \frac{1}{2}$ yields the family of Legendre polynomials. If $\alpha = 1$, we obtain the Chebyshev polynomials of the second kind. For general $\alpha > 0$, the functions $(C_k^{(\alpha)}(x))_{k \geq 0}$ are the *Gegenbauer/ultraspherical polynomials*, defined via their generating function

$$(1 - 2xt + t^2)^{-\alpha} = \sum_{k=0}^{\infty} C_k^{(\alpha)}(x) t^k.$$

For all $\alpha \geq 0$, the polynomials $(C_k^{(\alpha)}(x))_{k \geq 0}$ form a complete orthogonal set in the Hilbert space $L^2([-1,1], w_\alpha)$, where w_α is the weight function

$$w_\alpha(x) := (1 - x^2)^{\alpha - \frac{1}{2}}, \qquad x \in (-1, 1).$$

Thus, another definition of $C_k^{(\alpha)}(x)$ is that it is a polynomial of degree k, with $C_0^{(\alpha)}(x) = 1$, and such that the $C_k^{(\alpha)}$ are orthogonal with respect to the bilinear form

$$\langle f, g \rangle := \int_{-1}^{1} f(x) g(x) w_\alpha(x)\, dx, \qquad f, g \in L^2([-1, 1], w_\alpha),$$

and satisfy

$$\langle C_k^{(\alpha)}, C_k^{(\alpha)} \rangle = \frac{\pi 2^{1-2\alpha} \Gamma(k + 2\alpha)}{k!\,(k + \alpha)(\Gamma(\alpha))^2}.$$

Yet another definition is that the Gegenbauer polynomials $C_k^{(\alpha)}(x)$ for $\alpha > 0$ satisfy the differential equation

$$(1 - x^2) y'' - (2\alpha + 1) x y' + k(k + 2\alpha) y = 0.$$

We also have a direct formula

$$C_k^{(\alpha)}(x) := \sum_{j=0}^{\lfloor k/2 \rfloor} (-1)^j \frac{\Gamma(k - j + \alpha)}{\Gamma(\alpha) j!\,(k - 2j)!} (2x)^{k-2j},$$

as well as a recursion

$$C_0^{(\alpha)}(x) := 1, \qquad C_1^{(\alpha)}(x) := 2\alpha x,$$

$$C_k^{(\alpha)}(x) := \frac{1}{k} \left(2x(k + \alpha - 1) C_{k-1}^{(\alpha)}(x) - (k + 2\alpha - 2) C_{k-2}^{(\alpha)}(x) \right) \quad \forall k \geq 2.$$

11.7 Sketch of Proof of Theorem 11.16

Schoenberg's theorem 11.16 has subsequently been studied by many authors, and in a variety of settings over the years. This includes classifying the positive definite functions on different kinds of spaces: locally compact groups, spheres, and products of these. We next give a proof-sketch of this result. In what follows, we use without reference the observation (akin to Lemma 2.1) that the set of functions f, such that $f \circ \cos$ is positive definite on S^{r-1}, also forms a closed convex cone, which is, moreover, closed under entrywise products.

We first outline why (2) $\implies$ (1) in Theorem 11.16. By the above observation, it suffices to show that $C_k^{(\alpha)} \circ \cos$ is positive definite on S^{r-1}. The proof is by induction on r. For the base case $r = 2$, let $\theta_1, \theta_2, \ldots, \theta_n \in S^1 = [0, 2\pi)$. Up to the sign, their distance matrix has (i, j) entry $d(\theta_i, \theta_j) = \theta_i - \theta_j$ (or a suitable translate modulo 2π). Now, by Lemma 1.15, the matrix $(\cos(k(\theta_i - \theta_j)))_{i,j=1}^n$ is positive semidefinite. But this is precisely the matrix obtained by applying $C_k^{(0)} \circ \cos$ to the distance matrix above. This proves one implication for $d = 2$. The induction step (for general $r \geq 2$) follows from addition formulas for $C_k^{(\alpha)}$.

For the converse implication, set $\alpha := \frac{r-2}{2}$ and note that $f \in L^2([-1, 1], w_\alpha)$. Hence, f has a series expansion $\sum_{k=0}^\infty c_k C_k^{(\alpha)}(x)$, with $c_k \in \mathbb{R}$. Now recover the c_k via

$$c_k = \int_{-1}^1 f(x) C_k^{(\alpha)}(x) w_\alpha(x)\, dx,$$

since the $C_k^{(\alpha)}$ form an orthonormal family. Note that $C_k^{(\alpha)}$ and f are both positive definite (upon precomposing with the cosine function), hence so is their product by the Schur product theorem. A result of W.H. Young now shows that $c_k \geq 0$ for all $k \geq 0$. $\square$

11.8 Entrywise Preservers in a Fixed Dimension

We conclude by discussing a natural mathematical refinement of Schoenberg's theorem:

*"Which functions entrywise preserve positivity in a **fixed** dimension?"*

This turns out to be a challenging, yet important question from the point of view of applications (see Section 7.1 for more on this.) In particular, note that there exist functions which preserve positivity on $\mathbb{P}_n$ but not on $\mathbb{P}_{n+1}$: the power functions x^α with $\alpha \in (n - 3, n - 2)$ for $n \geq 3$, by Theorem 5.3. By Vasudeva's theorem 11.4, it follows that these "noninteger" power functions cannot be absolutely monotonic.

Surprisingly, while Schoenberg's theorem is classical and provides a complete description in the dimension-free case, not much is known about the fixed-dimension case: namely, the classification of functions $f : I \to \mathbb{R}$, such that $f[-] : \mathbb{P}_n(I) \to \mathbb{P}_n(\mathbb{R})$ for a fixed integer $n \geq 1$.

- If $n = 1$, then clearly, any function $f : [0, \infty) \to [0, \infty)$ works.
- For $n = 2$ and $I = (0, \infty)$, these are precisely the functions $f : (0, \infty) \to \mathbb{R}$ that are nonnegative, nondecreasing, and multiplicatively midconvex. This was shown by Vasudeva (see Theorem 6.7), and it implies similar results for $I = [0, \infty)$ and $I = \mathbb{R}$.
- For every integer $n \geq 3$, the question is *open* to date.

Given the scarcity of results in this direction, a promising line of attack has been to study refinements of the problem. These can involve restricting the test set of matrices in fixed dimension (say, under rank or sparsity constraints) or the test set of functions (say, to only the entrywise powers) as was studied in Chapters 5–9; or to use both restrictions. See Section 7.2 for more on this discussion, as well as Part III, where we study *polynomial* preservers in a fixed dimension.

To conclude: while the general problem in fixed dimension $n \geq 3$ is open to date, there is a known result: a necessary condition satisfied by positivity preservers on $\mathbb{P}_n$, shown by R.A. Horn in his 1969 paper [126] and attributed to his advisor, Loewner. The result is more than 50 years old; yet even today, it remains essentially the only known result for general preservers f on $\mathbb{P}_n$. In Chapters 12 and 13, we will state and prove this result – in fact, a stronger version. We will then show (stronger versions of) Vasudeva's and Schoenberg's theorems, via a different approach than the one by Schoenberg, Rudin, or others: we crucially use the fixed-dimension theory via the result of Horn and Loewner.

12

Loewner's Determinant Calculation in Horn's Thesis

As mentioned in Chapter 11, the goal in Part II is to prove a stronger form of Schoenberg's theorem 11.2, in the spirit of Rudin's theorem 11.3 but replacing the word "Toeplitz" by "Hankel." In order to do so, we will first prove a stronger version of Vasudeva's theorem 11.4, in which the test set is once again reduced to only low-rank Hankel matrices.

In turn, our proof of this version of Vasudeva's theorem relies on a fixed-dimension result, alluded to at the end of Chapter 11. Namely, we state and prove a stronger version of a 1969 theorem of Horn [126] (attributed by him to Loewner), in this chapter and Chapter 13.

Theorem 12.1 (Horn–Loewner theorem, stronger version)   *Let $I = (0, \infty)$, and fix $u_0 \in (0, 1)$ and an integer $n \geq 1$. Define $\mathbf{u} := (1, u_0, \dots, u_0^{n-1})^T$. Suppose $f : I \to \mathbb{R}$ is such that $f[-]$ preserves positivity on the set $\{a\mathbf{1}_{n \times n} + b\mathbf{u}\mathbf{u}^T : a, b > 0\}$ as well as on the rank-1 matrices in $\mathbb{P}_2(I)$ and the Toeplitz matrices in $\mathbb{P}_2(I)$. Then*

(1) *$f \in C^{n-3}(I)$ and $f, f', \dots, f^{(n-3)}$ are nonnegative on I. Moreover, $f^{(n-3)}$ is convex and nondecreasing on I.*

(2) *If, moreover, $f \in C^{n-1}(I)$, then $f^{(n-2)}, f^{(n-1)}$ are also nonnegative on I.*

All test matrices here are Hankel of rank ≤ 2 – and are, moreover, totally nonnegative (TN) by Corollary 3.8 since they arise as the truncated moment matrices of the measures $a\delta_1 + b\delta_{u_0}$. This is used later in Part II, to prove stronger versions of Vasudeva's and Schoenberg's theorems (see Chapter 14), with similarly reduced test sets of low-rank Hankel matrices.

Remark 12.2   In the original result by Horn (and Loewner), f was assumed to be continuous and to preserve positivity on all of $\mathbb{P}_N((0, \infty))$. In Theorem 12.1, we have removed the continuity hypothesis, in the spirit of Rudin's work, and also greatly reduced the test set of matrices.

Remark 12.3 We also observe that Theorem 12.1 is the "best possible," in that the number of nonzero derivatives that must be positive is sharp. For example, let $n \geq 2$, $I = (0, \infty)$, and $f : I \to \mathbb{R}$ be given by: $f(x) := x^{\alpha}$, where $\alpha \in (n - 2, n - 1)$. Using Theorem 5.3, $f[-]$ preserves positivity on the test sets $\{a\mathbf{1}_{n \times n} + b\mathbf{u}\mathbf{u}^T : a, b > 0\}$ and $\mathbb{P}_2(I)$ Moreover, $f \in C^{n-1}(I)$ and f, $f', \ldots, f^{(n-1)}$ are strictly positive on I. However, $f^{(n)}$ is negative on I.

This low-rank Hankel example (and more generally, Theorem 5.3) further shows that there exist (power) functions that preserve positivity on $\mathbb{P}_n$ but not on $\mathbb{P}_{n+1}$. In Part III, we will show that there also exist polynomial preservers with the same property.

We now proceed toward the proof of Theorem 12.1 for general functions. A major step is the next calculation, which essentially proves the result for smooth functions. In the sequel, define the *Vandermonde determinant* of a vector $\mathbf{u} = (u_1, \ldots, u_n)^T$ to be 1 if $n = 1$, and

$$V(\mathbf{u}) := \prod_{1 \leq j < k \leq n} (u_k - u_j) = \det \begin{pmatrix} 1 & u_1 & \cdots & u_1^{n-1} \\ 1 & u_2 & \cdots & u_2^{n-1} \\ \vdots & \vdots & \ddots & \vdots \\ 1 & u_n & \cdots & u_n^{n-1} \end{pmatrix}, \quad \text{if } n > 1.$$

$$(12.1)$$

Proposition 12.4 *Fix an integer $n > 0$ and define $N := \binom{n}{2}$. Suppose $a \in \mathbb{R}$ and let a function $f : (a - \epsilon, a + \epsilon) \to \mathbb{R}$ be N-times differentiable for some fixed $\epsilon > 0$. Now fix vectors $\mathbf{u}, \mathbf{v} \in \mathbb{R}^n$, and define $\Delta : (-\epsilon', \epsilon') \to \mathbb{R}$ via $\Delta(t) := \det f[a\mathbf{1}_{n \times n} + t\mathbf{u}\mathbf{v}^T]$ for a sufficiently small $\epsilon' \in (0, \epsilon)$. Then $\Delta(0) = \Delta'(0) = \cdots = \Delta^{(N-1)}(0) = 0$, and*

$$\Delta^{(N)}(0) = \binom{N}{0, 1, \ldots, n - 1} V(\mathbf{u}) V(\mathbf{v}) \prod_{k=0}^{n-1} f^{(k)}(a),$$

where the first factor on the right is the multinomial coefficient.

This computation was originally due to Loewner. While the result seemingly involves (higher) derivatives, it is in fact a completely algebraic phenomenon, valid over any ground ring. For the interested reader, we isolate this phenomenon in Proposition 12.6; its proof is more or less the same as the one provided for Proposition 12.4. To gain some feel for the computations, the reader may wish to work out the $N = 3$ case first.

Proof Let $\mathbf{w}_k$ denote the kth column of $a\mathbf{1}_{n \times n} + t\mathbf{u}\mathbf{v}^T$; thus, $\mathbf{w}_k$ has jth entry $a + tu_j v_k$. To differentiate $\Delta(t)$, we will use the multilinearity of the determinant and the Laplace expansion of $\Delta(t)$ into a linear combination of $n!$ "monomials," each of which is a product of n terms $f(\cdot)$. Using the product rule, taking the derivative yields n terms from each monomial, and we may rearrange all of these terms into n "clusters" of terms (grouping by the column which gets differentiated), and regroup back using the Laplace expansion to obtain

$$\Delta'(t) = \sum_{k=1}^{n} \det(f[\mathbf{w}_1] \mid \cdots \mid f[\mathbf{w}_{k-1}] \mid v_k \mathbf{u} \circ f'[\mathbf{w}_k] \mid f[\mathbf{w}_{k+1}] \mid \cdots \mid f[\mathbf{w}_n]).$$

Now apply the derivative repeatedly, using this principle. Using the chain rule, for $M \geq 0$ the derivative $\Delta^{(M)}(t)$ – evaluated at $t = 0$ – is an integer linear combination of terms of the form

$$\det\left(v_1^{m_1}\mathbf{u}^{\circ m_1} \circ f^{(m_1)}[a\mathbf{1}] \mid \cdots \mid v_n^{m_n}\mathbf{u}^{\circ m_n} \circ f^{(m_n)}[a\mathbf{1}]\right)$$
$$= \det\left(f^{(m_1)}(a)v_1^{m_1}\mathbf{u}^{\circ m_1} \mid \cdots \mid f^{(m_n)}(a)v_n^{m_n}\mathbf{u}^{\circ m_n}\right), \quad m_1 + \cdots + m_n = M,$$
$$(12.2)$$

where $\mathbf{1} = (1, \ldots, 1)^T \in \mathbb{R}^n$ and all $m_j \geq 0$. Notice that if any $m_j = m_k$ for $j \neq k$, then the corresponding determinant (12.2) vanishes. Thus, the lowest degree derivative $\Delta^{(M)}(0)$ whose expansion contains a nonvanishing determinant is when $M = 0 + 1 + \cdots + (n-1) = N$. This proves the first part of the result.

To show the second part, consider $\Delta^{(N)}(0)$. Once again, the only determinant terms that do not vanish in its expansion correspond to applying $0, 1, \ldots, n-1$ derivatives to the columns in some order. We first compute the integer multiplicity of each such determinant, noting by symmetry that these multiplicities are all equal. As we are applying N derivatives to Δ (before evaluating at 0), the derivative applied to get f' in some column can be any of $\binom{N}{1}$; now the two derivatives applied to get f'' in a (different) column can be chosen in $\binom{N-1}{2}$ ways; and so on. Thus, the multiplicity is precisely

$$\binom{N}{1}\binom{N-1}{2}\binom{N-3}{3}\cdots\binom{2n-3}{n-2} = \prod_{k=0}^{n-1}\binom{N - \binom{k}{2}}{k}$$
$$= \frac{N!}{\prod_{k=0}^{n-1} k!} = \binom{N}{0, 1, \ldots, n-1}.$$

We next compute the sum of all determinant terms. Each term corresponds to a unique permutation of the columns $\sigma \in S_n$, with say $\sigma_k - 1$ the order of the derivative applied to the kth column $f[\mathbf{w}_k]$. Using (12.2), the determinant corresponding to σ equals

$$\prod_{k=0}^{n-1} f^{(k)}(a) v_k^{\sigma_k - 1} \cdot (-1)^\sigma \cdot \det \left(\mathbf{u}^{\circ 0} \mid \mathbf{u}^{\circ 1} \mid \cdots \mid \mathbf{u}^{\circ(n-1)} \right)$$

$$= V(\mathbf{u}) \prod_{k=0}^{n-1} f^{(k)}(a) \cdot (-1)^\sigma \prod_{k=0}^{n-1} v_k^{\sigma_k - 1}.$$

Summing this term over all $\sigma \in S_n$ yields precisely

$$V(\mathbf{u}) \prod_{k=0}^{n-1} f^{(k)}(a) \sum_{\sigma \in S_n} (-1)^\sigma \prod_{k=0}^{n-1} v_k^{\sigma_k - 1} = V(\mathbf{u}) \prod_{k=0}^{n-1} f^{(k)}(a) \cdot V(\mathbf{v}).$$

Now, multiply by the (common) integer multiplicity computed above, to finish the proof. □

We next present the promised algebraic formulation of Proposition 12.4. For this, some notation is required. Fix a commutative (unital) ring R and an R-algebra S. The first step is to formalize the notion of the derivative, on a subclass of S-valued functions. This involves more structure than the more common notion of a *derivation*, so we give it a different name.

Definition 12.5   Given a commutative ring R, a commutative R-algebra S (with $R \subset S$), and an R-module X, a *differential calculus* is a pair (A, ∂), where A is an R-subalgebra of functions: $X \to S$ (under pointwise addition and multiplication and R-action) which contains the constant functions, and $\partial \colon A \to A$ satisfies the following properties:

(1) ∂ is R-linear, i.e., $\partial \sum_j r_j f_j = \sum_j r_j \partial f_j$ for all $r_j \in R$, $f_j \in A$ (and all j).
(2) ∂ is a derivation (product rule): $\partial(fg) = f \cdot (\partial g) + (\partial f) \cdot g$ for $f, g \in A$.
(3) ∂ satisfies a variant of the chain rule for composing with linear functions. Namely, if $x' \in X, r \in R$, and $f \in A$, then the function $g \colon X \to S$, $g(x) := f(x' + rx)$ also lies in A, and moreover, $(\partial g)(x) = r \cdot (\partial f)(x' + rx)$.

With this definition at hand, we can now state the desired algebraic generalization of Proposition 12.4; the proof is essentially the same.

Proposition 12.6   *Suppose R, S, and X are as in Definition 12.5, with an associated differential calculus (A, ∂). Now, fix an integer $n > 0$, two vectors $\mathbf{u}$, $\mathbf{v} \in R^n$, a vector $a \in X$, and a function $f \in A$. Define $N \in \mathbb{N}$ and $\Delta \colon X \to R$ via*

$$N := \binom{n}{2}, \qquad \Delta(t) := \det f\left[a \mathbf{1}_{n \times n} + t \mathbf{u} \mathbf{v}^T \right], \quad t \in X.$$

Then $\Delta(0_X) = (\partial \Delta)(0_X) = \cdots = (\partial^{N-1} \Delta)(0_X) = 0_R$, and

$$\left(\partial^N \Delta\right)(0_X) = \begin{pmatrix} N \\ 0, 1, \ldots, n-1 \end{pmatrix} V(\mathbf{u}) V(\mathbf{v}) \prod_{k=0}^{n-1} \left(\partial^k f\right)(a).$$

Notice that the algebra A is supposed to remind the reader of "smooth functions," and is used here for ease of exposition. One can instead work with an appropriate algebraic notion of "N-times differentiable functions" in order to truly generalize Proposition 12.4; we leave the details to the interested reader.

Remark 12.7   Note that Proposition 12.4 is slightly more general than the original argument of Horn and Loewner, which involved the special case $\mathbf{u} = \mathbf{v}$. As the above proof (and Proposition 12.6) shows, the argument is essentially algebraic, hence holds for any $\mathbf{u}, \mathbf{v}$.

Finally, we use Proposition 12.4 to prove the Horn–Loewner theorem 12.1 for smooth functions. The remainder of the proof – for arbitrary functions – will be discussed in Chapter 13.

Proof of Theorem 12.1 for smooth functions   Suppose f is smooth – or more generally, C^N where $N = \binom{n}{2}$. Then the result is shown by induction on n. For $n = 1$ the result says that f is nonnegative if it preserves positivity on the given test set, which is obvious. For the induction step, we know that f, $f', \ldots, f^{(n-2)} \geq 0$ on I, since the given test set of $(n-1) \times (n-1)$ matrices can be embedded into the test set of $n \times n$ matrices. (Here we do not use the test matrices in $\mathbb{P}_2(I)$.) Now define $f_\epsilon(x) := f(x) + \epsilon x^n$ for each $\epsilon > 0$, and note by the Schur product theorem 2.10 (or Lemma 11.1) that f_ϵ also satisfies the hypotheses.

Given $a, t > 0$ and the vector $\mathbf{u} = \left(1, u_0, \ldots, u_0^{n-1}\right)^T$ as in the theorem, define

$$\Delta(t) := \det f_\epsilon\left[a\mathbf{1}_{n \times n} + t\mathbf{u}\mathbf{u}^T\right]$$

as in Proposition 12.4, but replacing $f, \mathbf{v}$ by $f_\epsilon, \mathbf{u}$ respectively. Then $\Delta(t) \geq 0$ for $t > 0$ by assumption, so

$$0 \leq \lim_{t \to 0^+} \frac{\Delta(t)}{t^N}, \qquad \text{where } N = \binom{n}{2}.$$

On the other hand, by Proposition 12.4 and applying L'Hôpital's rule,

$$\lim_{t \to 0^+} \frac{\Delta(t)}{t^N} = \frac{\Delta^{(N)}(0)}{N!} = \frac{1}{N!} \begin{pmatrix} N \\ 0, 1, \ldots, n-1 \end{pmatrix} V(\mathbf{u})^2 \prod_{k=0}^{n-1} f_\epsilon^{(k)}(a)$$

$$= V(\mathbf{u})^2 \prod_{k=0}^{n-1} \frac{f_\epsilon^{(k)}(a)}{k!}.$$

Thus, the right-hand side here is nonnegative. Since $\mathbf{u}$ has distinct coordinates, we can cancel all positive factors to conclude that

$$\prod_{k=0}^{n-1} f_\epsilon^{(k)}(a) \geq 0, \quad \forall \epsilon, a > 0.$$

But $f_\epsilon^{(k)}(a) = f^{(k)}(a) + \epsilon n(n-1) \cdots (n-k+1)a^{n-k}$, and this is positive for $k = 0, \ldots, n-2$ by the induction hypothesis. Hence,

$$f_\epsilon^{(n-1)}(a) = f^{(n-1)}(a) + \epsilon\, n! \geq 0, \quad \forall \epsilon, a > 0.$$

It follows that $f^{(n-1)}(a) \geq 0$, hence $f^{(n-1)}$ is nonnegative on $(0, \infty)$, as desired. $\square$

We conclude this line of proof by mentioning that the Horn–Loewner theorem, as well as Proposition 12.4 and its algebraic avatar in Proposition 12.6 afford generalizations; the latter results reveal a surprising and novel application to Schur polynomials and to symmetric function identities. For more details, the reader is referred to Question 23.10, and to the recent paper by Khare [142].

The final remark is that there is a *different*, simpler proof of Theorem 12.1 for smooth functions, essentially by Vasudeva (1979, [230]) and along the lines of FitzGerald–Horn's 1977 argument ([84] and see also the proof of Theorem 5.3). Vasudeva's proof is direct, so does not lead to the connections to Schur polynomials mentioned in the preceding paragraph.

Simpler proof of Theorem 12.1 for smooth functions   This proof in fact works for $f \in C^{n-1}(I)$. Akin to the previous proof, this argument also works more generally than for $\mathbf{u} = (1, u_0, \ldots, u_0^{n-1})^T$: choose arbitrary distinct real scalars $v_1, \ldots, v_n$ and write $v := (v_1, \ldots, v_n)^T$. Then for $a > 0$ and small $t > 0$, $a\mathbf{1}_{n \times n} + tvv^T \in \mathbb{P}_n(I)$. Now, given $0 \leq m \leq n-1$, choose a vector $u \in R^n$ which is orthogonal to the vectors $\mathbf{1}, v, v^{\circ 2}, v^{\circ(m-1)}$ but not to $v^{\circ m}$, and compute using the hypotheses and the Taylor expansion of f at a

$$0 \leq u^T f\big[a\mathbf{1}_{n \times n} + tvv^T\big]u = u^T \left(\sum_{l=0}^{m-1} f^{(l)}(a)\frac{t^l}{l!}v^{\circ l}(v^{\circ l})^T + \frac{t^m}{m!}C\right)u$$

$$= \frac{t^m}{m!}u^T Cu,$$

where $C_{n \times n}$ has (j, k) entry $(v_j v_k)^m f^{(m)}(a + \theta_{jk}tv_j v_k)$ with all $\theta_{j,k} \in (0, 1)$. Divide by $t^m/m!$ and let $t \to 0^+$; since f is $C^{(m)}$, we obtain $0 \leq \left(u^T v^{\circ m}\right)^2 f^{(m)}(a)$. As this holds for all $0 \leq m \leq n-1$, the proof is complete. $\square$

13
The Stronger Horn–Loewner Theorem via Mollifiers

We continue with the proof of the Horn–Loewner theorem 12.1. This is in three steps:

(1) Theorem 12.1 holds for smooth functions. This was proved in Chapter 12.
(2) If Theorem 12.1 holds for smooth functions, then it holds for continuous functions. Here, we need to assume $n \geq 3$.
(3) If f satisfies the hypotheses in Theorem 12.1, then it is continuous. This follows from Vasudeva's 2×2 result – see (the proof of) Theorem 6.7 and Remark 6.8.

To carry out the second step – as well as a similar step in proving Schoenberg's theorem, see Section 17.2 – we will use a standard tool in analysis called *mollifiers*.

13.1 An Introduction to (One-Variable) Mollifiers

In this section, we examine some basic properties of mollifiers of one variable: the theory extends to $\mathbb{R}^n$ for all $n > 1$, but that is not required in what follows.

First, recall that one can construct smooth functions $g : \mathbb{R} \to \mathbb{R}$, such that g and all its derivatives vanish on $(-\infty, 0]$: for instance, $g(x) = \exp(-1/x) \cdot \mathbf{1}(x > 0)$. Indeed, one shows that $g^{(n)}(x) = p_n(1/x)g(x)$ for some polynomial p_n; hence, $g^{(n)}(x) \to 0$ as $x \to 0$. Hence:

Lemma 13.1 *Given scalars $-1 < a < b < 0$, there exists a smooth function ϕ that vanishes outside $[a, b]$, is positive on (a, b), and is a probability distribution on $\mathbb{R}$.*

Of course, the assumption $[a,b] \subset (-1,0)$ is completely unused in the proof of the lemma but is included for easy reference since we will require it in what follows.

Proof   The function $\varphi(x) := g(x - a)g(b - x)$ is nonnegative, smooth, and supported precisely on (a,b). In particular, $\int_{\mathbb{R}} \varphi > 0$, so the normalization $\phi := \varphi / \int_{\mathbb{R}} \varphi$ has the desired properties. $\square$

We now introduce mollifiers.

Definition 13.2   A *mollifier* is a one-parameter family of functions (in fact probability distributions)

$$\left\{ \phi_\delta(x) := \tfrac{1}{\delta} \phi \left(\tfrac{x}{\delta} \right) : \delta > 0 \right\},$$

with real domain and range, corresponding to any function ϕ that satisfies Lemma 13.1.

A continuous, real-valued function f (with suitable domain inside $\mathbb{R}$) is said to be *mollified* by convolving with the family ϕ_δ. In this case, we define

$$f_\delta(x) := \int_{\mathbb{R}} f(t) \phi_\delta(x - t)\, dt,$$

where one extends f outside its domain by zero. (This is called *convolution*: $f_\delta = f * \phi_\delta$.)

Remark 13.3   Mollifiers, or *Friedrichs mollifiers*, were used by Horn and Loewner in the late 1960s, as well as previously by Rudin in his 1959 proof of Schoenberg's theorem. They were a relatively modern tool at the time, having been introduced by Friedrichs in his seminal 1944 paper on PDEs [91], as well as slightly earlier by Sobolev in his famous 1938 paper [224] (which contained the proof of the Sobolev embedding theorem).

Returning to the definition of a mollifier, notice by the change of variables $u = x - t$ and Lemma 13.1 that

$$f_\delta(x) = \frac{1}{\delta} \int_{\mathbb{R}} f(x - u) \phi \left(\frac{u}{\delta} \right) du = \int_{-\delta}^{0} f(x - u) \phi_\delta(u)\, du. \tag{13.1}$$

In particular, f_δ is a "weighted average" of the image set $f([x, x + \delta])$, since ϕ is a probability distribution. Now it is not hard to see that f_δ is continuous and converges to f pointwise as $\delta \to 0^+$. In fact, more is true:

Proposition 13.4   *If $I \subset \mathbb{R}$ is a right-open interval and $f : I \to R$ is continuous, then for all $\delta > 0$, the mollified functions f_δ are smooth on $\mathbb{R}$ (where we* *extend f outside I by zero), and converge uniformly to f on compact subsets of I as $\delta \to 0^+$.*

To prove this result, we show two lemmas in somewhat greater generality. First, some notation: a (Lebesgue measurable) function $f : \mathbb{R} \to \mathbb{R}$ is said to be *locally* L^1 if it is L^1 on each compact subset of $\mathbb{R}$.

Lemma 13.5   *If $f : \mathbb{R} \to \mathbb{R}$ is locally L^1, and $\psi : \mathbb{R} \to \mathbb{R}$ is continuous with compact support, then $f * \psi : \mathbb{R} \to \mathbb{R}$ is also continuous.*

Proof   Suppose $x_n \to x$ in $\mathbb{R}$; without loss of generality $|x_n - x| < 1$ for all $n > 0$. Also choose $r, M > 0$, such that ψ is supported on $[-r, r]$ and $M = \|\psi\|_{L^\infty(\mathbb{R})} = \max_{\mathbb{R}} |\psi(x)|$. Then for each $t \in \mathbb{R}$, we have

$$f(t)\psi(x_n - t) \to f(t)\psi(x - t),$$

$$|f(t)\psi(x_n - t)| \leq M|f(t)| \cdot \mathbf{1}(|x - t| \leq r + 1)$$

(the second inequality follows by considering separately the cases $|x - t| \leq r + 1$ and $|x - t| > r + 1$). Since the right-hand side is integrable, Lebesgue's dominated convergence theorem applies

$$\lim_{n \to \infty} (f * \psi)(x_n) = \lim_{n \to \infty} \int_{\mathbb{R}} f(t)\psi(x_n - t)\, dt = \int_{\mathbb{R}} \lim_{n \to \infty} f(t)\psi(x_n - t)\, dt$$

$$= \int_{\mathbb{R}} f(t)\psi(x - t)\, dt = (f * \psi)(x),$$

so $f * \psi$ is continuous on $\mathbb{R}$. $\square$

Lemma 13.6   *If $f : \mathbb{R} \to \mathbb{R}$ is locally L^1, and $\psi : \mathbb{R} \to \mathbb{R}$ is C^1 with compact support, then $f * \psi : \mathbb{R} \to \mathbb{R}$ is also C^1 and $(f * \psi)' = f * \psi'$ on $\mathbb{R}$.*

Proof   We compute

$$(f * \psi)'(x) = \lim_{h \to 0} \frac{1}{h} \int_{\mathbb{R}} f(y)\psi(x + h - y)\, dy - \frac{1}{h} \int_{\mathbb{R}} f(y)\psi(x - y)\, dy$$

$$= \lim_{h \to 0} \int_{\mathbb{R}} f(y) \frac{\psi(x + h - y) - \psi(x - y)}{h}\, dy$$

$$= \lim_{h \to 0} \int_{\mathbb{R}} f(y)\psi'(x - y + c(h, y))\, dy,$$

where for each $y \in \mathbb{R}$, $c(h, y) \in [0, h]$ is chosen using the Mean Value Theorem. While $c(h, y) \to 0$ as $h \to 0$, the problem is that y is not fixed inside the integral. Thus, to proceed, we argue as follows: suppose ψ is supported inside $[-r, r]$ as above, hence so is ψ'. Choose any sequence $h_n \to 0$. Now **claim** that the last integral above, evaluated at h_n, converges to $(f * \psi')(x)$ as

$n \to \infty$ – hence so does the limit of the last integral above. Indeed, we may first assume all $h_n \in (-1, 1)$; and then check for each n that the above integral equals

$$\int_{\mathbb{R}} f(y)\psi'(x - y + c(h_n, y))\, dy = \int_{x-(r+1)}^{x+(r+1)} f(y)\psi'(x - y + c(h_n, y))\, dy$$

by choice of r. Now the integrand on the right-hand side is bounded above by $M_1|f(y)|$ in absolute value, where $M_1 := \|\psi'\|_{L^\infty(\mathbb{R})}$. Hence, by the dominated convergence theorem,

$$\lim_{n \to \infty} \int_{x-(r+1)}^{x+(r+1)} f(y)\psi'(x - y + c(h_n, y))\, dy$$

$$= \int_{x-(r+1)}^{x+(r+1)} f(y)\psi'(x - y)\, dy = (f * \psi')(x),$$

where the first equality also uses that ψ is C^1 and the second is by choice of r. Since this happens for every sequence $h_n \to 0$, it follows that $(f * \psi)'(x) = (f * \psi')(x)$. Moreover, $f * \psi'$ is continuous by Lemma 13.5, since ψ is C^1. This shows $f * \psi$ is C^1 as claimed.   □

Finally, we show the claimed properties of mollified functions.

Proof of Proposition 13.4   Extending f by zero outside I, it follows that f is locally L^1 on $\mathbb{R}$. Repeatedly applying Lemma 13.6 to $\psi = \phi_\delta, \phi_\delta', \phi_\delta'', \ldots$, we conclude that $f_\delta \in C^\infty(\mathbb{R})$.

To prove local uniform convergence, let K be a compact subset of I and $\epsilon > 0$. Denote $b := \sup K$ and $a := \inf K$. Since I is right open, there is a number $l > 0$, such that $J := [a, b + l] \subset I$. Since f is uniformly continuous on J, given $\epsilon > 0$ there exists $\delta \in (0, l)$, such that $|x - y| < \delta$, $x, y \in J \implies |f(x) - f(y)| < \epsilon$.

Now claim that if $0 < \xi < \delta$ then $\|f_\xi - f\|_{L^\infty(K)} \leq \epsilon$; note this proves the uniform convergence of the family f_δ to f on K. To show the claim, compute using (13.1) for $x \in K$

$$|f_\xi(x) - f(x)| = \left| \int_{-\xi}^{0} (f(x - u) - f(x))\phi_\xi(u)\, du \right|$$

$$\leq \int_{-\xi}^{0} |f(x - u) - f(x)|\phi_\xi(u)\, du \leq \epsilon \int_{-\xi}^{0} \phi_\xi(u)\, du = \epsilon.$$

This is true for all $x \in K$ by the choice of $\xi < \delta < l$, and hence, proves the claim.   □

13.2 Completing the Proof of the Horn–Loewner Theorem

With mollifiers in hand, we finish the proof of Theorem 12.1. As mentioned at the start of this chapter, the proof can be divided into three steps, and two are now already worked out. It remains to show the second step, that is, if $n \geq 3$ and if the result holds for smooth functions, then it holds for continuous functions.

Thus, suppose $I = (0, \infty)$ and $f : I \to \mathbb{R}$ is continuous. Define the mollified functions f_δ, $\delta > 0$ as above; note each f_δ is smooth. Moreover, given $a, b > 0$, by (13.1) the function f_δ satisfies

$$f_\delta\big[a\mathbf{1}_{n \times n} + b\mathbf{u}\mathbf{u}^T\big] = \int_{-\delta}^{0} \phi_\delta(y) \cdot f\big[(a + |y|)\mathbf{1}_{n \times n} + b\mathbf{u}\mathbf{u}^T\big]\,dy, \qquad (13.2)$$

and this is positive semidefinite by the assumptions for f. Thus, $f_\delta[-]$ preserves positivity on the given test set in $\mathbb{P}_n(I)$; a similar argument shows that $f_\delta[-]$ preserves positivity on $\mathbb{P}_2(I)$. Hence, by the proof in Chapter 12, f_δ, $f'_\delta, \ldots, f_\delta^{(n-1)}$ are nonnegative on I.

Observe that the theorem amounts to deducing a similar statement for f; however, as f is a priori known only to be continuous, we can only deduce nonnegativity for a discrete version of the derivatives – namely, divided differences:

Definition 13.7 Suppose I is a real interval and a function $f : I \to \mathbb{R}$. Given $h > 0$ and an integer $k \geq 0$, the *kth order forward differences with step size $h > 0$* are defined as follows:

$$\big(\Delta_h^0 f\big)(x) := f(x),$$

$$\big(\Delta_h^k f\big)(x) := \big(\Delta_h^{k-1} f\big)(x + h) - \big(\Delta_h^{k-1}\big)(x) = \sum_{j=0}^{k} \binom{k}{j}(-1)^{k-j} f(x + jh),$$

whenever $k > 0$ and $x, x + kh \in I$. Similarly, the *kth order divided differences with step size $h > 0$* are

$$\big(D_h^k f\big)(x) := \frac{1}{h^k}\big(\Delta_h^k f\big)(x), \qquad \forall k \geq 0, \ x, x + kh \in I.$$

The key point is that if a function is differentiable to some order, and its derivatives of that order are nonnegative on an open interval, then using the mean value theorem for divided differences, one shows the corresponding divided differences are also nonnegative, hence so are the corresponding forward differences. Remarkably, the converse also holds, *including differentiability!* This is a classical result by Boas and Widder:

Theorem 13.8   *Suppose $I \subset \mathbb{R}$ is an open interval, bounded or not, and $f : I \to \mathbb{R}$.*

(1) *(Cauchy's mean value theorem for divided differences: special case.) If f is k-times differentiable in I for some integer $k > 0$, and $x, x + kh \in I$ for $h > 0$, then there exists $y \in (x, x + kh)$, such that $(D_h^k f)(x) = f^{(k)}(y)$.*
(2) *(Boas–Widder, [44]) Suppose $k \geq 2$ is an integer, and $f : I \to \mathbb{R}$ is continuous and has all forward differences of order k nonnegative on I:*

$$(\Delta_h^k f)(x) \geq 0, \quad \text{whenever } h > 0 \text{ and } x, x + kh \in I.$$

Then on all of I, the function $f^{(k-2)}$ exists, is continuous and convex, and has nondecreasing left- and right-hand derivatives.

We make a few remarks on Boas and Widder's result. First, for $k = 2$ the result seems similar to Ostrowski's theorem 6.2, except for the local boundedness being strengthened to continuity. Second, note that while $f_\pm^{(k-1)}$ is nondecreasing by the theorem, one cannot claim here that the lower-order derivatives $f, \ldots, f^{(k-2)}$ are nondecreasing on I. Indeed, a counterexample for such an assertion for $f^{(l)}$, where $0 \leq l \leq k-2$, is $f(x) = -x^{l+1}$ on $I \subset \mathbb{R}$. Finally, we refer the reader to Section 21.1 for additional related observations and results.

Proof   The second part will be proved in detail in Chapter 21. For the first, consider the Newton form of the Lagrange interpolation polynomial $P(X)$ for $f(X)$ at $X = x, x + h, \ldots, x + kh$. The highest term of $P(X)$ is

$$(D_h^k f)(x) \cdot (X - x_{k-1}) \cdots (X - x_1)(X - x_0),$$

$$\text{where } x_j = x_0 + jh \; \forall j \geq 0.$$

Writing $g(X) := f(X) - P(X)$ to be the remainder function, note that g vanishes at $x, x + h, \ldots, x + kh$. Successively applying Rolle's theorem to g, $g', \ldots, g^{(k-1)}$, it follows that $g^{(k)}$ has a root in $(x, x + kh)$, say y. But then,

$$0 = g^{(k)}(y) = f^{(k)}(y) - (D_h^k f)(x)k!,$$

which concludes the proof.   □

Returning to our proof of the stronger Horn–Loewner theorem 12.1, since $f_\delta, f_\delta', \ldots, f_\delta^{(n-1)} \geq 0$ on I, by the above theorem the divided differences of

f_δ up to order $n-1$ are nonnegative on I, hence the same holds for the forward differences of f_δ. Applying Proposition 13.4, the forward differences of f of orders $k = 0, \ldots, n-1$ are also nonnegative on I. Finally, invoke the Boas–Widder theorem for $k = 2, \ldots, n-1$ to conclude the proof of the (stronger) Horn–Loewner theorem – noting for "low orders" that f is nonnegative and nondecreasing on I by using forward differences of orders $k = 0, 1$ respectively, and hence $f, f' \geq 0$ on I as well.   □

14

Stronger Vasudeva and Schoenberg Theorems via Bernstein's Theorem

14.1 The Theorems of Vasudeva and Bernstein

Having shown (the stronger form of) the Horn–Loewner theorem 12.1, we use it to prove the following strengthening of Vasudeva's theorem 11.4. In it, recall from Definition 6.10 that HTN_n denotes the set of $n \times n$ totally nonnegative (*TN*) Hankel matrices. (These are automatically positive semidefinite.)

Theorem 14.1 (Vasudeva's theorem, stronger version – see Remark 14.15) *Suppose $I = (0,\infty)$ and $f : I \to \mathbb{R}$. The following are equivalent:*

(1) *The entrywise map $f[-]$ preserves positivity on $\mathbb{P}_n(I)$ for all $n \geq 1$.*
(2) *The entrywise map $f[-]$ preserves positivity on all matrices in HTN_n with positive entries and rank at most 2 for all $n \geq 1$.*
(3) *The function f equals a convergent power series $\sum_{k=0}^{\infty} c_k x^k$ for all $x \in I$, with the Maclaurin coefficients $c_k \geq 0$ for all $k \geq 0$.*

To show the theorem, we require the following well-known classical result by Bernstein:

Definition 14.2 If $I \subset \mathbb{R}$ is open, we say that $f : I \to \mathbb{R}$ is *absolutely monotonic* if f is smooth on I and $f^{(k)} \geq 0$ on I for all $k \geq 0$.

Theorem 14.3 (Bernstein [34]) *Suppose $-\infty < a < b \leq \infty$. If $f : [a,b) \to \mathbb{R}$ is continuous at a and absolutely monotonic on (a,b), then f can be extended analytically to the complex disk $D(a, b - a)$.*

With Bernstein's theorem at hand, the "stronger Vasudeva theorem" follows easily:

Proof of Theorem 14.1 That (3) $\implies$ (1) follows from the Schur product theorem or Lemma 11.1; and clearly (1) $\implies$ (2). Now suppose (2) holds.

By the stronger Horn–Loewner theorem 12.1, $f^{(k)} \geq 0$ on I for all $k \geq 0$, i.e., f is absolutely monotonic on I. In particular, f is nonnegative and non-decreasing on $I = (0, \infty)$, so it can be continuously extended to the origin via $f(0) := \lim_{x \to 0^+} f(x) \geq 0$. Now apply Bernstein's theorem with $a = 0$ and $b = \infty$ to deduce that f agrees on $[0, \infty)$ with an entire function $\sum_{k=0}^{\infty} c_k x^k$. Moreover, since $f^{(k)} \geq 0$ on I for all k, it follows that $f^{(k)}(0) \geq 0$, i.e., $c_k \geq 0 \; \forall k \geq 0$. Restricting to I, we obtain (3), as desired.   □

On a related note, recall Theorems 5.3 and 6.11, which showed that when studying entrywise powers preserving the two closed convex cones $\mathbb{P}_n([0, \infty))$ and HTN_n, the answers were identical. This is perhaps not surprising, given Theorem 3.6. In this vein, we observe that such an equality of preserver sets also holds when classifying the entrywise maps preserving Hankel TN matrices with positive entries:

Corollary 14.4   *With $I = (0, \infty)$ and $f : I \to \mathbb{R}$, the three assertions in Theorem 14.1 are further equivalent to:*

(4) The entrywise map $f[-]$ preserves total nonnegativity on all matrices in HTN_n with positive entries for all $n \geq 1$.

(5) The entrywise map $f[-]$ preserves total nonnegativity on the matrices in HTN_n with positive entries and rank at most 2 for all $n \geq 1$.

Proof   Clearly (4) $\implies$ (5) $\implies$ (2), where (1)–(3) are as in Theorem 14.1. That (3) $\implies$ (4) follows from Lemma 11.1 and Theorem 3.6.   □

Remark 14.5   Here is one situation where the two sets of preservers – of positivity on $\mathbb{P}_n$ for all n, and of total nonnegativity on HTN_n for all n – differ: if we also allow zero entries, as opposed to only positive entries as in the preceding corollary and Theorem 14.1. In this case, one shows that the preservers of HTN_n for all n are the functions f, such that $f|_{(0, \infty)}$ is absolutely monotonic, and hence a power series with nonnegative Maclaurin coefficients; and, such that $0 \leq f(0) \leq \lim_{x \to 0^+} f(x)$, since the only Hankel TN matrices with a zero entry arise as truncated moment matrices of measures $a\delta_0$. On the other hand, by considering the rank-2 Hankel positive semidefinite matrix

$$A := \begin{pmatrix} 2 & 1 & 1 \\ 1 & 1 & 0 \\ 1 & 0 & 1 \end{pmatrix}, \text{ and considering the inequality}$$

$$\lim_{x \to 0^+} \det f[xA] \geq 0,$$

it follows that f is continuous at 0^+. (Note, A is not TN.) In particular, the entrywise preservers of positivity on $\bigcup_{n \geq 1} \mathbb{P}_n([0, \infty))$ are precisely the functions $\sum_{k \geq 0} c_k x^k$, with all $c_k \geq 0$.

Remark 14.6 In contrast, if one instead tries to classify the entrywise preservers of total nonnegativity on all (possibly symmetric) TN matrices, then one obtains only the constant or linear functions $f(x) = c, cx$ for $c, x \geq 0$. See the 2020 preprint [22] by Belton–Guillot–Khare–Putinar.

To complete the proof of the stronger Vasudeva theorem 14.1 as well as its corollary (Corollary 14.4), it remains to show Bernstein's theorem.

Proof of Bernstein's theorem 14.3 We first claim $f^{(k)}(a^+)$ exists and equals $\lim_{x \to a^+} f^{(k)}(x)$ for all $k \geq 0$. The latter limit here exists because $f^{(k+1)} \geq 0$ on (a, b), so $f^{(k)}(x)$ is nonnegative and nondecreasing on $[a, b)$.

It suffices to show the claim for $k = 1$. But here we compute

$$f'(a^+) = \lim_{h \to 0^+} \frac{f(a+h) - f(a)}{h} = \lim_{h \to 0^+} f'(a + c(h)),$$

where $c(h) \in [0, h]$ exists and goes to zero as $h \to 0^+$, by the Mean Value Theorem. The claim now follows from the previous paragraph. In particular, $f^{(k)}$ exists and is continuous, nonnegative, and nondecreasing on $[a, b)$.

Applying Taylor's theorem, we have

$$f(x) = f(a^+) + f'(a^+)(x - a) + \cdots + f^{(n)}(a^+)\frac{(x-a)^n}{n!} + R_n(x),$$

where R_n is the Taylor remainder

$$R_n(x) = \int_a^x \frac{(x-t)^n}{n!} f^{(n+1)}(t) \, dt. \tag{14.1}$$

By the assumption on f, we see that $R_n(x) \geq 0$. Changing variables to $t = a + y(x - a)$, the limits for y change to $0, 1$, and we have

$$R_n(x) = \frac{(x-a)^{n+1}}{n!} \int_0^1 (1-y)^n f^{(n+1)}(a + y(x-a)) \, dy.$$

Since $f^{(n+2)} \geq 0$ on $[a, b)$, if $a \leq x \leq c$ for some $c < b$, then uniformly in $[a, c]$ we have $0 \leq f^{(n+1)}(a + y(x-a)) \leq f^{(n+1)}(a + y(c-a))$. Therefore, using Taylor's remainder formula once again, we obtain

$$0 \leq R_n(x) \leq \frac{(x-a)^{n+1}}{n!} \int_0^1 (1-y)^n f^{(n+1)}(a + y(c-a)) \, dy$$

$$= R_n(c)\frac{(x-a)^{n+1}}{(c-a)^{n+1}}$$

$$= \frac{(x-a)^{n+1}}{(c-a)^{n+1}} \left(f(c) - \sum_{k=0}^{n} f^{(k)}(a^+)\frac{(c-a)^k}{k!} \right)$$

$$\leq f(c)\frac{(x-a)^{n+1}}{(c-a)^{n+1}}.$$

From this it follows that $\lim_{n\to\infty} R_n(x) = 0$ for all $x \in [a,c)$. Since this holds for all $c \in (a,b)$, the Taylor series of f converges to f on $[a,b)$. In other words,

$$f(x) = \sum_{k=0}^{\infty} \frac{f^{(k)}(a^+)}{k!}(x-a)^k, \qquad x \in [a,b).$$

Now if $z \in D(a, b-a)$, then clearly $a + |z-a| < a + (b-a) = b$. Choosing any $c \in (a + |z-a|, b)$, we check that the Taylor series converges (absolutely) at z

$$\left| \sum_{k=0}^{\infty} \frac{f^{(k)}(a^+)}{k!}(z-a)^k \right| \leq \sum_{k=0}^{\infty} \frac{f^{(k)}(a^+)}{k!}|z-a|^k$$

$$\leq \sum_{k=0}^{\infty} \frac{f^{(k)}(a^+)}{k!}|c-a|^k$$

$$= f(c) < \infty.$$

This completes the proof of Bernstein's theorem – and with it, the stronger form of Vasudeva's theorem.   □

Remark 14.7   We mention for completeness that Bernstein's theorem admits an extension, which was already shown by Bernstein in 1926 [33], and which say that even if $f^{(k)} \geq 0$ in (a,b) only for $k \geq 0$ *even*, then f is necessarily analytic in (a,b). (In fact, Bernstein worked only with divided differences – see Theorem 21.2.) This was further extended by Boas as follows:

Theorem 14.8 (Boas, 1941 [14])   *Let $\{n_p : p \geq 1\}$ be an increasing sequence of positive integers, such that n_{p+1}/n_p is uniformly bounded. Let $(a,b) \subset \mathbb{R}$ and $f : (a,b) \to \mathbb{R}$ be smooth. If for each $p \geq 1$, the derivative $f^{(n_p)}$ does not change sign in (a,b), then f is analytic in (a,b).*

14.2 The Stronger Version of Schoenberg's Theorem

We now come to the main result of this part of the text: the promised strengthening of Schoenberg's theorem.

Theorem 14.9 (Schoenberg's theorem, stronger version)   *Given $f : \mathbb{R} \to \mathbb{R}$, the following are equivalent:*

(1) *The entrywise map $f[-]$ preserves positivity on $\mathbb{P}_n(\mathbb{R})$ for all $n \geq 1$.*
(2) *The entrywise map $f[-]$ preserves positivity on the Hankel matrices in $\mathbb{P}_n(\mathbb{R})$ of rank at most 3 for all $n \geq 1$.*

(3) *The function f equals a convergent power series $\sum_{k=0}^{\infty} c_k x^k$ for all $x \in \mathbb{R}$, with the Maclaurin coefficients $c_k \geq 0$ for all $k \geq 0$.*

See also Theorem 14.13 for two a priori weaker, yet equivalent, assertions.

Remark 14.10   Recall the two definitions of positive definite functions, from Definition 11.12 and the discussion preceding Lemma 11.13. The metric-space version for Euclidean and Hilbert spheres was connected by Schoenberg to functions of the form $f \circ \cos$, by requiring that $\left(f(\langle x_j, x_k \rangle) \right)_{j,k \geq 0}$ be positive semidefinite for all choices of vectors $x_j \in S^{r-1}$ (for $2 \leq r \leq \infty$). A *third* notion of positive definite kernels on Hilbert spaces $\mathcal{H}$ arises from here, and is important in machine learning among other areas (see, e.g., [182, 225, 229]): one says $f : \mathbb{R} \to \mathbb{R}$ is *positive definite on $\mathcal{H}$* if, for any choice of finitely many points $x_j, j \geq 0$, the matrix $(f(\langle x_j, x_k \rangle))_{j,k \geq 0}$ is positive semidefinite. Since Gram matrix ranks are bounded above by $\dim \mathcal{H}$, this shows that Rudin's 1959 theorem 11.3 classifies the positive definite kernels/functions on $\mathcal{H}$ for any real Hilbert space of dimension ≥ 3. The stronger Schoenberg theorem 14.9 provides a second proof.

Returning to the stronger Schoenberg theorem 14.9, clearly (1) $\Longrightarrow$ (2), and (3) $\Longrightarrow$ (1) by the Pólya–Szegő observation 11.1. Thus, the goal over the next few sections is to prove (2) $\Longrightarrow$ (3). The proof is simplified when some of the arguments below are formulated in the language of *moment sequences* and their preservers. We begin by defining these and explaining the dictionary between moment sequences and positive-semidefinite Hankel matrices, due to Hamburger (among others).

Definition 14.11   Recall that given an integer $k \geq 0$ and a real measure μ supported on a subset of $\mathbb{R}$, μ has kth moment equal to the following (if it converges)

$$s_k(\mu) := \int_{\mathbb{R}} x^k \, d\mu.$$

Henceforth, we only work with *admissible* measures, i.e., such that μ is nonnegative on $\mathbb{R}$ and $s_k(\mu)$ converges for all $k \geq 0$. The *moment sequence* of such a measure μ is the sequence

$$\mathbf{s}(\mu) := (s_0(\mu), s_1(\mu), \ldots).$$

We next define *transforms* of moment sequences: a function $f : \mathbb{R} \to \mathbb{R}$ acts entrywise, to take moment sequences to real sequences

$$f[\mathbf{s}(\mu)] := (f(s_k(\mu)))_{k \geq 0}. \tag{14.2}$$

We are interested in examining when the transformed sequence (14.2) is also the moment sequence of an admissible measure supported on $\mathbb{R}$. This connects to the question of positivity preservers via the following classical result.

Theorem 14.12 (Hamburger [112])   *A real sequence* $(s_k)_{k \geq 0}$ *is the moment sequence of an admissible measure, if and only if the semi-infinite Hankel matrix* $H := (s_{j+k})_{j,k \geq 0}$ *is positive semidefinite.*

Recall that the easy half of this result was proved early on, in Lemma 1.19.

Thus, entrywise functions preserving positivity on Hankel matrices are intimately related to moment-sequence preservers. Also note that if a measure μ has finite support in the real line, then by examining, for example (1.3), the Hankel moment matrix H_μ (i.e., every submatrix) has rank at most the size of the support set. From this and Hamburger's theorem, we deduce all but the last sentence of the following result:

Theorem 14.13   *Theorem 14.9(2) implies the following a priori weaker statement:*

(4) *For each measure*

$$\mu = a\delta_1 + b\delta_{u_0} + c\delta_{-1}, \quad with\ u_0 \in (0,1),\ a,b,c \geq 0, \tag{14.3}$$

there exists an admissible (nonnegative) measure $\sigma = \sigma_\mu$ *on* $\mathbb{R}, i$ *such that* $f(s_k(\mu)) = s_k(\sigma)\ \forall k \geq 0.$

In turn, this implies the still-weaker statement:

(5) *For each measure* μ *as in* (14.3), *with semi-infinite Hankel moment matrix* H_μ, *the matrix* $f[H_\mu]$ *is positive semidefinite.*

In fact, these statements are equivalent to the assertions in Theorem 14.9.

Remark 14.14   In this text, we do not prove Hamburger's theorem; but we have used it to state Theorem 14.13(4) – i.e., in working with the admissible measure $\sigma = \sigma_\mu$. A closer look reveals that the use of Hamburger's theorem and moment sequences is not required to prove Schoenberg's theorem, or even its stronger form in Theorem 14.13, which is explained in (proving) Theorem 14.13(5). Our workaround is explained in Chapter 15, via a "positivity-certificate trick" involving limiting sum-of-squares representations of polynomials. That said, moment sequences help simplify the presentation of the proof, and hence we will continue to use them in the proof, Chapters 16 and 17.

The next three chapters are devoted to proving (5) $\implies$ (3) (in Theorems 14.9 and 14.13). Here is an outline of the steps in the proof:

(1) All matrices $A = \begin{pmatrix} a & b \\ b & c \end{pmatrix} \in \mathbb{P}_2((0,\infty))$ with $a \geq c$ occur as leading principal submatrices of the Hankel moment matrices H_μ, where μ is as in (14.3).

(2) Apply the stronger Horn–Loewner theorem 12.1 and Bernstein's theorem to deduce that $f|_{(0,\infty)} = \sum_{k=0}^\infty c_k x^k$ for some $c_k \geq 0$.

(3) If f satisfies assertion (5) in Theorem 14.13, then f is continuous on $\mathbb{R}$.

(4) If, moreover, f is smooth and satisfies assertion (5) in Theorem 14.13, then f is real analytic.

(5) Real analytic functions satisfy the desired implication above:
$$(5) \implies (3).$$

(6) Using mollifiers and complex analysis, one can go from smooth functions to continuous functions.

Notice that Steps 3, 4–5, and 6 resemble the three steps in the proof of the stronger Horn–Loewner theorem 12.1.

In this chapter, we complete the first two steps in the proof.

Step 1   For the first step, suppose $A = \begin{pmatrix} a & b \\ b & c \end{pmatrix} \in \mathbb{P}_2((0,\infty))$ with $a \geq c$. There are three cases. First, if $b = \sqrt{ac}$ then use $\mu = a\delta_{b/a}$, since $0 < b/a \leq 1$.

Henceforth, assume $0 < b < \sqrt{ac} \leq a$. (In particular, $2b < 2\sqrt{ac} \leq a + c$.) The second case is if $b > c$; we then find $t > 0$, such that $A - t\mathbf{1}_{2\times2}$ is singular. This condition amounts to a linear equation in t, with solution (to be verified by the reader)

$$t = \frac{ac - b^2}{a + c - 2b} > 0.$$

Then $c - t = \dfrac{(b - c)^2}{a + c - 2b} > 0$, so $a - t, b - t > 0$ and $A = \begin{pmatrix} s_0(\mu) & s_1(\mu) \\ s_1(\mu) & s_2(\mu) \end{pmatrix}$, where

$$\mu = \frac{ac - b^2}{a + c - 2b}\delta_1 + \frac{(a - b)^2}{a + c - 2b}\delta_{\frac{b-c}{a-b}}, \quad \text{with } \frac{b - c}{a - b} \in (0, 1).$$

The third case is when $0 < b \leq c \leq \sqrt{ac} \leq a$, with $b < \sqrt{ac}$. Now find $t > 0$, such that the matrix $\begin{pmatrix} a - t & b + t \\ b + t & c - t \end{pmatrix} \in \mathbb{P}_2((0,\infty))$ and is singular. To do so requires solving a linear equation, which yields

$$t = \frac{ac - b^2}{a + c + 2b}, \qquad c - t = \frac{(b + c)^2}{a + c + 2b},$$

$$a - t = \frac{(a+b)^2}{a+c+2b}, \qquad b + t = \frac{(a+b)(b+c)}{a+c+2b},$$

and all of these are strictly positive. So $a, b, c > 0$ are the first three moments of

$$\mu = \frac{ac - b^2}{a+c+2b}\delta_{-1} + \frac{(a+b)^2}{a+c+2b}\delta_{\frac{b+c}{a+b}}, \quad \text{with } \frac{b+c}{a+b} \in (0, 1].$$

Step 2   Observe that the hypotheses of the stronger Horn–Loewner theorem 12.1 (for all n) can be rephrased as saying that $f[-]$ sends the rank-1 matrices in $\mathbb{P}_2(I)$ and the Toeplitz matrices in $\mathbb{P}_2(I)$ to $\mathbb{P}_2(\mathbb{R})$, and that assertion (4) in Theorem 14.13 holds for all measures $a\delta_1 + b\delta_{u_0}$, where $u_0 \in (0, 1)$ is fixed and $a, b > 0$. By Step 1 and the hypotheses, we can apply the stronger Horn–Loewner theorem in our setting for each $n \geq 3$, hence $f|_{(0,\infty)}$ is smooth and absolutely monotonic. As in the proof of the stronger Vasudeva theorem 14.1, extend f continuously to the origin, say to a function $\widetilde{f}$, and apply Bernstein's theorem 14.3. It follows that $\widetilde{f}|_{[0,\infty)}$ is a power series with nonnegative Maclaurin coefficients, and Step 2 follows by restricting to $\widetilde{f}|_{(0,\infty)} = f|_{(0,\infty)}$.   □

Remark 14.15   From Step 2, it follows that assertions (1) and (2) in the stronger Vasudeva theorem 14.1 can be further weakened, to deal only with the rank-1 matrices in $\mathbb{P}_2(I)$, the Toeplitz matrices in $\mathbb{P}_2(I)$, and with the (Hankel *TN* moment matrices of) measures $a\delta_1 + b\delta_{u_0}$, for a *single* fixed $u_0 \in (0, 1)$ and all $a, b \geq 0$ with $a + b > 0$.

15

Proof of the Stronger Schoenberg Theorem (Part I) – Positivity Certificates

We continue with the proof of the stronger Schoenberg theorem 14.9. Previously, we have shown the first two of the six steps in the proof (these are listed following Theorem 14.13).

Step 3   The next step is to show that if assertion (4) (or (5)) in Theorem 14.13 holds, then f is continuous on $\mathbb{R}$. Notice from Steps 1 and 2 of the proof that f is absolutely monotonic, hence continuous, on $(0, \infty)$.

15.1 Integration Trick and Proof of Continuity

At this stage, we transition to moment-sequence preservers, via Hamburger's theorem 14.12. The following "integration trick" will be used repeatedly in what follows: Suppose $p(t)$ is a real polynomial that takes nonnegative values for $t \in [-1, 1]$. Write $p(t) = \sum_{k=0}^{\infty} a_k t^k$ (with only finitely many a_k nonzero, but not necessarily all positive, note). If $\mu \geq 0$ is an admissible measure – in particular, nonnegative by Definition 14.11 – then by assumption and Hamburger's theorem we have $f(s_k(\mu)) = s_k(\sigma_\mu) \ \forall k \geq 0$, for some admissible measure $\sigma_\mu \geq 0$ on $\mathbb{R}$, where $f \colon \mathbb{R} \to \mathbb{R}$ satisfies Theorem 14.13(4) or (5). Now **assuming** σ_μ is supported on $[-1, 1]$ (which is not a priori clear from the hypotheses), we have

$$0 \leq \int_{-1}^{1} p(t)\, d\sigma_\mu = \sum_{k=0}^{\infty} \int_{-1}^{1} a_k t^k \, d\sigma_\mu = \sum_{k=0}^{\infty} a_k s_k(\sigma_\mu) = \sum_{k=0}^{\infty} a_k f(s_k(\mu)).$$
(15.1)

Example 15.1   Suppose $p(t) = 1 - t^d$ on $[-1, 1]$, for some integer $d \geq 1$. Then $f(s_0(\mu)) - f(s_d(\mu)) \geq 0$. As a further special case, if $\mu = a\delta_1 + b\delta_{u_0} + c\delta_{-1}$ as in Theorem 14.13(4), if σ_μ is supported on $[-1, 1]$ then this would imply

135

$$f(a+b+c) \geq f(a+bu_0^d + c(-1)^d), \qquad \forall u_0 \in (0,1),\ a,b,c \geq 0.$$

It is not immediately clear how the preceding inequalities can be obtained by considering only the preservation of matrix positivity by $f[-]$ (or more involved such assertions). As we will explain in Section 15.2, this has connections to real algebraic geometry; in particular, to a well-known program of Hilbert.

Returning to the proof of continuity in Schoenberg's theorem, we suppose without further mention that f satisfies only Theorem 14.13(5) – and hence, is absolutely monotonic on $(0, \infty)$. We begin by showing two preliminary lemmas, which are used in the proof of continuity.

Lemma 15.2 f *is bounded on compact subsets of* $\mathbb{R}$.

Proof   If $K \subset \mathbb{R}$ is compact, say $K \subset [-M, M]$ for some $M > 0$, then note that $f|_{(0,\infty)}$ is nondecreasing, hence $0 \leq |f(x)| \leq f(M)$, $\forall x \in (0, M]$. Now apply $f[-]$ to the matrix $B := \begin{pmatrix} x & -x \\ -x & x \end{pmatrix}$, arising from $\mu = x\delta_{-1}$, with $x > 0$. The positivity of $f[B]$ implies $|f(-x)| \leq f(x) \leq f(M)$. Similarly considering $\mu = \frac{M}{2}\delta_1 + \frac{M}{2}\delta_{-1}$ shows that $|f(0)| \leq f(M)$.   □

Now say $\mu = a\delta_1 + b\delta_{u_0} + c\delta_{-1}$ as above, or more generally, μ is any nonnegative measure supported in $[-1, 1]$. It is easily seen that its moments $s_k(\mu)$, $k \geq 0$ are all uniformly bounded in absolute value – in fact, by the mass $s_0(\mu)$. Our next lemma shows that the converse is also true.

Lemma 15.3   *Given an admissible measure* σ *on* $\mathbb{R}$, *the following are equivalent:*

(1) *The moments of* σ *are all uniformly bounded in absolute value.*
(2) *The measure* σ *is supported on* $[-1, 1]$.

Proof   As discussed above, (2) $\implies$ (1). To show the converse, suppose (1) holds but (2) fails. Then σ has positive mass in $(1, \infty) \sqcup (-\infty, -1)$. We obtain a contradiction in the first case; the proof is similar in the other case. Thus, suppose σ has positive mass on

$$(1, \infty) = \left[1 + \tfrac{1}{1}, 1 + \tfrac{1}{0}\right) \sqcup \left[1 + \tfrac{1}{2}, 1 + \tfrac{1}{1}\right) \sqcup \cdots \sqcup \left[1 + \tfrac{1}{n+1}, 1 + \tfrac{1}{n}\right) \sqcup \cdots,$$

where $1/0 := \infty$. Then $\sigma(I_n) > 0$ for some $n \geq 0$, where we denote $I_n := \left[1 + \tfrac{1}{n+1}, 1 + \tfrac{1}{n}\right)$ for convenience. But now we obtain the desired contradiction

$$s_{2k}(\sigma) = \int_{\mathbb{R}} x^{2k} \, d\sigma \geq \int_{1+\frac{1}{n+1}}^{1+\frac{1}{n}} x^{2k} \, d\sigma$$

$$\geq \int_{1+\frac{1}{n+1}}^{1+\frac{1}{n}} \left(1 + \tfrac{1}{n+1}\right)^k \, d\sigma \geq \sigma(I_n) \left(1 + \tfrac{1}{n+1}\right)^k,$$

and this is not uniformly bounded over all $k \geq 0$. □

With these basic lemmas at hand, we have:

Proof of Step 3 for the stronger Schoenberg theorem: continuity Suppose $f : \mathbb{R} \to \mathbb{R}$ satisfies Theorem 14.13(4). (We explain in Section 15.3, how to weaken the hypotheses to Theorem 14.13(5).) Given a measure $\mu = a\delta_1 + b\delta_{u_0} + c\delta_{-1}$ for $u_0 > 0$ and $a, b, c \geq 0$, note that $|s_k(\mu)| \leq s_0(\mu) = a + b + c$. Hence, by Lemma 15.2, the moments $s_k(\sigma_\mu)$ are uniformly bounded over all k. By Lemma 15.3, it follows that σ_μ must be supported in $[-1, 1]$. In particular, we can apply the integration trick (15.1) above.

We use this trick to prove continuity at $-\beta$ for $\beta \geq 0$. (By Step 2, this proves the continuity of f on $\mathbb{R}$.) Thus, fix $\beta \geq 0$, $u_0 \in (0, 1)$, and $b > 0$, and define

$$\mu := (\beta + bu_0)\delta_{-1} + b\delta_{u_0}.$$

Let $p_{\pm, 1}(t) := (1 \pm t)(1 - t^2)$; note that these polynomials are nonnegative on $[-1, 1]$. By the integration trick (15.1),

$$\int_{-1}^{1} p_{\pm, 1}(t) \, d\sigma_\mu(t) \geq 0$$

$$\implies \quad s_0(\sigma_\mu) - s_2(\sigma_\mu) \geq \pm(s_1(\sigma_\mu) - s_3(\sigma_\mu))$$

$$\implies \quad f(s_0(\mu)) - f(s_2(\mu)) \geq |f(s_1(\mu)) - f(s_3(\mu))|$$

$$\implies \quad f(\beta + b(1 + u_0)) - f\big(\beta + b(u_0 + u_0^2)\big)$$

$$\geq \left| f(-\beta) - f\big(-\beta - bu_0(1 - u_0^2)\big) \right|.$$

Now let $b \to 0^+$. Then the left-hand side goes to zero by Step 2 (in Chapter 14), hence so does the right-hand side. This implies f is left continuous at $-\beta$ for all $\beta \geq 0$. To show f is right continuous at $-\beta$, use $\mu' := (\beta + bu_0^3)\delta_{-1} + b\delta_{u_0}$ instead of μ. □

Remark 15.4 Akin to its use in proving the continuity of f, the integration trick (15.1) can also be used to prove the boundedness of f on compact

sets $[-M, M]$, as in Lemma 15.2. To do so, work with the polynomials
$p_{\pm,0}(t) := 1 \pm t$, which are also nonnegative on $[-1, 1]$. Given $0 \le x < M$,
applying (15.1) to $\mu = M\delta_{x/M}$ and $\mu' = x\delta_{-1}$ shows Lemma 15.2.

15.2 The Integration Trick Explained: Semialgebraic Geometry

In Section 15.1, we used the following integration trick: if $\sigma \ge 0$ is a real
measure supported in $[-1, 1]$ with all moments finite, i.e., the Hankel moment
matrix $H_\sigma := (s_{j+k}(\sigma))_{j,k=0}^\infty$ is positive semidefinite; and if a polynomial
$p(t) = \sum_{k \ge 0} a_k t^k$ is nonnegative on $[-1, 1]$, then

$$0 \le \int_{-1}^1 p(t)\, d\sigma = \sum_{k=0}^\infty \int_{-1}^1 a_k t^k \, d\sigma = \sum_{k=0}^\infty a_k s_k(\sigma).$$

This integration trick is at the heart of the link between moment problems
and (Hankel) matrix positivity. This trick is now explained; namely, how
this integral inequality can be understood purely in terms of the positive
semidefiniteness of H_σ. This also has connections to real algebraic geometry
and Hilbert's seventeenth problem.

The basic point is as follows: if a d-variate polynomial (in one or several
variables) is a sum of squares of real polynomials – also called a *sum-of-squares polynomial* – then it is automatically nonnegative on $\mathbb{R}^d$. However,
Hilbert showed in his 1888 paper [121] – following the doctoral dissertation
of Hermann Minkowski – that for $d \ge 2$, there exist polynomials that are
not sums of squares, yet are nonnegative on $\mathbb{R}^d$. The first such example was
constructed in 1967, and is the well-known Motzkin polynomial $M(x, y) = x^4 y^2 + x^2 y^4 - 3x^2 y^2 + 1$ [178].[1,2,3] Such phenomena are also studied on

---

[1] Indeed, as explained, e.g., in [198], by the AM–GM inequality we have $M(x, y) \ge 2t^3 - 3t^2 + 1 = (2t^2 - t - 1)(t - 1)$, where $t = x^2 y^2$. Now either $t \in [0, 1]$, so both factors on the right are nonpositive; or $t > 1$, so both factors are positive. Next, suppose $M(x, y) = \sum_j f_j(x, y)^2$ is a sum of squares; since $M(x, 0) = M(0, y) = 1$, it follows that $f_j(x, 0), f_j(0, y)$ are constants, and hence the f_j are of the form $f_j(x, y) = a_j + b_j xy + c_j x^2 y + d_j xy^2$. Now equating the coefficient of $x^2 y^2$ in $M = \sum_j f_j^2$ gives $-3 = \sum_j b_j^2$, a contradiction.

[2] Hilbert then showed in [122] that every nonnegative polynomial on $\mathbb{R}^2$ is a sum of four squares of rational functions; e.g., the Motzkin polynomial equals $\frac{x^2 y^2 (x^2 + y^2 - 2)^2 (x^2 + y^2 + 1) + (x^2 - y^2)^2}{(x^2 + y^2)^2}$. For more on this problem, see, e.g., [198].

[3] The reader may recall the name of Motzkin from Theorem 3.4, in an entirely different context. As a historical digression, we mention several relatively "disconnected" areas of mathematics, in *all* of which, remarkably, Motzkin made fundamental contributions. His thesis [174] was a

polytopes (results of Farkas, Pólya, and Handelman), and on more general "semialgebraic sets" including compact ones (results of Stengle, Schmüdgen, Putinar, and Vasilescu, among others).

Now given, say, a one-variable polynomial that is nonnegative on a semialgebraic set such as $[-1, 1]$, one would like a *positivity certificate* for it, meaning a sum-of-squares representation mentioned above, or more generally, a *limiting sum-of-squares representation*. To make this precise, define the L^1-*norm*, or the *Wiener norm*, of a polynomial $p(t) = \sum_{k \geq 0} a_k t^k$ as

$$\| p(t) \|_{1,+} := \sum_{k \geq 0} |a_k|. \tag{15.2}$$

One would thus like to find a sequence p_n of sum-of-squares polynomials, such that $\| p_n(t) - p(t) \|_{1,+} \to 0$ as $n \to \infty$. Two simple cases are if there exist polynomials $q_n(t)$, such that (1) $p_n(t) = q_n(t)^2$ $\forall n$ or (2) $p_n(t) = \sum_{k=1}^{n} q_k(t)^2$ $\forall n$.

How does this connect to matrix positivity? It turns out that in our given situation, what is required is precisely a positivity certificate. For example, say $p(t) = (3 - t)^2 = 9 - 6t + t^2 \geq 0$ on $\mathbb{R}$. Then

$$\int_{-1}^{1} p \, d\sigma = 9s_0(\sigma) - 6s_1(\sigma) + s_2(\sigma) = (3, \, -1) \begin{pmatrix} s_0(\sigma) & s_1(\sigma) \\ s_1(\sigma) & s_2(\sigma) \end{pmatrix} (3, \, -1)^T$$

$$= (3e_0 - e_1)^T H_\sigma (3e_0 - e_1),$$

$$\tag{15.3}$$

where $e_0 = (1, 0, 0, \ldots)^T, e_1 = (0, 1, 0, 0, \ldots)^T, \ldots$ comprise the standard basis for $\mathbb{R}^{\mathbb{N} \sqcup \{0\}}$, and H_σ is the semi-infinite, positive semidefinite Hankel moment matrix for σ. From this calculation, it follows that $\int_{-1}^{1} p \, d\sigma$ is nonnegative – and this holds more generally, whenever there exists a (limiting) sum-of-squares representation for p.

<hr>

landmark work in the area of linear inequalities/linear programming, introducing in particular the *Motzkin transposition theorem* and the *Fourier–Motzkin Elimination (FME) algorithm*. Additionally, he proved in the same thesis the fundamental fact in geometric combinatorics that, a convex polyhedral set is the Minkowski sum of a compact (convex) polytope and a convex polyhedral cone. Third, in his thesis Motzkin also characterized the matrices that satisfy the variation diminishing property; see Theorem 3.4. Then in [175], Motzkin studied what is now called the *Motzkin number* in combinatorics: this is the number of different ways to draw nonintersecting chords between n marked points on a circle. In [176], he provided the first example of a principal ideal domain that is not a Euclidean domain: $\mathbb{Z}[(1 + \sqrt{-19})/2]$.

In [177], he provided an ideal-free short proof of Hilbert's Nullstellensatz, together with degree bounds. Motzkin also provided in [178] the aforementioned polynomial $M(x, y)$ in connection to Hilbert's seventeenth problem.

We now prove the existence of such a limiting sum-of-squares representation in two different ways for general polynomials $p(t)$ that are nonnegative on $[-1, 1]$, and in a constructive third way for the special family of polynomials

$$p_{\pm,n}(t) := (1 \pm t)(1 - t^2)^n, \qquad n \geq 0.$$

(Note, we used $p_{\pm,0}$ and $p_{\pm,1}$ to prove the local boundedness and continuity of f on $\mathbb{R}$, respectively; and Chapter 16 uses $p_{\pm,n}$ to prove that smoothness implies real analyticity.)

Proof 1　We claim more generally that for any dimension $d \geq 1$, every polynomial that is nonnegative on $[-1, 1]^d$ has a limiting sum-of-squares representation. This is proved at the end of the 1976 paper of Berg, Christensen, and Ressel [29].

Proof 2　Here is a constructive proof of a positivity certificate for the polynomials $p_{\pm,n}(t) = (1 \pm t)(1 - t^2)^n$, $n \geq 0$. (It turns out, we only need to work with these in order to show the stronger Schoenberg theorem.) First, notice that

$$
\begin{aligned}
p_{+,0}(t) &= (1 + t), & p_{-,0}(t) &= (1 - t), \\
p_{+,1}(t) &= (1 - t)(1 + t)^2, & p_{-,1}(t) &= (1 + t)(1 - t)^2, & (15.4) \\
p_{+,2}(t) &= (1 + t)(1 - t^2)^2, & p_{-,2}(t) &= (1 - t)(1 - t^2)^2,
\end{aligned}
$$

and so on. Thus, if we show that $p_{\pm,0}(t) = 1 \pm t$ are limits of sum-of-squares polynomials, then so are $p_{\pm,n}(t)$ for all $n \geq 0$ (where limits are taken in the Wiener norm). But we have

$$
\begin{aligned}
\frac{1}{2}(1 \pm t)^2 &= \frac{1}{2} \pm t + \frac{t^2}{2}, \\
\frac{1}{4}(1 - t^2)^2 &= \frac{1}{4} - \frac{t^2}{2} + \frac{t^4}{4}, & (15.5) \\
\frac{1}{8}(1 - t^4)^2 &= \frac{1}{8} - \frac{t^4}{4} + \frac{t^8}{8},
\end{aligned}
$$

and so on. Adding the first k of these equations shows that the partial sum

$$p_k^{\pm}(t) := (1 - \frac{1}{2^k}) \pm t + \frac{t^{2^k}}{2^k} = (1 \pm t) + \frac{t^{2^k} - 1}{2^k}$$

is a sum-of-squares polynomial, for every $k \geq 1$. This provides a positivity certificate for $1 \pm t$, as desired. It also implies the sought-for interpretation of the integration trick in Step 3 above

$$\left| \int_{-1}^{1} [p_k^{\pm}(t) - (1 \pm t)] \, d\sigma \right| \leq \int_{-1}^{1} \left| p_k^{\pm}(t) - (1 \pm t) \right| d\sigma$$

$$\leq \int_{-1}^{1} \frac{1}{2^k} \, d\sigma + \int_{-1}^{1} \frac{t^{2^k}}{2^k} \, d\sigma \leq \frac{1}{2^k} \cdot 2s_0(\sigma),$$

which goes to 0 as $k \to \infty$. Hence, using the notation following (15.3),

$$\int_{-1}^{1} (1 \pm t) \, d\sigma = \lim_{k \to \infty} \int_{-1}^{1} p_k^{\pm}(t) \, d\sigma$$

$$= \frac{1}{2}(e_0 \pm e_1)^T H_\sigma (e_0 \pm e_1)$$

$$+ \sum_{j=2}^{\infty} \frac{1}{2^j} (e_0 - e_{2^{j-1}})^T H_\sigma (e_0 - e_{2^{j-1}}),$$

and this is nonnegative because H_σ is positive semidefinite.

Proof 3   If we only want to interpret the integration trick (15.1) in terms of the positivity of the Hankel moment matrix H_σ, then the restriction of using the Wiener norm $\| \cdot \|_{1,+}$ can be relaxed, and one can work instead with the weaker notion of the uniform norm. With this metric, we claim more generally that *every continuous function* $f(t_1, \ldots, t_d)$ that is nonnegative on a compact subset $K \subset \mathbb{R}^d$ has a limiting sum-of-squares representation on K. (Specialized to $d = 1$ and $K = [-1, 1]$, this proves the integration trick.)

To see the claim, observe that $\sqrt{f(t_1, \ldots, t_d)} \colon K \to [0, \infty)$ is continuous, so by the Stone–Weierstrass theorem, there exists a polynomial sequence q_n converging uniformly to $\sqrt{f}$ in $L^\infty(K)$. Thus, $q_n^2 \to f$ in $L^\infty(K)$, as desired. Explicitly, if $d = 1$ and $q_n(t) = \sum_{k=0}^{\infty} c_{n,k} t^k$, then define the semi-infinite vectors

$$\mathbf{u}_n := (c_{n,0}, c_{n,1}, \ldots)^T, \qquad n \geq 1.$$

Now compute for any admissible measure σ supported in K

$$\int_K f \, d\sigma = \lim_{n \to \infty} \int_K q_n^2(t) \, d\sigma = \lim_{n \to \infty} \mathbf{u}_n^T H_\sigma \mathbf{u}_n \geq 0, \qquad (15.6)$$

which is a positivity certificate for all continuous, nonnegative functions on compact $K \subset \mathbb{R}$.

This reasoning extends to all dimensions $d \geq 1$ and compact $K \subset \mathbb{R}^d$, by Lemma 1.21.

15.3 From the Integration Trick to the Positivity-Certificate Trick

Proof 2 in Section 15.2 is the key to understanding why Hamburger's theorem is not required to prove the stronger Schoenberg theorem 14.13 (namely, (5) $\implies$ (3)). Specifically, we only need to use the following fact:

> For each fixed $n \geq 0$, if $\sum_k a_k t^k$ is the expansion of $p_{\pm,n}(t) = (1 \pm t)$ $(1 - t^2)^n \geq 0$ on $[-1,1]$, then $\sum_k a_k f(s_k(\mu)) \geq 0$.

This was derived above using the integration trick (15.1) via the auxiliary admissible measure σ_μ, which exists by Theorem 14.13(4). We now explain a workaround via a related "positivity-certificate trick" that requires using only that $f[H_\mu]$ is positive semidefinite, hence allowing us to work with the weaker hypothesis, Theorem 14.13(5) instead. In particular, one can avoid using Hamburger's theorem and requiring the existence of σ_μ.

The positivity-certificate trick is as follows:

Theorem 15.5   *Fix a semi-infinite real Hankel matrix $H = (f_{j+k})_{j,k \geq 0}$ that is positive semidefinite (i.e., its principal minors are positive semidefinite), with all entries f_j uniformly bounded. If a polynomial $p(t) = \sum_{j \geq 0} a_j t^j$ has a positivity certificate – i.e., a Wiener-limiting sum-of-squares representation – then $\sum_{j \geq 0} a_j f_j \geq 0$.*

According to Proof 2 in Section 15.2 above (see the discussion around (15.5)), Theorem 15.5 applies to $p = p_{\pm,n}$ for all $n \geq 0$ and $H = f[H_\mu]$, where f, μ are as in Theorem 14.13(5). This implies the continuity of the entrywise positivity preserver f in the above discussion, and also suffices to complete the proof in the next two chapters, of the stronger Schoenberg theorem (Theorem 14.9).

Proof of Theorem 15.5   As an illustrative special case, if $p(t)$ is the square of a polynomial $q(t) = \sum_{j \geq 0} c_j t^j$, then as in (15.3),

$$\sum_{j \geq 0} a_j f_j = \sum_{j,k \geq 0} c_j c_k f_{j+k} = \mathbf{u}^T H \mathbf{u} \geq 0, \quad \text{where } \mathbf{u} = (c_0, c_1, \ldots)^T.$$

By additivity, the result therefore also holds for a sum of squares of polynomials. The subtlety in working with a limiting sum-of-squares representation is that the degrees of each sum-of-squares polynomial in the limiting sequence need not be uniformly bounded. Nevertheless, suppose in the Wiener norm (15.2) that

$$p(t) = \lim_{n \to \infty} q_n(t), \quad \text{where } q_n(t) = \sum_{k=0}^{K_n} q_{n,k}(t)^2$$

is a sum of squares of polynomials for each n.

Define the linear functional Ψ_H (given the Hankel matrix H) that sends a polynomial $p(t) = \sum_{j \geq 0} a_j t^j$ to the scalar $\Psi_H(p) := \sum_{j \geq 0} a_j f_j$. Now define the vectors $\mathbf{u}_{n,k}$ via

$$q_{n,k}(t) = \sum_{j \geq 0} q_{n,k}^{[j]} t^j \in \mathbb{R}[t], \quad \mathbf{u}_{n,k} := \left(q_{n,k}^{[0]}, q_{n,k}^{[1]}, \ldots \right)^T.$$

Similarly, define $q_n(t) = \sum_{j \geq 0} q_n^{[j]} t^j$. Then for all $n \geq 1$,

$$\sum_{j \geq 0} q_n^{[j]} f_j = \Psi_H(q_n) = \sum_{k=0}^{K_n} \Psi_H(q_{n,k}^2) = \sum_{k=0}^{K_n} \mathbf{u}_{n,k}^T H \mathbf{u}_{n,k} \geq 0.$$

Finally, taking the limit as $n \to \infty$, and writing $p(t) = \sum_{j \geq 0} a_j t^j$, we claim that

$$\sum_{j \geq 0} a_j f_j = \lim_{n \to \infty} \sum_{j \geq 0} q_n^{[j]} f_j \geq 0.$$

Indeed, the (first) equality holds because if $M \geq \sup_j |f_j|$ is a uniform (and finite) upper bound, then

$$\left| \sum_{j \geq 0} q_n^{[j]} f_j - \sum_{j \geq 0} a_j f_j \right| \leq \sum_{j \geq 0} |q_n^{[j]} - a_j| \, |f_j| \leq M \|q_n - p\|_{1,+},$$

and this goes to zero as $n \to \infty$. □

16

Proof of the Stronger Schoenberg Theorem (Part II) – Real Analyticity

Having explained the positivity-certificate trick, we return to the proof of the stronger Schoenberg theorem. The present goal is to prove that if a smooth function $f : \mathbb{R} \to \mathbb{R}$ satisfies assertion (5) in Theorem 14.13, then f is real analytic and hence satisfies assertion (3) in Theorem 14.9. (See Steps (4) and (5) in the list following Remark 14.14.) To show these results, we first discuss the basic properties of real analytic functions that are required in the proofs.

16.1 Preliminaries on Real Analytic Functions

Definition 16.1 Suppose $I \subset \mathbb{R}$ is an open interval, and $f : I \to \mathbb{R}$ is smooth, denoted $f \in C^\infty(I)$. Recall that the *Taylor series* of f at a point $x \in I$ is

$$(Tf)_x(y) := \sum_{j=0}^{\infty} \frac{f^{(j)}(x)}{j!} (y - x)^j, \qquad y \in I,$$

if this sum converges at y. Notice that this sum is not equal to $f(y)$ in general.

Next, we say that f is *real analytic on* I, denoted $f \in C^\omega(I)$, if $f \in C^\infty(I)$ and for all $x \in I$ there exists $\delta_x > 0$, such that the Taylor series of f at x converges to f on $(x - \delta_x, x + \delta_x)$.

Clearly, real analytic functions on I form a real vector space. Less obvious is the following useful property, which is stated without proof:

Proposition 16.2 *Real analytic functions are closed under composition. More precisely, if $I \xrightarrow{f} J \xrightarrow{g} \mathbb{R}$, and f, g are real analytic on their domains, then so is $g \circ f$ on I.*

We also develop a few preliminary results on real analytic functions, which are needed to prove the stronger Schoenberg theorem. We begin with an example of real analytic functions, which depicts what happens in our setting.

Lemma 16.3   *Suppose $I = (0, R)$ for $0 < R \leq \infty$, and $f(x) = \sum_{k=0}^{\infty} c_k x^k$ on I, where $c_k \geq 0 \ \forall k$. Then $f \in C^{\omega}(I)$, and $(Tf)_a(x)$ converges whenever $a, x \in I$ are such that $|x - a| < R - a$.*

In particular, if $R = \infty$ and $a > 0$, then $(Tf)_a(x) \to f(x)$ on the domain of f.

Proof   Note that $\sum_{k=0}^{\infty} c_k x^k$ converges on $(-R, R)$. Thus, we denote this extension to $(-R, R)$ also by f, and show more generally that $(Tf)_a(x)$ converges to $f(x)$ for $|x - a| < R - a$, $a \in [0, R)$ (whenever f is defined at x). Indeed,

$$f(x) = \sum_{k=0}^{\infty} c_k((x - a) + a)^k = \sum_{k=0}^{\infty} \sum_{j=0}^{k} \binom{k}{j} c_k (x - a)^j a^{k-j}.$$

Notice that this double sum is absolutely convergent, since

$$\sum_{k=0}^{\infty} \sum_{j=0}^{k} \binom{k}{j} c_k |x - a|^j a^{k-j} = f(a + |x - a|) < \infty.$$

Hence, we can rearrange the double sum (e.g., by Fubini's theorem), to obtain

$$f(x) = \sum_{j=0}^{\infty} \left(\sum_{m=0}^{\infty} \binom{m + j}{j} c_{m+j} a^m \right) (x - a)^j$$

$$= \sum_{j=0}^{\infty} \frac{f^{(j)}(a)}{j!} (x - a)^j = (Tf)_a(x)$$

using standard properties of power series. In particular, f is real analytic on I. $\square$

We also require the following well-known result on zeros of real analytic functions:

Theorem 16.4 (Identity theorem)   *Suppose $I \subset \mathbb{R}$ is an open interval and f, $g : I \to \mathbb{R}$ are real analytic. If the subset of I where $f = g$ has an accumulation point in I, then $f \equiv g$ on I.*

In other words, the zeros of a nonzero (real) analytic function form a discrete set.

Proof   Without loss of generality, we may suppose $g \equiv 0$. Suppose $c \in I$ is an accumulation point of the zero set of f. Expand f locally at c into its Taylor series, and claim that $f^{(k)}(c) = 0$ for all $k \geq 0$. Indeed, suppose for contradiction that

$$f^{(0)}(c) = \cdots = f^{(k-1)}(c) = 0 \neq f^{(k)}(c)$$

for some $k \geq 0$. Then,

$$\frac{f(x)}{(x-c)^k} = \frac{f^{(k)}(c)}{k!} + o(x-c),$$

so f is nonzero close to c, and this contradicts the hypotheses. Thus, $f^{(k)}(c) = 0 \; \forall k \geq 0$, which in turn implies that $f \equiv 0$ on an open interval around c.

Now consider the set $I_0 := \{x \in I : f^{(k)}(x) = 0 \; \forall k \geq 0\}$. Clearly I_0 is a closed subset of I. Moreover, if $c_0 \in I_0$ then $f \equiv (Tf)_{c_0} \equiv 0$ near c_0, hence the same happens at any point near c_0 as well. Thus, I_0 is also an open subset of I. Since I is connected, $I_0 = I$, and $f \equiv 0$.   □

16.2 Proof of the Stronger Schoenberg Theorem for Smooth Functions

We continue with the proof of the stronger Schoenberg theorem ((5) $\implies$ (2) in Theorems 14.9 and 14.13).

Akin to the proof of the stronger Horn–Loewner theorem 12.1, we have shown that any function satisfying the hypotheses in Theorem 14.13(5) must be continuous. Hence, by the first two steps in the proof – listed after Remark 14.14 – we have that $f(x) = \sum_{k=0}^{\infty} c_k x^k$ on $[0, \infty)$, with all $c_k \geq 0$.

Again, similar to the proof of the stronger Horn–Loewner theorem 12.1, we next prove the stronger Schoenberg theorem for smooth functions. The key step here is:

Theorem 16.5   *Let $f \in C^{\infty}(\mathbb{R})$ be as in the preceding discussion, and define the family of smooth functions*

$$H_a(x) := f(a + e^x), \qquad a, x \in \mathbb{R}.$$

Then H_a is real analytic on $\mathbb{R}$, for all $a \in \mathbb{R}$.

For ease of exposition, we break the proof into several steps.

Lemma 16.6   *For all $n \geq 1$, we have*

$$H_a^{(n)}(x) = a_{n,1} f'(a + e^x)e^x + a_{n,2} f''(a + e^x)e^{2x}$$
$$+ \cdots + a_{n,n} f^{(n)}(a + e^x)e^{nx},$$

where $a_{n,j}$ is a positive integer for all $1 \leq j \leq n$.

Proof and remarks   One shows by induction on $n \geq 1$ (with the base case of $n = 1$ immediate) that the array $a_{n,j}$ forms a weighted variant of Pascal's triangle, in that

$$a_{n,j} = \begin{cases} 1, & \text{if } j = 1, n, \\ a_{n-1,j-1} + j a_{n-1,j}, & \text{otherwise.} \end{cases}$$

This concludes the proof. Notice that some of the entries of the array $a_{n,j}$ are easy to compute inductively:

$$a_{n,1} = 1, \qquad a_{n,2} = 2^{n-1} - 1, \qquad a_{n,n-1} = \binom{n}{2}, \qquad a_{n,n} = 1.$$

An interesting combinatorial exercise may be to seek a closed-form expression and a combinatorial interpretation for the other entries.   □

Lemma 16.7   *We have the following bound:*

$$|H_a^{(n)}(x)| \leq H_{|a|}^{(n)}(x), \qquad \forall a, x \in \mathbb{R}, \ n \in \mathbb{Z}^{\geq 0}. \tag{16.1}$$

Proof   By Lemma 16.6 we have that $H_{|a|}^{(n)}(x) \geq 0$ for all a, x, n as in (16.1), so it remains to show the inequality. For this, we assume $a < 0$, and use the positivity-certificate trick from Chapter 15 – i.e., Theorem 15.5, applied to the polynomials

$$p_{\pm,n}(t) := (1 \pm t)(1 - t^2)^n, \qquad n \geq 0$$

and the admissible measure

$$\mu := |a|\delta_{-1} + e^x \delta_{e^{-h}}, \qquad a, h > 0, \ x \in \mathbb{R}.$$

Notice that $p_{\pm,n} \geq 0$ on $[-1, 1]$. Hence, by Theorem 15.5 – and akin to the calculation in Chapter 15 to prove continuity – we get

$$\sum_{k=0}^{n} \binom{n}{k} (-1)^{n-k} f(|a| + e^{x-2kh}) \geq \left| \sum_{k=0}^{n} \binom{n}{k} (-1)^{n-k} f(a + e^{x-(2k+1)h}) \right|.$$

Dividing both sides by $(2h)^n$ and sending $h \to 0^+$, we obtain

$$H_{|a|}^{(n)}(x) \geq |H_a^{(n)}(x)|. \qquad \qquad \square$$

Remark 16.8   In this computation, we do not need to use the measures $\mu = |a| \delta_{-1} + e^x \delta_{e^{-h}}$ for all $h > 0$. It suffices to fix a single $u_0 \in (0, 1)$ and consider the sequence $h_n := -\log(u_0)/n$, so we work with $\mu = |a| \delta_{-1} + e^x \delta_{u_0^{1/n}}$ (supported at $1, u_0^{1/n}$) for $a > 0, x \in \mathbb{R}, n \geq 1$.

Lemma 16.9   *For all integers $n \geq 0$, the assignment $(a, x) \mapsto H_a^{(n)}(x)$ is nondecreasing in both $a \geq 0$ and $x \in \mathbb{R}$. In particular if $a \geq 0$, then H_a is absolutely monotonic on $\mathbb{R}$, and its Taylor series at $b \in \mathbb{R}$ converges absolutely at all $x \in \mathbb{R}$.*

Proof   The monotonicity in $a \geq 0$ follows from the absolute monotonicity of $f|_{[0, \infty)}$ mentioned at the start of Section 16.2. The monotonicity in x for a fixed $a \geq 0$ follows because $H_a^{(n+1)}(x) \geq 0$ by Lemma 16.6.

To prove the (absolute) convergence of $(T H_a)_b$ at $x \in \mathbb{R}$, notice that

$$|(T H_a)_b(x)| = \left| \sum_{n=0}^{\infty} H_a^{(n)}(b) \frac{(x-b)^n}{n!} \right| \leq \sum_{n=0}^{\infty} H_a^{(n)}(b) \frac{|x-b|^n}{n!}$$

$$= (T H_a)_b(b + |x - b|).$$

We claim that this final (Taylor) series is bounded above by $H_a(b + |x - b|)$, which would complete the proof. Indeed, by Taylor's theorem, the nth Taylor remainder term for $H_a(b + |x - b|)$ can be written as (see, e.g., (14.1))

$$\int_b^{b+|x-b|} \frac{(b + |x-b| - t)^n}{n!} H_a^{(n+1)}(t) \, dt,$$

which is nonnegative from above. Taking $n \to \infty$ shows the claim and completes the proof. $\square$

Now we can prove the real analyticity of H_a:

Proof of Theorem 16.5   Fix scalars $a, \delta > 0$. We show that for all $b \in [-a, a]$ and $x \in \mathbb{R}$, the nth remainder term for the Taylor series $T H_b$ around the point x converges to zero as $n \to \infty$, uniformly near x. More precisely, define

$$\Psi_n(x) := \sup_{y \in [x-\delta, x+\delta]} \left| R_n((T H_b)_x)(y) \right|.$$

We then claim $\Psi_n(x) \to 0$ as $n \to \infty$ for all x. This will imply that at all $x \in \mathbb{R}$, $(TH_b)_x$ converges to H_b on a neighborhood of radius δ. Moreover, this holds for all $\delta > 0$ and at all $b \in [-a,a]$ for all $a > 0$.

Thus, it remains to prove for each $x \in \mathbb{R}$ that $\Psi_n(x) \to 0$ as $n \to \infty$. By the above results, we have for all $n \in \mathbb{Z}^{\geq 0}$

$$|H_b^{(n)}(y)| \leq H_{|b|}^{(n)}(y) \leq H_a^{(n)}(y) \leq H_a^{(n)}(x + \delta),$$

$$\forall b \in [-a,a], \ y \in [x - \delta, x + \delta].$$

Using a standard estimate for the Taylor remainder, for all b, y, n as above, it follows that

$$|R_n((TH_b)_x)(y)| \leq H_a^{(n+1)}(x+\delta)\frac{|y-x|^{n+1}}{(n+1)!} \leq H_a^{(n+1)}(x+\delta)\frac{\delta^{n+1}}{(n+1)!}.$$

But the right-hand term goes to zero by the calculation in Lemma 16.9, since

$$0 \leq \sum_{n=-1}^{\infty} H_a^{(n+1)}(x+\delta)\frac{\delta^{n+1}}{(n+1)!} \leq H_a(x+\delta+\delta) = f(a + e^{x+2\delta}) < \infty.$$

Hence, we obtain

$$\lim_{n \to \infty} \sup_{y \in [x-\delta, x+\delta]} |R_n((TH_b)_x)(y)| \to 0, \quad \forall x \in \mathbb{R}, \ \delta > 0, \ b \in [-a,a], \ a > 0.$$

From above, this shows that the Taylor series of H_b converges locally to H_b at all $x \in \mathbb{R}$, for all b as desired. (In fact, the "local" neighborhood of convergence around x is all of $\mathbb{R}$.) □

With the above analysis in hand, we can prove Steps 4 and 5 of the proof of the stronger Schoenberg theorem (see the list after Remark 14.14):

Suppose $f : \mathbb{R} \to \mathbb{R}$ satisfies assertion (5) of Theorem 14.13.

(4) *If f is smooth on $\mathbb{R}$, then f is real analytic on $\mathbb{R}$.*
(5) *If f is real analytic on $\mathbb{R}$, then $f(x) = \sum_{k=0}^{\infty} c_k x^k$ on $\mathbb{R}$, with $c_k \geq 0 \ \forall k$.*

Proof of Step 4 for the stronger Schoenberg theorem Given $x \in \mathbb{R}$, we want to show that the Taylor series $(Tf)_x$ converges to f locally around x. Choose $a > |x|$ and define

$$L_a(y) := \log(a + y) = \log(a) + \log(1 + y/a), \quad y \in (-a, a).$$

This is real analytic on $(-a, a)$ (e.g., akin to Lemma 16.3). Hence, by Proposition 16.2 and Theorem 16.5, the composite

$$y \xrightarrow{L_a} \log(a + y) \xrightarrow{H_{-a}} H_{-a}(L_a(y)) = f(-a + \exp(\log(a + y))) = f(y)$$

is also real analytic on $(-a, a)$, so around $x \in \mathbb{R}$.   □

Proof of Step 5 for the stronger Schoenberg theorem   By Step 4, f is real analytic on $\mathbb{R}$; and, by Steps 1 and 2 $f(x) = \sum_{k=0}^{\infty} c_k x^k$ on $(0, \infty)$, with $c_k \geq 0 \; \forall k$. Let $g(x) := \sum_{k=0}^{\infty} c_k x^k \in C^{\omega}(\mathbb{R})$. Since $f \equiv g$ on $(0, \infty)$, it follows by the Identity Theorem 16.4 that $f \equiv g$ on $\mathbb{R}$.   □

17

Proof of the Stronger Schoenberg Theorem (Part III) – Complex Analysis

We can now complete the proof of the final Step 6 (listed after Remark 14.14) of the stronger Schoenberg theorem. Namely, suppose $f : \mathbb{R} \to \mathbb{R}$ is such that for each measure

$$\mu = a\delta_1 + b\delta_{u_0} + c\delta_{-1}, \quad \text{with } u_0 \in (0,1), \ a,b,c \geq 0,$$

with semi-infinite Hankel moment matrix H_μ, the matrix $f[H_\mu]$ is positive semidefinite.

Under these assumptions, we have previously shown (in Steps 1, 2; 3; 4, 5 respectively):

- There exist real scalars $c_0, c_1, \ldots \geq 0$ such that $f(x) = \sum_{k=0}^\infty c_k x^k$ for all $x \in (0, \infty)$.
- f is continuous on $\mathbb{R}$.
- If f is smooth, then $f(x) = \sum_{k=0}^\infty c_k x^k$ on $\mathbb{R}$.

We now complete the proof by showing that one can pass from smooth functions to continuous functions. The tools we will use are the "three Ms": Montel, Morera, and Mollifiers. We first discuss some basic results in complex analysis that are required.

17.1 Tools from Complex Analysis

Definition 17.1   Suppose $D \subset \mathbb{C}$ is open and $f : D \to \mathbb{C}$ is a continuous function.

(1) (Holomorphic) A function f is *holomorphic at a point* $z \in D$ if the limit $\lim_{y \to z} \frac{f(y)-f(z)}{y-z}$ exists. A function f is *holomorphic on D* if it is holomorphic at every point of D.

(2) (Complex analytic) f is said to be *complex analytic around* $c \in D$ if f
    can be expressed as a power series locally around c, which converges to
    $f(z)$ for every z sufficiently close to c. Similarly, f is *analytic on* D if it
    is analytic at every point of D.
(3) (Normal) Let $\mathcal{F}$ be a family of holomorphic functions: $D \to \mathbb{C}$. Then $\mathcal{F}$ is
    normal if given any compact $K \subset D$ and a sequence $\{f_n : n \geq 1\} \subset \mathcal{F}$,
    there exists a subsequence f_{n_k} and a function $f : K \to \mathbb{C}$, such that
    $f_{n_k} \to f$ uniformly on K.

Remark 17.2   Note that it is not specified that the limit function f
be holomorphic. However, this will turn out to be the case, as we shall see
presently.

We use without proof the following results (and Cauchy's theorem, which
we do not state):

Theorem 17.3   *Let $D \subset \mathbb{C}$ be an open subset.*

(1) *A function $f : D \to \mathbb{C}$ is holomorphic if and only if f is complex analytic.*
(2) *(Montel) Let $\mathcal{F}$ be a family of holomorphic functions on D. If $\mathcal{F}$ is
    uniformly bounded on D, then $\mathcal{F}$ is normal on D.*
(3) *(Morera) Suppose that for every closed oriented piecewise C^1 curve γ in
    D, we have $\oint_\gamma f \, dz = 0$. Then f is holomorphic on D.*

17.2 Proof of the Stronger Schoenberg Theorem: Conclusion

Let $f : \mathbb{R} \to \mathbb{R}$ be as described above; in particular, f is continuous on $\mathbb{R}$
and absolutely monotonic on $(0, \infty)$. As discussed in the proof of the stronger
Horn–Loewner theorem 12.1, we mollify f with the family $\phi_\delta(u) = \phi(u/\delta)$
for $\delta > 0$ as in Proposition 13.4. As shown in (13.2), f_δ satisfies assertion (5)
in Theorem 14.13, so (e.g., by the last bulleted point above, and Steps 4 and 5)

$$f_\delta(x) = \sum_{k=0}^{\infty} c_{k,\delta} x^k \ \forall x \in \mathbb{R}, \quad \text{with } c_{k,\delta} \geq 0 \ \forall k \geq 0, \ \delta > 0.$$

Since f_δ is a power series with infinite radius of convergence, it extends
analytically to an entire function on $\mathbb{C}$ (see, e.g., Lemma 16.3). Let us call
this f_δ as well; now define

$$\mathcal{F} := \{f_{1/n} : n \geq 1\}.$$

We claim that for any $0 < r < \infty$, the family $\mathcal{F}$ is uniformly bounded on the complex disk $D(0, r)$. Indeed, since $f_\delta \to f$ uniformly on $[0, r]$ by Proposition 13.4, we have that $|f_{1/n} - f|$ is uniformly bounded over all n and on $[0, r]$, say by $M_r > 0$. Now if $z \in D(0, r)$, then

$$|f_{1/n}(z)| \le \sum_{k=0}^{\infty} c_{k,1/n}|z|^k = f_{1/n}(|z|) \le M_r + f(|z|) \le M_r + f(r) < \infty,$$

and this bound (uniform over $z \in D(0, r)$) does not depend on n.

By Montel's theorem, the previous claim implies that $\mathcal{F}$ is a normal family on $D(0, r)$ for each $r > 0$. Hence, on the closed disk $\overline{D}(0, r)$, there is a subsequence f_{1/n_l} with n_l increasing, which converges uniformly to some (continuous) $g = g_r$. Since f_{1/n_l} is holomorphic for all $l \ge 1$, by Cauchy's theorem we obtain for every closed oriented piecewise C^1 curve $\gamma \subset D(0, r)$

$$\oint_\gamma g_r \, dz = \oint_\gamma \lim_{l \to \infty} f_{1/n_l} \, dz = \lim_{l \to \infty} \oint_\gamma f_{1/n_l} \, dz = 0.$$

It follows by Morera's theorem that g_r is holomorphic, hence analytic, on $D(0, r)$. Moreover, $g_r \equiv f$ on $(-r, r)$ by the properties of mollifiers; thus, f is real analytic on $(-r, r)$ for every $r > 0$. Now apply the previous step of the proof – i.e., the Identity Theorem 16.4, together with the power series representation of f on $(0, \infty)$, to conclude the proof. □

17.3 Concluding Remarks and Variations

We conclude with several generalizations of the above results. First, the results by Horn–Loewner, Vasudeva, and Schoenberg (more precisely, their stronger versions) that were shown in this part of the text, together with the proofs given above, can be refined to versions with *bounded domains* $(0, \rho)$ or $(-\rho, \rho)$ for $0 < \rho < \infty$. The small change is to use admissible measures with bounded mass

$$\mu = a\delta_1 + b\delta_{u_0} + c\delta_{-1}, \qquad \text{where } u_0 \in (0, 1), \ a, b, c \ge 0$$

as above, but moreover, one now imposes the condition that $s_0(\mu) = a + b + c < \rho$.

Second, all of these results, including for bounded domains (i.e., masses of the underlying measures), can be extended to studying functions of several variables. In this case, given a domain $I \subset \mathbb{R}$ and integers $m, n \ge 1$, a function $f : I^m \to \mathbb{R}$ acts entrywise on an *m-tuple of $n \times n$ matrices* $A_1 = (a_{jk}^{(1)}), \ldots, A_m = (a_{jk}^{(m)})$ in $I^{m \times m}$, via

$$f[A_1, \ldots, A_m] := (f(a_{jk}^{(1)}, \ldots, a_{jk}^{(m)}))_{j,k=1}^n. \qquad (17.1)$$

One can now ask the multivariable version of the same question as above:

"Which functions, when applied entrywise to m-tuples of positive (semidefinite) matrices, preserve positivity?"

Observe that the coordinate functions $f(x_1, \ldots, x_m) := x_l$ work for all $1 \le l \le m$. Hence, by the Schur product theorem and the Pólya–Szegő observation (Lemma 11.1, since $\mathbb{P}_n$ is a closed convex cone for all $n \ge 1$), every convergent multipower series of the form

$$f(\mathbf{x}) := \sum_{\mathbf{n} \ge 0} c_{\mathbf{n}} \mathbf{x}^{\mathbf{n}}, \qquad \text{with } c_{\mathbf{n}} \ge 0 \ \forall \mathbf{n} \ge 0 \qquad (17.2)$$

preserves positivity in all dimensions (where $\mathbf{x}^{\mathbf{n}} := x_1^{n_1} \cdots x_m^{n_m}$, etc.). Akin to the Schoenberg–Rudin theorem in the one-variable case, it was shown by Fitzgerald et al. in 1995 [85] that the functions (17.2) are the only such preservers.

One can ask if the same result holds when one restricts the test set to m-tuples of Hankel matrices of rank at most 3, as in the treatment above. While this does turn out to yield the same classification, the proofs get more involved and now require multivariable machinery. For these stronger multivariate results, we refer the reader to the paper *"Moment-sequence transforms"* by Belton et al. [25].

18

Preservers of Loewner Positivity on Kernels

In this chapter and Chapter 19, we study the transforms that preserve positive (semi)definiteness, and Loewner monotonicity and convexity, on kernels on infinite domains. We begin with preservers of positive semidefinite and positive definite kernels.

Definition 18.1 Let X, Y be nonempty sets, and $K : X \times Y \to \mathbb{R}$ a kernel.

(1) Given $\mathbf{x} \in X^m$ and $\mathbf{y} \in Y^n$ for integers $m, n \geq 1$, define $K[\mathbf{x}; \mathbf{y}]$ to be the $m \times n$ real matrix, with (j, k) entry $K(x_j, y_k)$.

(2) Given an integer $n \geq 1$, define $X^{n, \neq}$ to be the set of all n-tuples in X with pairwise distinct coordinates.

(3) A kernel $K : X \times X \to \mathbb{R}$ is said to be *positive semidefinite* (respectively, *positive definite*) if K is symmetric – i.e., $K(x, y) = K(y, x) \, \forall x, y \in X$ – and for all $n \geq 1$ and tuples $\mathbf{x} \in X^{n, \neq}$, the matrix $K[\mathbf{x}; \mathbf{x}]$ is positive semidefinite (respectively, positive definite).

By "padding principal submatrices by the identity kernel," it is easily seen that given subsets $X \subset Y$ and a positive (semi)definite kernel K on $X \times X$, we can embed K into a kernel $\widetilde{K} : Y \times Y \to \mathbb{R}$ that is also positive (semi)definite: define $\widetilde{K}(x, y)$ to be $\mathbf{1}_{x=y}$ if $(x, y) \notin X \times X$, and $K(x, y)$ otherwise.

Now given a set X and a domain $I \subset \mathbb{R}$, we will study the inner transforms

$$\mathscr{F}_X^{\text{psd}}(I) := \big\{ F : I \to \mathbb{R} \mid \text{if } K : X \times X \to I \text{ is positive}$$
$$\text{semidefinite, so is } F \circ K \big\},$$

$$\mathscr{F}_X^{\text{pd}}(I) := \{ F : I \to \mathbb{R} \mid \text{if } K : X \times X \to I \text{ is positive definite, so is } F \circ K \}.$$

(see the beginning of this text). Here, $F \circ K$ sends $X \times X$ to $\mathbb{R}$.

Notice that if X is finite then $\mathscr{F}_X^{\text{psd}}(I)$ is precisely the set of entrywise maps preserving positivity on $\mathbb{P}_{|X|}(I)$; as mentioned in Section 11.8, this question remains open for all $|X| \geq 3$. If instead X is infinite, then the answer follows from Schoenberg and Rudin's results:

Theorem 18.2   *Fix* $0 < \rho \leq \infty$, *and suppose* I *is any of* $(0, \rho)$, $[0, \rho)$, *or* $(-\rho, \rho)$. *If* X *is an infinite set, then* $\mathscr{F}_X^{\mathrm{psd}}(I)$ *consists of all power series with nonnegative Maclaurin coefficients, which are convergent on* I.

This observation is useful in the study of positive definite kernels in computer science.

Proof   For $I = (0, \rho)$ or $(-\rho, \rho)$ with $\rho = \infty$, the result follows by embedding every positive semidefinite matrix into a kernel on $X \times X$, and applying Theorems 11.4 and 11.3, respectively. If $I = [0, \rho)$, then from above we have the desired power series expansion on $(0, \infty)$, and it remains to show that any preserver F is right continuous at 0. To see why, first note that $F(0) \geq 0$, and F is nondecreasing and nonnegative on $(0, \infty)$, so $F(0^+) := \lim_{x \to 0^+} F(x)$ exists. Now consider a three-point subset $\{x_1, x_2, x_3\}$ of X, with complement X', and define the kernel

$$K_0(x, y) := \begin{cases} 3, & \text{if } x = y, \\ 1, & \text{if } (x, y) = (x_1, x_2), (x_2, x_3), (x_3, x_2), (x_2, x_1), \\ 0, & \text{otherwise.} \end{cases} \quad (18.1)$$

Thus, K_0 is the padding by the identity of a positive definite 3×3 matrix. It follows that $F \circ (cK_0)$ is positive semidefinite for $c > 0$, so its principal submatrix $\begin{pmatrix} F(3c) & F(0) \\ F(0) & F(3c) \end{pmatrix} \in \mathbb{P}_2$. It follows by taking determinants and then $c \to 0^+$ that $F(0^+) \geq F(0) \geq 0$.

Finally, let $\mathbf{x} := (x_1, x_2, x_3)$; then

$$0 \leq \lim_{c \to 0^+} \det F[cK_0[\mathbf{x}; \mathbf{x}]] = -F(0^+)(F(0^+) - F(0))^2.$$

Thus, either $F(0^+) > 0$, in which case $F(0^+) = F(0)$, or else $F(0^+) = 0$, hence $F(0) = 0 = F(0^+)$ as well. This concludes the proof for $\rho = \infty$; for $\rho < \infty$, the result follows by the remarks in Section 17.3. □

Having worked out the preservers of positive semidefinite kernels, we move on to the positive definite case. If X is finite, then the fixed-dimension case is again open; but for infinite X we have:

Theorem 18.3   *Fix* $0 < \rho \leq \infty$, *and suppose* I *is any of* $(0, \rho)$, $[0, \rho)$, *or* $(-\rho, \rho)$. *If* X *is an infinite set, then* $\mathscr{F}_X^{pd}(I)$ *consists of all nonconstant power series with nonnegative Maclaurin coefficients, which are convergent on* I.

Proof   By the Schur product theorem, every monomial x^k for $k \geq 1$ preserves positive definiteness. This observation shows one implication.

Conversely, suppose first that $F \in \mathscr{F}_X^{pd}(I)$ is continuous. Now every positive semidefinite kernel $K: X \times X \to \mathbb{R}$ is the pointwise limit as $\epsilon \to 0^+$ of the family $K_\epsilon: X \times X \to \mathbb{R}$, given by $K_\epsilon(x, y) := K(x, y) + \epsilon \mathbf{1}_{x=y}$. Since F is continuous, it follows that F preserves positive semidefinite kernels as well, reducing the problem to the preceding result. Moreover, F is clearly not constant, e.g., by considering its action on the identity kernel.

The rest of the proof is devoted to showing that F is continuous on I. First suppose $I = (0, \rho)$ and $A \in \mathbb{P}_2(I)$ is positive definite. Then there exists $\epsilon \in (0, \rho/2)$ such that $A' := A - \epsilon \operatorname{Id}_{2 \times 2}$ is still positive definite. Choose $x_1, x_2 \in X$ and define the kernel

$$K: X \times X \to \mathbb{R}, \qquad (x, y) \mapsto \begin{cases} a_{jk}, & \text{if } x = x_j, y = x_k, \ 1 \leq j, k \leq 2; \\ \rho/2, & \text{if } x = y \notin \{x_1, x_2\}; \\ \epsilon, & \text{otherwise.} \end{cases}$$

Clearly,

$$K = \epsilon' \mathbf{1}_{X \times X} + \left(A' \oplus (\rho/2 - \epsilon) \operatorname{Id}_{X \setminus \{x_1, x_2\}} \right),$$

and so K is positive definite on X with all values in $I = (0, \rho)$. Hence, $F \circ K$ is also positive definite. It follows that the entrywise map $F[-]$ preserves positive definiteness on 2×2 matrices. Now invoke Lemma 6.9 to conclude that F is continuous on $(0, \rho)$.

This concludes the proof for $I = (0, \rho)$. Next, suppose $I = [0, \rho)$; by the preceding case, F is given by a nonconstant power series as asserted, and we just need to show F is right continuous at 0. Since F is increasing on $(0, \rho)$, the limit $F(0^+) := \lim_{x \to 0^+} F(x)$ exists and $F(0^+) \geq F(0) \geq 0$. Now use the kernel K_0 from (18.1) and repeat the subsequent arguments.

The final case is if $I = (-\rho, \rho)$. In this case we fix $u_0 \in (0, 1)$ and a countable subset $Y := \{x_0, x_1, \ldots\} \subset X$. Denote $Y^c := X \setminus Y$. Given $a, b > 0$ such that $a + b < \rho$, let the measure

$$\mu = \mu_{a,b} := a\delta_{-1} + b\delta_{u_0}.$$

The corresponding Hankel moment matrix is H_μ, with (j,k) entry $a(-1)^{j+k} + bu_0^{j+k}$, and this is positive semidefinite of rank 2. Now for each $\epsilon > 0$, define the kernel $K_\epsilon : X \times X \to \mathbb{R}$, via

$$K_\epsilon(x,y) := \begin{cases} H_\mu(j,j) + \epsilon, & \text{if } x = y = x_j, \ j \geq 0; \\ H_\mu(j,k), & \text{if } (x,y) = (x_j, x_k), \ j \neq k; \\ \epsilon, & \text{if } x = y \in Y^c; \\ 0, & \text{otherwise.} \end{cases}$$

Clearly, K_ϵ is positive definite, with entries in $I = (-\rho, \rho)$ for sufficiently small $\epsilon > 0$. It follows that $F \circ K_\epsilon$ is positive definite. Since F is continuous on $[0, \rho)$ by the previous cases, $\lim_{\epsilon \to 0^+} F \circ K_\epsilon = F[H_\mu \oplus 0_{Y^c \times Y^c}]$ is positive semidefinite, and so $F[-]$ preserves positivity on the Hankel moment matrices H_μ for all $\mu = \mu_{a,b}$ as above. It follows by the proof of Step 3 for the stronger Schoenberg theorem above (see the computations following Lemma 15.3) that F is continuous on $(-\rho, \rho)$, as desired. This concludes the proof in all cases. □

19

Preservers of Loewner Monotonicity and Convexity on Kernels

Thus far, we have studied the preservers of positive (semi)definiteness, with a brief look in Chapter 9 at entrywise powers preserving other Loewner properties. In this chapter, we return to these properties. Specifically, we classify all composition operators preserving Loewner monotonicity and convexity, on kernels on infinite domains. (The case of finite domains remains open, as for positivity preservers.)

The results for infinite domains will crucially use the finite versions; thus, we begin by reminding the reader of the definitions. Roughly speaking, a function is *Loewner monotone* (see Definition 8.5) if $f[A] \geq f[B]$ whenever $A \geq B \geq 0_{n \times n}$. Similarly, a function is *Loewner convex* (see Definition 9.7) if $f[\lambda A + (1 - \lambda)B] \leq \lambda f[A] + (1 - \lambda)f[B]$ whenever $A \geq B \geq 0$ and $\lambda \in [0, 1]$.

As explained in Remark 8.6, for $n = 1$ the usual notion of a monotonically nondecreasing function coincides with Loewner monotonicity. The same holds for convex functions vis-à-vis Loewner convex functions, for $n = 1$. Now for $n = 1$, a differentiable function $f : (0, \infty) \to \mathbb{R}$ is monotone (respectively, convex) if and only if f' is positive, i.e., has image in $[0, \infty)$ (respectively, monotone). The following result by Hiai in 2009 [119] extends this to the corresponding Loewner properties, in every dimension:

Theorem 19.1 (Hiai, fixed dimension) *Suppose $0 < \rho \leq \infty$, $I = (-\rho, \rho)$, and $f : I \to \mathbb{R}$.*

(1) *Given $n \geq 2$, the function f is Loewner convex on $\mathbb{P}_n(I)$ if and only if f is differentiable on I and f' is Loewner monotone on $\mathbb{P}_n(I)$. This result also holds if we restrict both test sets to rank $\leq k$ matrices in $\mathbb{P}_n(I)$ for every $2 \leq k \leq n$.*

(2) *Given $n \geq 3$, the function f is Loewner monotone on $\mathbb{P}_n(I)$ if and only if
f is differentiable on I and f' is Loewner positive on $\mathbb{P}_n(I)$.*

Recall the related but somewhat weaker variant in Proposition 9.9.

Here we show the first part and a weaker version of the second part of
Theorem 19.1 – see Hiai's 2009 paper for the complete proof. (Note: Hiai
showed the first part only for $k = n$; also, we do not use the second part in the
present text.) First, as a consequence of the first part and the previous results,
we obtain the following Schoenberg-type classification of the corresponding
preservers:

Theorem 19.2 (Dimension-free preservers of monotonicity and convexity)
*Suppose $0 < \rho \leq \infty$, $I = (-\rho, \rho)$, and $f : I \to \mathbb{R}$. The following are
equivalent:*

(1) *f is Loewner monotone on $\mathbb{P}_n(I)$ for all n.*
(2) *f is Loewner monotone on the rank ≤ 3 Hankel matrices in $\mathbb{P}_n(I)$
for all n.*
(3) *$f(x) = \sum_{k=0}^{\infty} c_k x^k$ on I, with $c_1, c_2, \cdots \geq 0$.*

Similarly, one has equivalent conditions characterizing Loewner convexity:

(1) *f is Loewner convex on $\mathbb{P}_n(I)$ for all n.*
(2) *f is Loewner convex on the rank ≤ 3 matrices in $\mathbb{P}_n(I)$ for all n.*
(3) *$f(x) = \sum_{k=0}^{\infty} c_k x^k$ on I, with $c_2, c_3, \cdots \geq 0$.*

Proof   We begin with the dimension-free Loewner monotone maps. Clearly,
(1) $\implies$ (2). To show (3) $\implies$ (1), note that $f(x) - c_0$ is also Loewner
monotone for any $c_0 \in \mathbb{R}$ if $f(x)$ is, so it suffices to consider $f(x) = x^k$
for $k \geq 1$. But such a function is clearly monotone, by the Schur product
theorem. This is an easy exercise, or see e.g., the proof of Theorem 8.7. Finally,
note from the definition of Loewner monotonicity that $f - f(0)$ entrywise
preserves positivity if f is Loewner monotone – on $\mathbb{P}_n(I)$ or on subsets of
these that contain the zero matrix. In particular, if (2) holds then $f - f(0)$ is
a dimension-free positivity preserver, hence of the form $\sum_{k \geq 0} c_k x^k$ with all
$c_k \geq 0$ by Theorem 14.9 – or more precisely, its variant for restricted domains
$(-\rho, \rho)$ as in Section 17.3. Since $f - f(0)$ also vanishes at the origin, we have
$c_0 = 0$, proving (3).

We next come to convexity preservers. Clearly, (1) $\implies$ (2). To show
(3) $\implies$ (1), note that $f(x) - c_0 - c_1 x$ is also Loewner convex for any
$c_0, c_1 \in \mathbb{R}$ if $f(x)$ is, so it again suffices to consider $f(x) = x^k$ for integers
$k \geq 2$. In fact, we claim by induction that x^k is Loewner convex for all $k \geq 0$.

The convexity of $1, x$ is immediate, and for the induction step, if x^k is convex, then for any integer $n \geq 1$, scalar $\lambda \in [0,1]$, and matrices $A \geq B \geq 0_{n \times n}$,

$$(\lambda A + (1 - \lambda)B)^{\circ(k+1)}$$

$$\leq (\lambda A + (1 - \lambda)B) \circ (\lambda A^{\circ k} + (1 - \lambda)B^{\circ k})$$

$$= \lambda A^{\circ(k+1)} + (1 - \lambda)B^{\circ(k+1)} - \lambda(1 - \lambda)(A - B) \circ (A^{\circ k} - B^{\circ k})$$

$$\leq \lambda A^{\circ(k+1)} + (1 - \lambda)B^{\circ(k+1)},$$

where the final inequality follows from the Loewner monotonicity of x^k and the Schur product theorem. Finally, if (2) holds, then by Theorem 19.1(1) for $k = 3$, f' exists and is Loewner monotone on rank ≤ 3 matrices in $\mathbb{P}_n(I)$ for all n, hence a power series as in the preceding set of equivalent statements. This immediately implies (3).   □

The remainder of this chapter is devoted to proving Theorem 19.1(1), beginning with some elementary properties of convex functions:

Lemma 19.3 (Convex functions)   *Suppose $I \subset \mathbb{R}$ is an interval and $f : I \to \mathbb{R}$ is convex.*

(1) *The function $(s,t) \mapsto \dfrac{f(t) - f(s)}{t - s}$, where $t > s$, is nondecreasing in both $t, s \in I$.*

(2) *If I is open, then $f'_{\pm}$ exist on I. In particular, f is continuous on I.*

(3) *If I is open and $z_1 < x < z_2$ in I, then $f'_+(z_1) \leq f'_-(x) \leq f'_+(x) \leq f'_-(z_2)$. In particular, $f'_{\pm}$ are nondecreasing in I, hence each continuous except at countably many points of jump discontinuity.*

(4) *If I is open, then for all $x \in I$,*

$$f'_+(x) = \lim_{z \to x^+} f'_{\pm}(z), \qquad f'_-(x) = \lim_{z \to x^-} f'_{\pm}(z).$$

(5) *If I is open, there exists a cocountable (therefore dense) subset $D \subset I$ on which f' exists. Moreover, f' is continuous and nondecreasing on D.*

Note that the assertions involving open intervals I may be carried over to the interiors of arbitrary intervals I on which f is convex.

Proof

(1) Suppose $s < t < u$ lie in I. One needs to show

$$\frac{f(t) - f(s)}{t - s} \leq \frac{f(u) - f(s)}{u - s} \leq \frac{f(u) - f(t)}{u - t}.$$

But both inequalities can be reformulated to say

$$f(t) \le \frac{u - t}{u - s} f(s) + \frac{t - s}{u - s} f(u),$$

which holds as f is convex.

(2) Given $t \in I$, choose $s < t < u$ in I, and note by the previous part that the ratio

$$\frac{f(x) - f(t)}{x - t}, \qquad x \in (t, u)$$

is nonincreasing in x as $x \to t^+$ and bounded below by $\dfrac{f(t) - f(s)}{t - s}$. Thus, $f'_+(t)$ exists; a similar argument works to show $f'_-(t)$ exists. In particular, the two limits $\lim_{x \to t^\pm} f(x) - f(t)$ are both zero, proving f is continuous at $t \in I$.

(3) The second sentence follows from the first, which in turn follows from the first part by taking limits and is left to the reader.

(4) The preceding part implies $f'_\pm(z)$ are nondecreasing as $z \to x^-$ and nonincreasing as $z \to x^+$, and shows "half" of the desired inequalities. We now show $f'_+(x) \ge \lim_{z \to x^+} f'_+(z)$; the remaining similar inequalities are shown similarly, and again left to the reader. Let $y \in I$, $y > x$; then the first part implies

$$\frac{f(y) - f(z)}{y - z} \ge \frac{f(y') - f(z)}{y' - z}, \qquad \forall x < z < y' < y.$$

Taking $y' \to z^+$, we have $f'_+(z) \le \frac{f(y) - f(z)}{y - z}$. From above, f is continuous on I, so

$$\frac{f(y) - f(x)}{y - x} = \lim_{z \to x^+} \frac{f(y) - f(z)}{y - z} \ge \lim_{z \to x^+} f'_+(z).$$

Finally, taking $y \to x^+$ concludes the proof.

(5) Let $D \subset I$ be the subset where f' exists, which is if and only if f'_+ is continuous (by the preceding part). In particular, D is cocountable from a previous part, and $f' = f'_+$ is continuous and nondecreasing on D by the same part.   □

The next preliminary result shows the continuity (respectively, differentiability) of monotone (respectively, convex) functions on 2×2 matrices:

Proposition 19.4   *Suppose $0 < \rho \le \infty$, $I = (-\rho, \rho)$, and $g : I \to \mathbb{R}$.*

(1) *If $g[-]$ is monotone on $\mathbb{P}_2(I)$, then g is continuous on I.*

(2) *If $g[-]$ is convex on $\mathbb{P}_2(I)$, then g is differentiable on I.*

Proof   We begin with the first assertion. It is easily verified that if g is monotone on $\mathbb{P}_2(I)$, then $g - g(0)$, when applied entrywise to $\mathbb{P}_2(I)$, preserves positivity. Hence, by (the bounded domain-variant of) Theorem 6.7, g is continuous on $(0, \rho)$. Moreover, we may assume without loss of generality that $g(0) = 0$.

Now let $0 < a < \rho$ and $0 < \epsilon < \rho - a$; then the monotonicity of g implies

$$\begin{pmatrix} a+\epsilon & a \\ a & a \end{pmatrix} \geq \begin{pmatrix} \epsilon & 0 \\ 0 & 0 \end{pmatrix} \geq 0_{2\times 2} \implies \begin{pmatrix} g(a+\epsilon) - g(\epsilon) & g(a) \\ g(a) & g(a) \end{pmatrix} \geq 0_{2\times 2}.$$

Pre- and post-multiplying this last matrix by $(1, -1)$ and $(1, -1)^T$ respectively, we have $g(a + \epsilon) - g(a) \geq g(\epsilon)$, and by the monotonicity of g (applied to $\epsilon 1_{2\times 2} \geq \epsilon' 1_{2\times 2}$ for $0 \leq \epsilon' < \epsilon < \rho$), it follows that g is nondecreasing on $[0, \rho)$. Now taking the limit as $\epsilon \to 0^+$, we have

$$0 = g(a^+) - g(a) \geq g(0^+) \geq 0,$$

where the first equality follows from the continuity of g. Hence, g is right continuous at 0.

Next, for the continuity of g on $(-\rho, 0)$, let

$$a \in (0, \rho), \quad 0 < \epsilon < \min(a, \rho - a), \quad A = \begin{pmatrix} 1 & -1 \\ -1 & 1 \end{pmatrix},$$

and deduce from the monotonicity of g

$$(a + \epsilon)A \geq aA \geq (a - \epsilon)A \geq 0$$
$$\implies \quad g[(a + \epsilon)A] \geq g[aA] \geq g[(a - \epsilon)A].$$

The positivity of the difference matrices implies, upon taking determinants

$$|g(a \pm \epsilon) - g(a)| \geq |g(-a \mp \epsilon) - g(-a)|.$$

Let $\epsilon \to 0^+$; then the continuity of g at a implies that at $-a$, as desired. A similar (one-sided) argument shows the left continuity of g at 0, via the step $g(\epsilon) - g(0) \geq |g(-\epsilon) - g(0)|$.

Next, we come to the second assertion. If g is convex on $\mathbb{P}_2(I)$, then restricting to the matrices $a 1_{2\times 2}$ for $a \in [0, \rho)$, it follows that g is convex on $[0, \rho)$. Hence, g'_+ exists on $(0, \rho)$ by Lemma 19.3. Now suppose $0 < s < t < \rho$ and $0 < \epsilon < \rho - t$. Then by the convexity of g, the matrix inequality (in the Loewner order)

$$\begin{pmatrix} t+\epsilon & t \\ t & t \end{pmatrix} \geq \begin{pmatrix} s+\epsilon & s \\ s & s \end{pmatrix} \geq 0_{2\times 2}$$

implies

$$\begin{pmatrix} g(\lambda(t+\epsilon)+(1-\lambda)(s+\epsilon)) & g(\lambda t+(1-\lambda)s) \\ g(\lambda t+(1-\lambda)s) & g(\lambda t+(1-\lambda)s) \end{pmatrix}$$

$$\leq \lambda \begin{pmatrix} g(t+\epsilon) & g(t) \\ g(t) & g(t) \end{pmatrix} + (1-\lambda) \begin{pmatrix} g(s+\epsilon) & g(s) \\ g(s) & g(s) \end{pmatrix},$$

for all $\lambda \in [0,1]$. Write this inequality in the following form: $\begin{pmatrix} \alpha & \beta \\ \beta & \beta \end{pmatrix} \geq 0_{2\times 2}$.

As above, pre- and post-multiplying this last matrix by $(1,-1)$ and $(1,-1)^T$ respectively yields

$$g(\lambda t+(1-\lambda)s+\epsilon)-g(\lambda t+(1-\lambda)s)$$

$$\leq \lambda(g(t+\epsilon)-g(t))+(1-\lambda)(g(s+\epsilon)-g(s)).$$

Divide by ϵ and let $\epsilon \to 0^+$; this shows g'_+ is convex, hence continuous by Lemma 19.3, on $(0,\rho)$.

Next, denote by g_0, g_1 the even and odd parts of g, respectively

$$g_0(t) := \frac{1}{2}(g(t)+g(-t)), \qquad g_1(t) := \frac{1}{2}(g(t)-g(-t)).$$

We claim that g_0, g_1 are convex on $[0,\rho)$. Indeed, by the convexity of g we deduce for $0 \leq s \leq t < \rho$ and $\lambda \in [0,1]$

$$\begin{pmatrix} t & -t \\ -t & t \end{pmatrix} \geq \begin{pmatrix} s & -s \\ -s & s \end{pmatrix} \geq 0$$

$$\implies \begin{pmatrix} g(c_\lambda) & g(-c_\lambda) \\ g(-c_\lambda) & g(c_\lambda) \end{pmatrix} \leq \lambda \begin{pmatrix} g(t) & g(-t) \\ g(-t) & g(t) \end{pmatrix}$$

$$+ (1-\lambda) \begin{pmatrix} g(s) & g(-s) \\ g(-s) & g(s) \end{pmatrix},$$

where $c_\lambda = \lambda t+(1-\lambda)s$. Pre- and post-multiplying this last inequality by $(1, \pm 1)$ and $(1, \pm 1)^T$ respectively, yields

$$g(\lambda t+(1-\lambda)s) \pm g(-(\lambda t+(1-\lambda)s))$$

$$\leq \lambda(g(t) \pm g(-t))+(1-\lambda)(g(s) \pm g(-s)).$$

This yields g_0, g_1 are convex on $[0,\rho)$. Next, note that if $0 \leq s < t < \rho$, and $0 < \epsilon \leq \min(t-s, \rho-t)$, then

$$\frac{g(s+\epsilon)-g(s)}{\epsilon} \leq \frac{g(t)-g(t-\epsilon)}{\epsilon},$$

by Lemma 19.3(1). Taking $\epsilon \to 0^+$ shows that $g'_+(s) \le g'_-(t)$ if $0 \le s < t < \rho$ and g is convex. Similarly, $g'_-(t) \le g'_+(t)$; therefore,

$$g'_+(s) \le g'_-(t) = (g_0)'_-(t) + (g_1)'_-(t) \le (g_0)'_+(t) + (g_1)'_+(t) = g'_+(t).$$

Since g'_+ is continuous, letting $s \to t^-$ shows $(g_j)'_-(t) = (g_j)'_+(t)$ for $j = 0, 1$. Thus, g_j is differentiable on $(0, \rho)$. Since g_0 is even and g_1 is odd, they are also differentiable on $(-\rho, 0)$. Hence, g_0, g_1, g are differentiable on $I \setminus \{0\}$. Finally, let $I' := (-2\rho/3, 2\rho/3)$ and define $h(x) := g(x + \rho/3)$. It is easy to check that h is convex on $\mathbb{P}_2(I')$, so it is differentiable at $-\rho/3$ by the above analysis, and hence g is differentiable at 0, as desired. □

With these preliminary results at hand, we now complete the remaining proof:

Proof of Theorem 19.1   We begin by showing the first assertion. First, suppose f is differentiable on I and f' is monotone on the rank $\le k$ matrices in $\mathbb{P}_n(I)$. Also assume $A \ge B \ge 0_{n \times n}$ are matrices of rank $\le k$. Now follow the proof of Proposition 9.9(3) to show that $f[-]$ is Loewner convex on rank $\le k$ matrices in $\mathbb{P}_n(I)$. Here we use the fact that since $A \ge B \ge 0$, we have the chain of Loewner inequalities

$$A \ge \lambda A + (1 - \lambda)B \ge \lambda \frac{A + B}{2} + (1 - \lambda)B \ge B, \qquad (19.1)$$

and hence the ranks of all matrices here are at most $\mathrm{rk}(A) \le k$.

The converse is shown in two steps; in fact, we will also prove that f is continuously differentiable on I. The first step is to show the result for $n = k = 2$. Note by Proposition 19.4 that f is differentiable on I. Now say $A \ge B \ge 0_{2 \times 2}$ with $A \ne B$ in $\mathbb{P}_2(I)$. Writing $A = \begin{pmatrix} a_1 & a \\ a & a_2 \end{pmatrix}$ and $B = \begin{pmatrix} b_1 & b \\ b & b_2 \end{pmatrix}$, we have $a_j \ge b_j \ge 0$ for $j = 1, 2$ and $(a - b)^2 \le (a_1 - b_1)(a_2 - b_2)$. Define $\delta \in [0, a_1 - b_1]$ and the matrix $C_{2 \times 2}$ via

$$(a - b)^2 = (a_1 - b_1 - \delta)(a_2 - b_2), \qquad C := \begin{pmatrix} b_1 + \delta & b \\ b & b_2 \end{pmatrix} \in \mathbb{P}_2(I).$$

Clearly, $A \ge C \ge B$, all matrices are in $\mathbb{P}_2(I)$, and $A - C, C - B$ have rank at most 1. Thus, we may assume without loss of generality that $A - B$ has rank 1; write $A - B = \begin{pmatrix} a & \sqrt{ab} \\ \sqrt{ab} & b \end{pmatrix} \in \mathbb{P}_2$. First, if $ab = 0$, then $f'[A] - f'[B]$ is essentially a scalar on the main diagonal. Now since f is convex on $[0, \rho)$ by considering $a\mathbf{1}_{2 \times 2}$ for $a \in [0, \rho)$, we have f' is nondecreasing on $(0, \rho)$, and hence $f'[A] \ge f'[B]$.

The other case is $a, b > 0$. In this case $A \geq B \geq 0$ and $A - B$ is rank 1 with no zero entries. Now follow the proof of Proposition 9.9(3) to infer $f'[A] \geq f'[B]$. Together, both cases show that f' is monotone on $\mathbb{P}_2(I)$, so f' is continuous on I by Proposition 19.4(1).

This shows the result for $n = k = 2$. Now suppose $n > 2$. First, f is convex on $\mathbb{P}_2(I)$, so f' is monotone on $\mathbb{P}_2(I)$ and hence continuous on I by the previous case. Second, to show that f' is monotone as asserted, suppose $A \geq B \geq 0_{n \times n}$ are matrices in $\mathbb{P}_n(I)$ of rank $\leq k$. Now claim that there is a chain of Loewner matrix inequalities

$$A = A_n \geq A_{n-1} \geq \cdots \geq A_0 = B,$$

satisfying: (1) $A_j \in \mathbb{P}_n(I)$ for all $0 \leq j \leq n$, and (2) $A_j - A_{j-1}$ has rank at most 1 for each $1 \leq j \leq n$. Note that such a chain of inequalities would already imply the reverse inclusions for the corresponding null spaces, so each A_j has rank at most k.

To show the claim, spectrally decompose $A - B = UDU^T$, where U is orthogonal and $D = \mathrm{diag}(\lambda_1, \ldots, \lambda_n)$ with $\lambda_j \geq 0$, and write

$$A_j := B + U \, \mathrm{diag}(\lambda_1, \ldots, \lambda_j, 0, \ldots, 0) U^T, \qquad 0 \leq j \leq n.$$

Note that $A_j \leq A$, so the same applies to each of their corresponding (nonnegative) diagonal entries. Thus, $0 \leq (A_j)_{ll} \leq a_{ll}$ for $1 \leq l \leq n$. Thus, the diagonal entries of each A_j lie in $I = (-\rho, \rho)$, hence so do the off-diagonal entries. This shows the claim.

Thus, to show $f'[A] = f'[A_n] \geq f'[A_0] = f'[B]$, it suffices to assume, as in the previous case of $n = k = 2$, that $A - B$ has rank 1. First, if $A - B$ has no zero entries, then $f'[A] \geq f'[B]$ by Proposition 9.9(3). Otherwise, suppose $A - B = uu^T$, with $u \in \mathbb{R}^n$ a nonzero vector having zero entries. Without loss of generality, write $u = \begin{pmatrix} v \\ 0 \end{pmatrix}$, with $v \in \mathbb{R}^l$ having no zero entries for some $1 \leq l \leq n - 1$. Accordingly, write $A = \begin{pmatrix} A_{11} & A_{12} \\ A_{21} & A_{22} \end{pmatrix}$, and similarly for B; it follows that $A_{ij} = B_{ij}$ for all $(i, j) \neq (1, 1)$, and $A_{11} = B_{11} + vv^T$. Now since f is Loewner convex on $[B, A]$, it is so on $[B_{11}, A_{11}]$, where all matrices are positive semidefinite and also have rank $\leq k$. Moreover, f' exists and is continuous on I from above. Again, by Proposition 9.9(3), it follows that f' is Loewner monotone on $[B_{11}, A_{11}]$ (if $k = 1$ then this assertion is true by one-variable calculus). But then,

$$f'[A] - f'[B] = \begin{pmatrix} f'[A_{11}] - f'[B_{11}] & 0 \\ 0 & 0 \end{pmatrix} \geq 0.$$

This proves the first assertion; we turn to the second. First, suppose $A \geq B \geq 0_{n \times n}$, and f' is Loewner positive on $\mathbb{P}_n(I)$. Then follow the proof of Proposition 9.9(2) to infer $f[A] \geq f[B]$. Conversely, we prove the result under a stronger hypothesis: namely, f is differentiable. Now the proof of Proposition 9.9(2) again applies: given $A \in \mathbb{P}_n(I)$, we have $A + \epsilon \mathbf{1}_{n \times n} \in \mathbb{P}_n(I)$ for small $\epsilon > 0$. By monotonicity, it follows that

$$\tfrac{1}{\epsilon}(f[A + \epsilon \mathbf{1}_{n \times n}] - f[A]) \in \mathbb{P}_n.$$

Taking $\epsilon \to 0^+$ proves $f'[A] \in \mathbb{P}_n$, as desired. $\qquad\qquad\square$

We now move to kernels.

Definition 19.5   Suppose X is a nonempty set, $I \subset \mathbb{R}$ a domain, and $\mathcal{V}$ is a set of (real symmetric) positive semidefinite kernels on $X \times X$, with values in I.

(1) The *Loewner order* on kernels on $X \times X$ is: $K \succeq L$ for K, L kernels on $X \times X$, if $K - L$ is a positive semidefinite kernel. (Note, if X is finite, this specializes to the usual Loewner ordering on real $|X| \times |X|$ matrices.)
(2) A function $F : I \to \mathbb{R}$ is *Loewner monotone on* $\mathcal{V}$ if $F \circ K \succeq F \circ L$ whenever $K \succeq L \succeq 0$ are kernels in $\mathcal{V}$.
(3) A function $F : I \to \mathbb{R}$ is *Loewner convex on* $\mathcal{V}$ (here I is assumed to be convex) if whenever $K \succeq L \succeq 0$ are kernels in $\mathcal{V}$, we have

$$\lambda F \circ K + (1 - \lambda) F \circ L \succeq F \circ (\lambda K + (1 - \lambda)L), \qquad \forall \lambda \in [0, 1].$$

The above results for matrices immediately yield the results for kernels:

Theorem 19.6   *Suppose* $0 < \rho \leq \infty$, $I = (-\rho, \rho)$, *and* $F : I \to \mathbb{R}$, *and* X *is an infinite set. The composition map* $F \circ -$ *is Loewner monotone (respectively, Loewner convex) on positive kernels on* $X \times X$, *if and only if* F *satisfies the respective equivalent conditions on matrices of all sizes, in Theorem 19.2.*

Proof   First, if F is Loewner monotone or convex on kernels on $X \times X$, then by restricting the defining inequalities to kernels on $Y \times Y$ (padded by zeros) for finite sets $Y \subset X$, it follows that F is, respectively, Loewner monotone or convex on $\mathbb{P}_n(I)$ for all $n \geq 1$.

To show the converse, suppose first that $F(y) = \sum_{k=0}^{\infty} c_k y^k$ on I, with all $c_k \geq 0$. To show that $F \circ -$ is Loewner monotone on kernels on $X \times X$, it suffices to do so on every "principal submatrix" of such kernels – i.e., for every finite indexing subset of X. But this is indeed true for F, by Theorem 19.2. A similar proof holds for Loewner convex maps. $\qquad\qquad\square$

20

Functions Acting Outside Forbidden Diagonal Blocks

We next explore a variant of the question of classifying the dimension-free preservers. Recall that Schoenberg's original motivation in proving his result was to classify the entrywise positivity preservers $f[-]$ on correlation/Gram matrices – with or without rankconstraints – since these are the matrices that arise as distance matrices on Euclidean spheres (after applying $\cos(\cdot)$ entrywise). In a sense, this is equivalent to applying $f/f(1)$ to the off-diagonal entries of correlation matrices and preserving positivity.

In a similar vein, motivated by modern applications via high-dimensional covariance estimation, Guillot and Rajaratnam in 2015 [110] classified entrywise maps that operate only on off-diagonal entries, and preserve positivity in all dimensions.

Theorem 20.1 (Guillot and Rajaratnam)   *Let* $0 < \rho \le \infty$ *and* $f \colon (-\rho, \rho) \to \mathbb{R}$*. Given a square matrix* $A \in \mathbb{P}_n((-\rho, \rho))$*, define* $f^*[A] \in \mathbb{R}^{n \times n}$ *to be the matrix with* (j,k)*-entry* $f(a_{jk})$ *if* $j \ne k$*, and* a_{jj} *otherwise. Then the following are equivalent:*

(1) $f^*[-]$ *preserves positivity on* $\mathbb{P}_n((-\rho, \rho))$ *for all* $n \ge 1$*.*
(2) *There exist scalars* $c_k \ge 0$*, such that* $f(x) = \sum_{k \ge 0} c_k x^k$ *and*
    $|f(x)| \le |x|$ *on all of* $(-\rho, \rho)$*. (Thus, if* $\rho = \infty$*, then* $f(x) \equiv cx$ *on* $\mathbb{R}$*, for some* $c \in [0, 1]$*.)*

Once again, the robust characterization of absolute monotonicity emerges out of this variant of entrywise operations.

The main result of this chapter provides – in a closely related setting – an example of a dimension-free preserver that is *not* absolutely monotonic. To elaborate: Theorem 20.1 was recently strengthened by Vishwakarma [231], where he introduced the more general model in which a different function $g(x)$ acts on the diagonal entries. Even more generally, Vishwakarma allowed $g[-]$

168

to act on prescribed principal submatrices/diagonal blocks and $f[-]$ to act on the remaining entries. To explain his results, we adopt the following notation throughout this chapter:

Definition 20.2   Fix $0 < \rho \leq \infty$, $I = (-\rho, \rho)$, and $f, g : I \to \mathbb{R}$. Also fix families of subsets $T_n \subset (2^{[n]}, \subset)$ for each $n \geq 1$, such that all elements in a fixed family T_n are pairwise incomparable. Now given $n \geq 1$ and a matrix $A \in I^{n \times n}$, define $(g, f)_{T_n}[A] \in \mathbb{R}^{n \times n}$ to be the matrix with (j, k)-entry $g(a_{jk})$ if there is some $E \in T_n$ containing j, k (here, j may equal k), and $f(a_{jk})$ otherwise.

Adopting this notation, Vishwakarma classifies the pairs (g, f) which preserve positivity according to a given sequence $\{T_n : n \geq 1\}$. Notice that if $T_n = \{[n]\}$ for $n > n_0$ and T_n is empty for $n \leq n_0$, this implies from Section 11.1 that $g(x)$ is absolutely monotonic as in Schoenberg–Rudin's results; and that $f[-]$ preserves positivity on $\mathbb{P}_{n_0}((-\rho, \rho))$. Such functions f do not admit a known characterization for $n_0 \geq 3$; and the following result will also not consider them. Thus, below we require $T_n \neq \{[n]\}$ for infinitely many $n \geq 1$.

Theorem 20.3 (Vishwakarma)   *Notation as in Definition 20.2. Suppose $\{T_n\}$ is such that $T_n \neq \{[n]\}$ for infinitely many $n \geq 1$. Then $(g, f)_{T_n}[-]$ preserves positivity on $\mathbb{P}_n(I)$ for all $n \geq 1$, if and only if exactly one of the following occurs:*

(1) *If T_n is the empty collection, i.e., $(g, f)_{T_n}[-] = f[-]$ for all $n \geq 1$, then $f(x) = \sum_{k \geq 0} c_k x^k$ on I, where $c_k \geq 0$ for all $k \geq 0$.*
(2) *If some T_n contains two nondisjoint subsets of $[n]$, then $g(x) = f(x)$, and $f(x)$ is a power series as in the preceding subcase.*
(3) *If $T_n \subset \{\{1\}, \dots, \{n\}\}$ for all $n \geq 2$, and some T_n is nonempty, then f is as in (1), and $0 \leq f \leq g$ on $[0, \rho)$.*
(4) *If $T_2 = \{\{1, 2\}\}$ and $T_n \subset \{\{1\}, \dots, \{n\}\}$ for all $n \geq 3$, then f is as in (1), $g(x)$ is nonnegative, nondecreasing, and multiplicatively midconvex on $[0, \rho)$, and $|g(x)| \leq g(|x|)$ for all x. If, moreover, some T_n is nonempty for $n \geq 3$, then $0 \leq f \leq g$ on $[0, \rho)$.*
(5) *Otherwise $T_n \not\subset \{\{1\}, \dots, \{n\}\}$ for some $n \geq 3$; and T_n is a partition of some subset of $[n]$ for each $n \geq 1$. In this case, with the additional assumption that $g(x) = \alpha x^k$ for some $\alpha \geq 0$ and $k \in \mathbb{Z}^{\geq 0}$:*

   (a) *If for all $n \geq 3$ we have $T_n = \{[n]\}$ or $\{\{1\}, \dots, \{n\}\}$, then f is as in (1) and $0 \leq f \leq g$ on $[0, \rho)$.*
   (b) *If T_n is not a partition of $[n]$ for some $n \geq 3$, then $f(x) = cg(x)$ for some $c \in [0, 1]$.*

(c) *If neither (a) nor (b) holds, then* $f(x) = cg(x)$ *for some* $c \in [-1/(K-1), 1]$, *where*

$$K := \max_{n \geq 1} |T_n| \in [2, +\infty].$$

In fact, the assertions in the above cases are equivalent to the weaker assertion (than above) that $(g, f)_{T_n}[-]$ *preserves positivity on the rank* ≤ 3 *matrices in* $\bigcup_{n \geq 1} \mathbb{P}_n(I)$.

We refer the reader to Vishwakarma's work for similar results with the domain I replaced by $(0, \rho), [0, \rho)$, or even the complex disk $D(0, \rho)$. As mentioned above, one interesting feature here is that in the final assertion (5)(c), we find the first example of a function that is not absolutely monotonic, yet is a dimension-free preserver, in this setting.

To prove Theorem 20.3, we require two well-known preliminaries, and a couple of additional results, shown below:

Proposition 20.4   *For a Hermitian matrix* $A_{n \times n}$, *denote its largest and smallest eigenvalues by* $\lambda_{\max}(A)$ *and* $\lambda_{\min}(A)$, *respectively.*

(1) *(Rayleigh–Ritz theorem) If* $A \in \mathbb{C}^{n \times n}$ *is Hermitian, then the ratio* $v^* A v / v^* v$, *as* v *runs over* $\mathbb{C}^n \setminus \{0\}$, *attains its maximum and minimum values, which equal* $\lambda_{\max}(A)$ *and* $\lambda_{\min}(A)$, *respectively.*

(2) *(Weyl's inequality, special case) If* $A, B \in \mathbb{C}^{n \times n}$ *are Hermitian, then*

$$\lambda_{\min}(A) + \lambda_{\min}(B) \leq \lambda_{\min}(A + B) \leq \lambda_{\min}(A) + \lambda_{\max}(B). \qquad (20.1)$$

The second assertion holds more generally; we do not state/prove/require it below.

Proof   For the first part, it suffices to show the minimum bound, since $\lambda_{\max}(A) = -\lambda_{\min}(-A)$. (That the bound is attained follows from the compactness of the unit complex sphere.) The matrix $A - \lambda_{\min}(A) \operatorname{Id}_{n \times n}$ is Hermitian with smallest eigenvalue zero, hence, is positive semidefinite. Thus, we compute for nonzero $v \in \mathbb{C}^n$

$$0 \leq \frac{v^* (A - \lambda_{\min}(A) \operatorname{Id}_{n \times n}) v}{v^* v} = \frac{v^* A v}{v^* v} - \lambda_{\min}(A).$$

This shows the first assertion. For the second, let $v \in \ker(A - \lambda_{\min}(A) \operatorname{Id}_{n \times n})$ be nonzero. Applying the previous part twice,

$$\lambda_{\min}(A + B) \leq \frac{v^* (A + B) v}{v^* v} = \frac{v^* A v}{v^* v} + \frac{v^* B v}{v^* v} \leq \lambda_{\min}(A) + \lambda_{\max}(B).$$

Similarly, if v is a nonzero eigenvector for $A + B$ with eigenvalue $\lambda_{\min}(A + B)$, then by the previous part applied twice,

$$\lambda_{\min}(A+B) = \frac{v^*(A+B)v}{v^*v} = \frac{v^*Av}{v^*v} + \frac{v^*Bv}{v^*v} \geq \lambda_{\min}(A) + \lambda_{\min}(B). \quad \square$$

We also require the following special case of the main result:

Lemma 20.5   Let $0 < \rho \leq \infty$, $I = (-\rho, \rho)$, and $f : I \to \mathbb{R}$. Let $g(x) = \alpha x^k$ for $\alpha \geq 0$ and $k \in \mathbb{Z}^{\geq 0}$. Finally, let $T_3 = \{\{1,2\}\}$ and $c_0 = 0$. The following are equivalent:

(1) $(g,f)_{T_3}[A] \in \mathbb{P}_3$ for all matrices $A \in \mathbb{P}_3(I)$.
(2) $(g,f)_{T_3}[A] \in \mathbb{P}_3$ for all rank-1 matrices $A \in \mathbb{P}_3(I)$.
(3) $f(x) \equiv cg(x)$ on I for some $c \in [c_0, 1]$.

The same equivalence holds if $T_3 = \{\{1,2\}, \{3\}\}$ and $c_0 = -1$.

Proof   First suppose $T_3 = \{\{1,2\}\}$ and $c_0 = 0$. Clearly (1) $\implies$ (2). Now suppose (2) holds. If $f \equiv 0$ or $g \equiv 0$, then the result is immediate, so suppose $f, g \not\equiv 0$ (hence, $\alpha > 0$). Now given $0 \leq |z| \leq w < \rho$ with $w \neq 0$, define

$$A(w,z) := \begin{pmatrix} z^2/w & z & z \\ z & w & w \\ z & w & w \end{pmatrix} = \frac{1}{w}uu^T, \quad \text{where} \quad u = (z, w, w)^T. \quad (20.2)$$

By choice of w, z, we have $A(w,z) \in \mathbb{P}_3(I)$, so $\det(g, f)_{T_3}[A(w,z)] \geq 0$.

There are now two cases. First, if $k = 0$, then $g(x) \equiv \alpha > 0$. Let $w > 0$ and expand the above determinant along the third row to compute

$$0 \leq \det \ldots A(w,z)] = -\alpha(f(z) - f(w))^2.$$

Using an increasing sequence $0 < w_n \to \rho^-$, this shows that f is constant on I, and by considering $(g, f)_{T_3}[0_{3\times3}]$, we have $f(x) \equiv c\alpha$ for some $c \in [0,1]$.

The other case is if $k > 0$, so that $g(0) = 0$. Now $f(0) = 0$ by considering $(g, f)_{T_3}[0_{3\times3}]$. Again expand the above determinant along the third row, to obtain

$$0 \leq \det(g, f)_{T_3}[A(w,z)] = -\frac{\alpha}{w^k}(w^k f(z) - z^k f(w))^2. \quad (20.3)$$

Thus, we have $f(z)/z^k = f(w)/w^k$ whenever $0 < |z| \leq w < \rho$. By using an increasing sequence $w_n \uparrow \rho^-$, this shows $f(z)/z^k$ is constant on $I \setminus \{0\}$, say $c \in \mathbb{R}$. By considering $A(w,w) = w1_{3\times3}$ for $w > 0$, it is not hard to see that $c \in [0,1]$, which proves (3).

Finally, if (3) holds, then $(g, f)_{T_3}[A]$ is the sum of $cg[A]$ and $(1-c)g[B]$ (padded by a zero row and column at the end), where B is the leading principal 2×2 submatrix of A. This shows (1) by the Schur product theorem.

The proof is similar if $T_3 = \{\{1,2\},\{3\}\}$ and $c_0 = -1$. Clearly, (1) $\Longrightarrow$ (2); similarly, the proof of (2) $\Longrightarrow$ (3) is unchanged (including the computation (20.3)) until the very last steps for both $k = 0$ and $k > 0$, at which points we can only conclude $c \in [-1,1]$. Finally, we assume (3) holds and show (1). The point is that for any scalar $c \in [-1,1]$ and any matrix $A \in \mathbb{P}_3(I)$, the principal minors of $(g, cg)_{T_3}[A]$ equal those of $(g, |c|g)_{T_3}[A]$, so that we may work with $|c| \in [0,1]$ instead of $c \in [-1,1]$. Now one shows (1) similarly as the previous case.   □

A final preliminary result – the second part easily follows from the first, and in turn strengthens Lemma 20.5.

Proposition 20.6   *Suppose for an integer $n \geq 3$ that $T_n \subset 2^{[n]}$ is a partition of $[n]$ into $k \geq 2$ nonempty subsets.*

(1) *Let $g(0) = 1$ and $f(0) = c$. Then $(g, f)_{T_n}[0_{n \times n}]$ is positive semidefinite if and only if $c \in [-1/(k-1), 1]$.*
(2) *Suppose $0 < \rho \leq \infty$, $I = (-\rho, \rho)$, and $f : I \to \mathbb{R}$. Also suppose $g : I \to \mathbb{R}$ is multiplicative and preserves positivity on $\mathbb{P}_n(I)$. If $T_n \neq \{\{1\}, \ldots, \{n\}\}$, then the following are equivalent:*

   (a) *$(g, f)_{T_n}[-]$ preserves positivity on $\mathbb{P}_n(I)$.*
   (b) *$(g, f)_{T_n}[-]$ preserves positivity on the rank-1 matrices in $\mathbb{P}_n(I)$.*
   (c) *$f(x) \equiv cg(x)$ on I for some $c \in [-1/(k-1), 1]$.*

The nonzero functions in part (2) include the powers $x^k, k \in \mathbb{Z}^{\geq 0}$ by the Schur product theorem; but also – as studied by Hiai in 2009 [119] – the "powers"

$$\phi_\alpha(x) := |x|^\alpha, \qquad \psi_\alpha(x) := \operatorname{sgn}(x)|x|^\alpha, \qquad \alpha \geq n - 2.$$

Proof   Let $T_n = \{J_1, \ldots, J_k\}$ with $\sqcup_j J_j = [n]$.

(1) Choose elements $j_1, \ldots, j_k$ with $j_i \in J_i$. By possibly relabeling the rows and columns, we may assume without loss of generality that $1 \leq j_1 < \cdots < j_k \leq n$. Now if $(g, f)_{T_n}[0_{n \times n}] \in \mathbb{P}_n$, then by considering the principal $k \times k$ submatrix corresponding to the indices $\{j_1, \ldots, j_k\}$, we obtain

$$C := c\mathbf{1}_{k \times k} + (1 - c)\operatorname{Id}_{k \times k} \in \mathbb{P}_k. \tag{20.4}$$

Since this matrix has eigenvalues $(1 - c)$ and $1 + (k - 1)c$, we get $c \in [-1/(k-1), 1]$, as desired.

For the converse, define the "decompression" of C, given by

$$\tilde{C} := c\mathbf{1}_{n \times n} + (1 - c)\sum_{j=1}^{k} \mathbf{1}_{J_j \times J_j} = (g, f)_{T_n}[0_{n \times n}] \in \mathbb{C}^{n \times n}. \tag{20.5}$$

We now show that if $c \in [-1/(k-1), 1]$, then $\widetilde{C} \in \mathbb{P}_n$. Indeed, given a vector $u \in \mathbb{C}^n$, define $u_{T_n} \in \mathbb{C}^k$ to have jth coordinate $\sum_{i \in J_j} u_i$. Then,

$$u^* \widetilde{C} u = u_{T_n}^* C u_{T_n} \geq 0, \qquad \forall u \in \mathbb{C}^n,$$

because the matrix C as in (20.4) is positive semidefinite as above.

(2) If $g \equiv 0$, then the result is easy to prove, so we suppose henceforth that $g \not\equiv 0$. Clearly, (a) implies (b). Next if (b) holds, then one can restrict to a suitable 3×3 submatrix – without loss of generality indexed by $1, 2, 3$, such that $T_n \cap \{1, 2, 3\} = \{\{1, 2\}, \{3\}\}$ by a slight abuse of notation. Hence, $f(x) \equiv cg(x)$ on I for some $c \in [-1, 1]$, by Lemma 20.5. Now if $g(x_0) \neq 0$, then $(g, f)_{T_n}[x_0 \mathbf{1}_{n \times n}]$ has as a principal submatrix, $g(x_0)C$, where C is as in (20.4). Hence, $c \in [-1/(k-1), 1]$ by the previous part, proving (c). Finally, given any matrix $A \in \mathbb{P}_n(I)$, we have

$$(g, cg)_{T_n}[A] = g[A] \circ \widetilde{C},$$

where $\widetilde{C}$ is as in (20.5). Now if (c) holds, then $\widetilde{C} \in \mathbb{P}_n$ by the previous part, and this shows (a) by the assumptions on g, f as well as the Schur product theorem. $\qquad \square$

With these results at hand, we are ready to proceed.

Proof of Theorem 20.3   Clearly, if $(g, f)_{T_n}[-]$ preserves positivity on $\mathbb{P}_n(I)$, then it does so on the rank ≤ 3 matrices in $\mathbb{P}_n(I)$. Thus, we will prove that this latter assertion implies the conclusions on (g, f) in the various cases; and that these conclusions imply in turn that $(g, f)_{T_n}[-]$ preserves positivity on $\mathbb{P}_n(I)$. This is done in each of the subcases (which place constraints on the family T_n). First if (1) all T_n are empty sets, then the result follows from the stronger Schoenberg–Rudin Theorem 11.3 (which holds over $(-\rho, \rho)$ instead of $\mathbb{R}$, as remarked in Section 17.3).

Next, suppose from (2) some T_n contains subsets $I_1, I_2 \subset [n]$ that are not disjoint. Clearly, if $g = f$ and f is as in (1), then $(g, f)_{T_n}[-] = f[-]$ preserves positivity by the Schur product theorem. Conversely, if $(g, f)_{T_n}[-]$ preserves positivity even on the rank-1 matrices in $\mathbb{P}_n((-\rho, \rho))$ for all $n \geq 3$, then there exist integers $n \geq 3$ and $a, b, c \in [n]$, such that

$$a, b \in I_1, \quad c \notin I_1, \qquad b, c \in I_2, \quad a \notin I_2.$$

By relabeling indices if needed, we will assume without loss of generality that $a = 1, b = 2$, and $c = 3$. Now let $x \in (-\rho, \rho)$ and define $A :=$
$\begin{pmatrix} |x| & x & x \\ x & |x| & |x| \\ x & |x| & |x| \end{pmatrix} \oplus 0_{(n-3) \times (n-3)} \in \mathbb{P}_n(I)$. If B denotes the leading principal 3×3 submatrix of $(g, f)_{T_n}[A]$, then

$$0 \le \det B = \det \begin{pmatrix} g(|x|) & g(x) & f(x) \\ g(x) & g(|x|) & g(|x|) \\ f(x) & g(|x|) & g(|x|) \end{pmatrix} = -g(|x|)(f(x) - g(x))^2.$$

If $g(|x|) = 0$, then by considering the 2×2 submatrices of B, we see that $f(x) = g(x) = 0$. If $g(|x|) \ne 0$, then it is positive, so we obtain $f(x) = g(x)$. This implies $f \equiv g$ on $(-\rho, \rho)$. Hence, $(g, f)_{T_n}[-] = f[-]$, and we reduced to case (1). This proves the equivalence for case (2).

Next suppose (3) holds. First assume f is as in (1) and $0 \le f \le g$ on $[0, \rho)$. If $A \in \mathbb{P}_n((-\rho, \rho))$, then $(g, f)_{T_n}[A]$ is the sum of $f[A]$ and a diagonal matrix with nonnegative entries. Hence, $(g, f)_{T_n}[A]$ is positive semidefinite by the Schur product theorem. The converse has two subcases. Let $s_n := \# \cup_{E \in T_n} E$, so $0 \le s_n \le n$, and hence, either $n - s_n$ or s_n is an unbounded sequence. If the former, then by restricting to the corresponding principal submatrices (padded by zeros), we are done by case (1) – considering the 2×2 matrix $\begin{pmatrix} g(x) & f(x) \\ f(x) & g(x) \end{pmatrix}$ or $\begin{pmatrix} g(x) & f(x) \\ f(x) & f(x) \end{pmatrix}$ for $x \in [0, \rho)$, we obtain $f(x) \le g(x)$, as desired.

Thus, we henceforth assume the latter holds, i.e., s_n is unbounded; restricting to these principal submatrices, we may assume without loss of generality that $T_n = \{\{1\}, \dots, \{n\}\}$ for all $n \ge 1$. We claim that $f[-]$ preserves positivity on rank ≤ 3 matrices in $\mathbb{P}_n(I)$ for all n. This would finish the proof in case (3), since now f is as in (1), and as above, this implies $0 \le f(x) \le g(x)$ for $x \in [0, \rho)$.

To prove the claim, let $A \in \mathbb{P}_n((-\rho, \rho))$, and let D_A be the diagonal matrix with (j, j)-entry $g(a_{jj}) - f(a_{jj})$. If $\mathbf{1}_{m \times m}$ denotes the all-ones $m \times m$ matrix, then $\mathbf{1}_{m \times m} \otimes A = \begin{pmatrix} A & \cdots & A \\ \vdots & \ddots & \vdots \\ A & \cdots & A \end{pmatrix}$, a matrix in $\mathbb{P}_{mn}(I)$. Also note that if A has rank ≤ 3, then by (2.3), so does $\mathbf{1}_{m \times m} \otimes A$. Now applying $(g, f)_{T_{mn}}[-]$ yields

$$(g, f)_{T_{mn}}[\mathbf{1}_{m \times m} \otimes A] = \mathbf{1}_{m \times m} \otimes f[A] + \mathrm{Id}_{m \times m} \otimes D_A \ge 0.$$

Hence, by (20.1),

$$0 \le \lambda_{\min}((g, f)_{T_{mn}}[\mathbf{1}_{m \times m} \otimes A]) \le \lambda_{\min}(\mathbf{1}_{m \times m} \otimes f[A]) + \lambda_{\max}(\mathrm{Id}_{m \times m} \otimes D_A)$$
$$= m\lambda_{\min}(f[A]) + \max_{1 \le j \le n} \{g(a_{jj}) - f(a_{jj})\},$$

where the equality holds because of (2.3) and since the eigenvalues of $\mathbf{1}_{m \times m}$ are $0, m$. From this it follows that $\lambda_{\min}(f[A]) \ge -\max_j (g(a_{jj}) - f(a_{jj}))/m$ for all $m \ge 1$. This shows $f[A]$ is positive semidefinite and concludes the proof in case (3).

If (4) holds, the proof in case (3) shows f is as in (1); and using $(g, f)_{T_2}[-] = g[-]$ via an argument similar to Theorem 6.7 shows the desired constraints on g. (This is left to the reader to work out.) The converse is shown using (variations of) the same proofs.

It remains to prove the equivalence in case (5); here we are also given that $g(x) = \alpha x^k$ for $\alpha, k \geq 0$ (and k an integer). If $\alpha = 0$, then the result is easy, so we suppose henceforth without loss of generality that $\alpha = 1$. In subcase (a), since $T_n = \{\{1\}, \ldots, \{n\}\}$ for infinitely many n by assumption, we can repeat the proof for case (3) to show that any preserver-pair (g, f) must satisfy $0 \leq f \leq g$ on $[0, \rho)$ and f is as in (1). Conversely, given such (g, f), if $T_n = \{[n]\}$, then $(g, f)_{T_n}[-] = g[-]$, which preserves positivity by the Schur product theorem. Otherwise, for $A \in \mathbb{P}_n(I)$, we compute

$$(g, f)_{T_n}[A] = f[A] + \text{diag}(g(a_{jj}) - f(a_{jj}))_{j=1}^n,$$

and both matrices are positive, hence so is $(g, f)_{T_n}[A]$, as desired.

Next for (b), we fix $n_1 \geq 3$ such that $T_{n_1} \not\subseteq \{\{1\}, \ldots, \{n_1\}\}$; also fix $n_0 \geq 3$ such that T_{n_0} is not a partition of $[n_0]$. If $f(x) = cg(x)$ for $c \in [0, 1]$, then $(g, f)_{T_{n_0}}[A]$ is the sum of $cA^{\circ k}$ and matrices of the form $(1 - c)B^{\circ k}$, where B is a principal submatrix of $A \in \mathbb{P}_{n_0}$, hence positive semidefinite. It follows by the Schur product theorem that $(g, f)_{T_{n_0}}[-]$ preserves positivity. Conversely, suppose $(g, f)_{T_n}[-]$ preserves positivity for all $n \geq 1$, on rank ≤ 3 matrices in $\mathbb{P}_n(I)$. At $n = n_1$, we can find three indices – labeled 1, 2, and 3 without loss of generality – such that for all $A \in \mathbb{P}_{n_1}(I)$, the leading 3×3 submatrix of $(g, f)_{T_{n_1}}[A]$ equals $(g, f)_{\{\{1,2\}\}}[A_{[3] \times [3]}]$ or $(g, f)_{\{\{1,2\}, \{3\}\}}[A_{[3] \times [3]}]$. Now using rank-1 matrices via Lemma 20.5 shows $f(x) = cg(x)$ for $c \in [-1, 1]$. Finally, considering matrices in $\mathbb{P}_{n_0}(I)$ yields $c \geq 0$, as desired.

The remaining subcase is (5)(c), in which case every T_n is a partition of $[n]$. Also note by the hypotheses that $K > 1$; and there exists $n_1 \geq 3$ and three indices – labeled 1, 2, and 3 without loss of generality – such that for all $A \in \mathbb{P}_{n_1}(I)$, the leading 3×3 submatrix of $(g, f)_{T_{n_1}}[A]$ equals $(g, f)_{\{\{1,2\}, \{3\}\}}[A_{[3] \times [3]}]$. Now using rank-1 matrices via Lemma 20.5 or Proposition 20.6 implies $f \equiv cg$, with $c \in [-1/(K - 1), 1]$. Conversely, if f and g are as specified and $T_n = \{[n]\}$, then $(g, f)_{T_n}[-] = g[-]$, which preserves positivity by the Schur product theorem. Else we are done by Proposition 20.6, since $k = |T_n| \leq K$.   □

21

The Boas–Widder Theorem on Functions with Positive Differences

In this chapter, we reproduce the complete proof of the theorem by Boas and Widder on functions with nonnegative forward differences [44]. This result was stated as Theorem 13.8(2), and we again write its statement here for convenience. In it and throughout this chapter, recall from just before Theorem 13.8 that given an interval $I \subset \mathbb{R}$ and a function $f : I \to \mathbb{R}$, the *kth order forward differences of f with step size $h > 0$* are defined as follows:

$$\left(\Delta_h^0 f\right)(x) := f(x),$$

$$\left(\Delta_h^k f\right)(x) := \left(\Delta_h^{k-1} f\right)(x+h) - \left(\Delta_h^{k-1} f\right)(x) = \sum_{j=0}^{k} \binom{k}{j} (-1)^{k-j} f(x+jh),$$

whenever $k > 0$ and $x, x + kh \in I$. It is easily seen that these difference operators commute

$$\Delta_\delta^m \left(\Delta_\epsilon^n f(x)\right) = \Delta_\epsilon^n \left(\Delta_\delta^m f(x)\right), \qquad \text{whenever } x, x + m\delta + n\epsilon \in I,$$

and so, we will omit parentheses and possibly permute these operators below, without further reference. Now we (re)state the theorem of interest:

Theorem 21.1 (Boas and Widder) *Suppose $k \geq 2$ is an integer, $I \subset \mathbb{R}$ is an open interval, bounded or not, and $f : I \to \mathbb{R}$ is a function that satisfies the following condition:*

$$\left(\Delta_h^k f\right)(x) \geq 0 \text{ whenever } h > 0 \text{ and } x, x + kh \in I, \quad \text{and } f \text{ is continuous on } I.$$
$$(H_k)$$

(In other words, f is continuous and has all forward differences of order k nonnegative on I.) Then on all of I, the function $f^{(k-2)}$ exists, is continuous and convex, and has nondecreasing left- and right-hand derivatives.

This is a "finite-order" result; for completeness, an order-free result along these lines was shown by Bernstein in his 1926 memoir [33]. In fact, Boas and Widder write that they were motivated to prove their Theorem 21.1 "in an effort to make more accessible" that result of Bernstein:

Theorem 21.2 (Bernstein, 1926)   *Given a subinterval $(a,b) \subset \mathbb{R}$ and a continuous function $f : (a,b) \to \mathbb{R}$, if the even-order forward differences*

$$(\Delta_\delta^{2n} f)(c) := \sum_{j=0}^{2n} \binom{2n}{j}(-1)^j f(c + j\delta), \qquad c \in (a,b), \ \delta \in (0, (b-c)/2n)$$

are all nonnegative, then f is analytic in (a,b).

21.1 Further Remarks and Results

Before writing down Boas and Widder's proof of Theorem 21.1, we make several additional observations beyond the result and its proof. The first observation (which was previously mentioned following Theorem 13.8(2)) is that while $f_\pm^{(k-1)}$ is nondecreasing by the above theorem, it is not always true that any other lower-order derivatives $f, \dots, f^{(k-2)}$ are nondecreasing on I. For example, let $0 \le l \le k - 2$ and consider $f(x) := -x^{l+1}$ on $I \subset \mathbb{R}$; then $f^{(l)}$ is strictly decreasing on I.

Second, it is natural to seek examples of *nonsmooth* functions satisfying the differentiability conditions of Theorem 21.1, but no more – in other words, to explore if Theorem 21.1 is indeed "sharp." This is now verified to be true:

Example 21.3   Let $I = (a,b) \subset \mathbb{R}$ be an open interval, where $-\infty \le a < b \le \infty$. Consider any function $g : I \to \mathbb{R}$ that is nondecreasing, hence Lebesgue integrable. For any interior point $c \in I$, the function $f_2(x) := \int_c^x g(t) \, dt$ satisfies (H_2)

$$\Delta_h^2 f_2(x) = \int_c^x g(t) \, dt - 2 \int_c^{x+h} g(t) \, dt + \int_c^{x+2h} g(t) \, dt$$

$$= \int_{x+h}^{x+2h} g(t) \, dt - \int_x^{x+h} g(t) \, dt$$

$$= \int_x^{x+h} (\Delta_h g)(t) \, dt \ge 0.$$

However, not every monotone g gives rise to an antiderivative that is differentiable on all of I.

Finally, to see that the condition (H_k) is sharp for all $k > 2$ as well, define f to be the $(k-1)$-fold indefinite integral of g. We claim that f satisfies (H_k). Continuity is obvious; and to study the kth order divided differences of f, first note by the fundamental theorem of calculus that f is $(k-2)$-times differentiable, with $f^{(k-2)}(x) \equiv f_2(x) = \int_c^x g(t)\, dt$. In particular, $\Delta_h^2 f \in C^{k-2}(a, b - kh)$ whenever $a < x < x + kh < b$ as in (H_k).

Now given such x, h, compute via the Cauchy mean-value theorem 13.8(1) for divided differences (and its notation)

$$\Delta_h^k f(x) = \Delta_h^{k-2}(\Delta_h^2 f)(x) = h^{k-2} D_h^{k-2}(\Delta_h^2 f)(x)$$

$$= \frac{h^{k-2}}{(k-2)!} (\Delta_h^2 f)^{(k-2)}(y),$$

for some $y \in (a, b - 2h)$. But this is easily seen to equal

$$= \frac{h^{k-2}}{(k-2)!} (\Delta_h^2 f^{(k-2)})(y) = \frac{h^{k-2}}{(k-2)!} \Delta_h^2 f_2(y),$$

and we just showed that this is nonnegative.   □

The final observation in this section is that there are natural analogues for $k = 0, 1$ of the Boas–Widder theorem (which is stated for $k \geq 2$). For this, we make the natural definition: *for $k < 0$, $f^{(k)}$ will denote the $|k|$-fold antiderivative of f*. Since f is assumed to be continuous, this is just the iterated indefinite Riemann integral starting at an(y) interior point of I. With this notation at hand:

Proposition 21.4   *The Boas–Widder theorem 21.1 also holds for $k = 0, 1$.*

Proof   In both cases, the continuity of $f^{(k-2)}$ is immediate by the fundamental theorem of calculus. Next, suppose $k = 1$ and choose $c \in I$. Now claim that if f is continuous and nondecreasing (i.e., (H_1)), then $f^{(-1)}(x) := \int_c^x f(t)\, dt$ is convex on I. Indeed, given $x_0 < x_1 \in I$, define $x_\lambda := (1 - \lambda)x_0 + \lambda x_1$ for $\lambda \in [0, 1]$, and compute

$$(1 - \lambda) f^{(-1)}(x_0) + \lambda f^{(-1)}(x_1) - f^{(-1)}(x_\lambda)$$

$$= (1 - \lambda) \int_c^{x_1} \mathbf{1}(t \leq x_0) f(t)\, dt$$

$$+ \lambda \int_c^{x_1} \mathbf{1}(t \leq x_1) f(t)\, dt - \int_c^{x_1} \mathbf{1}(t \leq x_\lambda) f(t)\, dt$$

$$= -(1 - \lambda) \int_{x_0}^{x_\lambda} f(t)\, dt + \lambda \int_{x_\lambda}^{x_1} f(t)\, dt.$$

But since f is nondecreasing, each integral – together with the accompanying sign – is bounded below by the corresponding expression where $f(t)$ is replaced by $f(x_\lambda)$. An easy computation now yields

$$(1 - \lambda) f^{(-1)}(x_0) + \lambda f^{(-1)}(x_1) - f^{(-1)}(x_\lambda)$$

$$\geq f(x_\lambda) \left(\lambda(x_1 - x_\lambda) - (1 - \lambda)(x_\lambda - x_0) \right) = 0;$$

therefore, $f^{(-1)}$ is convex, as desired.

This shows the result for $k = 1$. Next, if $k = 0$ then f is continuous and nonnegative on I, hence $f^{(-1)}$ is nondecreasing on I. Now the above computation shows that $f^{(-2)}$ is convex; the remaining assertions are obvious. $\square$

21.2 Proof of the Main Result

In this section, we reproduce Boas and Widder's proof of Theorem 21.1. We first make a few clarifying remarks about this proof.

(1) As Boas and Widder mention, Theorem 21.1 was previously shown by T. Popoviciu in 1934 [186] via an alternate argument using divided differences involving unequally spaced points. Here we will only explain Boas and Widder's proof.

(2) There is a minor error in the arguments of Boas and Widder, which is resolved by adding one word. See Remark 21.7 and the proof of Lemma 13 for more details. (There are a few other minor typos in the writing of Lemmas 6 and 10 and in some of the proofs; these are corrected without elaboration in the exposition in this chapter.)

(3) Boas and Widder do not explicitly write out a proof of the convexity of f (in the case $k = 2$). This is addressed below as well – see the paragraph following Proposition 21.8.

Notice that Theorem 21.1 follows for the case of unbounded domain I from that for bounded domains, so *we assume henceforth that*

$$I = (a,b), \qquad \text{with } -\infty < a < b < \infty.$$

We now reproduce a sequence of 14 lemmas shown by Boas and Widder, which culminate in the above theorem. These lemmas are numbered Lemma 1, ..., Lemma 14, and will be referred to only in this chapter. The rest of the results, equations, and remarks – starting from Theorem 21.1 and ending with

Proposition 21.8 – are numbered using the default counter in this text. None of the results in this chapter are cited elsewhere in the text.

The first of the 14 lemmas by Boas and Widder says that if the kth order "equi-spaced" forward differences are nonnegative, then so are the kth order "possibly nonequi-spaced" differences (the converse is immediate):

Lemma 1   *If $f(x)$ satisfies (H_k) in (a,b) for some $k \geq 2$, then for any k positive numbers $\delta_1, \ldots, \delta_k > 0$,*

$$\Delta_{\delta_1} \Delta_{\delta_2} \cdots \Delta_{\delta_k} f(x) \geq 0, \qquad \text{whenever } a < x < x + \delta_1 + \delta_2 + \cdots + \delta_k < b.$$

Proof   The key step is to prove using (H_k) that

$$\Delta_h^{k-1} \Delta_{\delta_1} f(x) \geq 0, \qquad \text{whenever } a < x < x + (k-1)h + \delta_1 < b. \quad (21.1)$$

After this, the lemma is proved using induction on $k \geq 2$. Indeed, (21.1) is precisely the assertion in the base case $k = 2$; and using (21.1) we can show the induction step as follows: for a fixed $\delta_1 \in (0, b-a)$, it follows that $\Delta_{\delta_1} f$ satisfies (H_{k-1}) in the interval $(a, b - \delta_1)$. Therefore,

$$\Delta_{\delta_2} \cdots \Delta_{\delta_k} (\Delta_{\delta_1} f(x)) \geq 0, \qquad \text{whenever } a < x < x + \delta_1 + \cdots + \delta_k < b.$$

Since the Δ_{δ_j} commute, and since δ_1 was arbitrary, the induction step follows.

Thus, it remains to show (21.1). Let $h > 0$ and $n \in \mathbb{N}$ be such that $a < x < x + h/n + (k-1)h < b$. One can check using an easy telescoping computation that

$$\Delta_h f(x) = \sum_{i=0}^{n-1} \Delta_{h/n} f(x + ih/n);$$

and iterating this procedure, we obtain

$$\Delta_h^{k-1} f(x) = \sum_{i_1=0}^{n-1} \cdots \sum_{i_{k-1}=0}^{n-1} \Delta_{h/n}^{k-1} f\big(x + [i_1 + \cdots + i_{k-1}]h/n\big). \quad (21.2)$$

(This works by induction on $k \geq 2$: the previous telescoping identity is the base case for $k = 2$, and for the induction step we evaluate the innermost sum using the base case.)

From the above computations, it further follows that

$$\Delta_{h/n} \Delta_h^{k-1} f(x) = \sum_{i_1=0}^{n-1} \cdots \sum_{i_{k-1}=0}^{n-1} \Delta_{h/n}^k f(x + [i_1 + \cdots + i_{k-1}]h/n) \geq 0,$$

where the final inequality uses the assumption (H_k). From this it follows that $\Delta_h^{k-1} f(x) \leq \Delta_h^{k-1} f(x + h/n)$.

Now suppose x is such that $a < x < x + mh/n + (k-1)h < b$. Applying the preceding inequality to $x, x + h/n, \ldots, x + (m-1)h/n$, we obtain

$$\Delta_h^{k-1} f(x) \le \Delta_h^{k-1} f(x + h/n) \le \cdots \le \Delta_h^{k-1} f(x + mh/n). \qquad (21.3)$$

We can now prove (21.1). As in it, choose $\delta_1 > 0$, such that $a < x < x + \delta_1 + (k-1)h < b$; and choose sequences m_j, n_j of positive integers, such that $m_j/n_j \to \delta_1/h$ and $x + m_j h/n_j + (k-1)h < b$ for all $j \ge 1$.

Since $f(x)$ is continuous, $f(x + m_j h/n_j)$ converges to $f(x + \delta_1)$, and $\Delta_h^{k-1} f(x + m_j h/n_j)$ to $\Delta_h^{k-1} f(x + \delta_1)$, as $j \to \infty$. Hence, using (21.3) with m_j, n_j in place of m, n respectively, we obtain by taking limits

$$\Delta_h^{k-1} f(x) \le \Delta_h^{k-1} f(x + \delta_1).$$

But this is equivalent to (21.1), as desired. □

Lemma 2 *If $f(x)$ satisfies (H_k) in (a, b) for some $k \ge 2$, then $\Delta_\epsilon^{k-1} f(x)$ and $\Delta_\epsilon^{k-1} f(x - \epsilon)$ are nondecreasing functions of x in $(a, b - (k-1)\epsilon)$ and $(a + \epsilon, b - (k-2)\epsilon)$, respectively.*

Proof   For the first part, suppose $y < z$ are points in $(a, b - (k-1)\epsilon)$, and set

$$\delta_1 := z - y, \qquad \delta_2 = \cdots = \delta_k := \epsilon.$$

Then by Lemma 1 – or simply (21.1) – it follows that

$$\Delta_\epsilon^{k-1} f(z) - \Delta_\epsilon^{k-1} f(y) = \Delta_{\delta_1} \Delta_\epsilon^{k-1} f(y) \ge 0,$$

which is what was asserted.

Similarly, for the second part we suppose $y < z$ are points in $(a + \epsilon, b - (k-2)\epsilon)$. Then $y - \epsilon < z - \epsilon$ are points in $(a, b - (k-1)\epsilon)$, so we are done by the first part. (Remark: Boas and Widder repeat the computations of the first part in this second part; but this is not required.) □

We **assume for the next four lemmas** that f satisfies (H_2) in the interval $x \in (a, b)$.

Lemma 3   *Suppose f satisfies (H_2) in (a, b) and $x \in (a, b)$. Then $h^{-1}\Delta_h f(x)$ is a nondecreasing function of h in $(a - x, b - x)$.*

Remark 21.5   Notice that $h = 0$ lies in $(a - x, b - x)$, and at this point the expression $h^{-1}\Delta_h f(x)$ is not defined. Hence, the statement of Lemma 3 actually says that $h \mapsto h^{-1}\Delta_h f(x)$ is nondecreasing for h in $(0, b - x)$ and separately for h in $(a - x, 0)$. The latter can be reformulated as follows: since $\Delta_{-h} f(x) = -\Delta_h f(x - h)$, Lemma 3 asserts that the map $h \mapsto h^{-1}\Delta_h f(x - h)$ is a nonincreasing function of h in $(0, x - a)$.

Proof of Lemma 3   We first prove the result for $h \in (0, b-x)$. Thus, suppose $0 < \epsilon < \delta < b - x$. By condition ($H_2$), for all integers $n \geq 2$ we have

$$\Delta^2_{\delta/n} f(x) \geq 0, \quad \Delta^2_{\delta/n} f(x + \delta/n) \geq 0, \quad \dots,$$
$$\Delta^2_{\delta/n} f(x + (n-2)\delta/n) \geq 0$$
$$\implies \Delta_{\delta/n} f(x) \leq \Delta_{\delta/n} f(x + \delta/n) \leq \dots \leq \Delta_{\delta/n} f(x + (n-1)\delta/n).$$

If $0 < m < n$, then the average of the first m terms here cannot exceed the average of all n terms. Therefore,

$$\frac{f(x + m\delta/n) - f(x)}{m\delta/n} \leq \frac{f(x + \delta) - f(x)}{\delta}.$$

Now since $\epsilon \in (0, \delta)$, choose integer sequences $0 < m_j < n_j$, such that $m_j/n_j \to \epsilon/\delta$ as $j \to \infty$. Applying the preceding inequality (with m, n replaced respectively by m_j, n_j) and taking limits, it follows that $\epsilon^{-1} \Delta_\epsilon f(x) \leq \delta^{-1} \Delta_\delta f(x)$, since f is continuous. This proves the first part of the lemma, for positive h.

The proof for negative $h \in (a - x, 0)$ is similar, and is shown using the reformulation of the assertion in Remark 21.5. Given $0 < \epsilon < \delta < x - a$, by condition ($H_2$) it follows for all integers $0 < m < n$ that

$$\Delta_{\delta/n} f(x - \delta) \leq \Delta_{\delta/n} f(x - (n-1)\delta/n) \leq \dots \leq \Delta_{\delta/n} f(x - \delta/n)$$
$$\implies \frac{f(x) - f(x - \delta)}{\delta} \leq \frac{f(x) - f(x - m\delta/n)}{m\delta/n},$$

this time using the last m terms instead of the first. Now work as above: using integer sequences $0 < m_j < n_j$, such that $m_j/n_j \to \epsilon/\delta$, it follows from the continuity of f that $\delta^{-1} \Delta_\delta f(x - \delta) \leq \epsilon^{-1} \Delta_\epsilon f(x - \epsilon)$, as desired. □

We next define the one-sided derivatives of functions.

Definition 21.6   Let f be a real-valued function on (a, b). Define

$$f'_+(x) := \lim_{\delta \to 0^+} \frac{\Delta_\delta f(x)}{\delta}, \qquad f'_-(x) := \lim_{\delta \to 0^-} \frac{\Delta_\delta f(x)}{\delta} = \lim_{\delta \to 0^+} \frac{\Delta_\delta f(x - \delta)}{\delta}.$$

Lemma 4   *Suppose f satisfies (H_2) in (a, b). Then f'_+, f'_- exist and are finite and nondecreasing on all of (a, b).*

Proof   That $f'_\pm$ exist on (a, b) follows from Lemma 3, though the limits may possibly be infinite. Now fix scalars $\delta, \epsilon, x, y, z$ satisfying

$$0 < \delta < \epsilon \quad \text{and} \quad a < z - \epsilon < x - \epsilon < x < x + \epsilon < y + \epsilon < b,$$

which implies that $a < z < x < y < b$. Then we have

$$\frac{\Delta_\epsilon f(z - \epsilon)}{\epsilon} \le \frac{\Delta_\epsilon f(x - \epsilon)}{\epsilon} \le \frac{\Delta_\delta f(x - \delta)}{\delta}$$

$$\le \frac{\Delta_\delta f(x)}{\delta} \le \frac{\Delta_\epsilon f(x)}{\epsilon} \le \frac{\Delta_\epsilon f(y)}{\epsilon},$$

where the five inequalities follow respectively using Lemma 2, Remark 21.5, Lemma 2, Lemma 3, and Lemma 2.

Now let $\delta \to 0^+$ keeping ϵ, x, y, z fixed, this yields

$$\frac{\Delta_\epsilon f(z - \epsilon)}{\epsilon} \le f'_-(x) \le f'_+(x) \le \frac{\Delta_\epsilon f(y)}{\epsilon},$$

which implies that $f'_\pm(x)$ are finite on (a, b). In turn, letting $\epsilon \to 0^+$ yields

$$f'_-(z) \le f'_-(x) \le f'_+(x) \le f'_+(y),$$

which shows that $f'_\pm$ are nondecreasing on (a, b). $\qquad\square$

Lemma 5   *If f satisfies (H_2) in (a, b) then f approaches a limit in $(-\infty, +\infty]$ as x goes to a^+ and x goes to b^-.*

Proof   Note by Lemma 2 that $\Delta_\delta f(x)$ is nondecreasing in $x \in (a, b - \delta)$. Hence, $\lim_{x \to a^+} \Delta_\delta f(x)$ exists and is finite, or equals $-\infty$. (The key point is that it is not $+\infty$.) Therefore, since f is continuous,

$$+\infty > \lim_{x \to a^+} \Delta_\delta f(x) = \lim_{x \to a^+} (f(x + \delta) - f(x)) = f(a + \delta) - f(a^+).$$

It follows that $f(a^+)$ exists and cannot equal $-\infty$.

By the same reasoning, the limit $\lim_{x \to (b-\delta)^-} \Delta_\delta f(x)$ exists and is finite, or equals $+\infty$, in which case

$$-\infty < \lim_{x \to (b-\delta)^-} \Delta_\delta f(x) = f(b^-) - f(b - \delta).$$

It follows that $f(b^-)$ exists and cannot equal $-\infty$. $\qquad\square$

Lemma 6   *Suppose $f(x)$ satisfies (H_2) in (a, b).*

(1) *If $f(a^+) < +\infty$, define $f(a) := f(a^+)$. Then $f'_+(a)$ exists and is finite or $-\infty$.*

(2) *If $f(b^-) < +\infty$, define $f(b) := f(b^-)$. Then $f'_-(b)$ exists and is finite or $+\infty$.*

Proof   First, if $f(a^+)$ or $f(b^-)$ are not $+\infty$ then they are finite by Lemma 5. To show (1), by Lemma 3, for $h \in (0, b - a)$ the map $h \mapsto h^{-1}\Delta_h f(x)$ is nondecreasing. Therefore, $h \mapsto h^{-1}\Delta_h f(a)$ is the limit of a set of nondecreasing functions in h, so it too is nondecreasing in h. This proves (1).

The second part is proved similarly, using that $h \mapsto h^{-1}\Delta_h f(b - h)$ is a nonincreasing function in h.   □

Common hypothesis for Lemmas 7–14 f **satisfies** (H_k) **in** (a, b)**, for some** $k \geq 3$.

(We use this hypothesis below without mention.)

Lemma 7   *For any $a < x < b$, the map $h \mapsto h^{-k+1}\Delta_h^{k-1} f(x)$ is a non-decreasing function of h in $(0, (b - x)/(k - 1))$.*

Proof   First, note that the given map is indeed well-defined. Now we prove the result by induction on $k \geq 2$; the following argument is similar in spirit to (for instance) computing by induction the derivative of x^{k-1}.

For $k = 2$ the result follows from Lemma 3. To show the induction step, given fixed $0 < h < (b - a)/(k - 2)$ and $\delta \in (0, b - a)$, it is clear by Lemma 1 that if f satisfies (H_k) in (a, b), then we have, respectively

$$\Delta_h^{k-2} f \text{ satisfies } (H_2) \text{ in } (a, b - (k - 2)h),$$
$$\Delta_\delta f \text{ satisfies } (H_{k-1}) \text{ in } (a, b - \delta). \tag{21.4}$$

In particular, if $0 < \delta < \epsilon < (b - x)/(k - 1)$, then we have

$$\frac{\Delta_\epsilon \Delta_\epsilon^{k-2} f(x)}{\epsilon^{k-2} \epsilon} \geq \frac{\Delta_\delta \Delta_\epsilon^{k-2} f(x)}{\epsilon^{k-2} \delta} = \frac{\Delta_\epsilon^{k-2} \Delta_\delta f(x)}{\epsilon^{k-2} \delta} \geq \frac{\Delta_\delta^{k-2} \Delta_\delta f(x)}{\delta^{k-2} \delta}.$$

Indeed, the first inequality is by the assertion for $k = 2$, which follows via Lemma 3 from the first condition in (21.4); and the second inequality is by the induction hypothesis (i.e., the assertion for $k - 1$) applied using the second condition in (21.4).

We saw in the preceding calculation that $\epsilon^{-k+1}\Delta_\epsilon^{k-1} f(x) \geq \delta^{-k+1}\Delta_\delta^{k-1} f(x)$. But this is precisely the induction step.   □

Lemma 8   *There is a point $c \in [a, b]$, such that $f(x)$ satisfies (H_{k-1}) in (c, b) and $-f(x)$ satisfies (H_{k-1}) in (a, c).*

Proof   Define subsets $A, B \subset (a, b)$ via

$$A := \left\{ x \in (a, b) : \Delta_\delta^{k-1} f(x) \geq 0 \text{ for all } \delta \in (0, (b - x)/(k - 1)) \right\},$$
$$B := (a, b) \setminus A.$$

If both A, B are nonempty, and $z \in A, y \in B$, then we claim that $y < z$. Indeed, since $y \notin A$, there exists $0 < \epsilon < (b - y)/(k - 1)$, such that $\Delta_\epsilon^{k-1} f(y) < 0$. By Lemma 2, if $z' \in (a, y]$, then $\Delta_\epsilon^{k-1} f(z') < 0$, and hence $z' \notin A$. Now conclude that $z > y$.

The above analysis implies the existence of $c \in [a,b]$, such that $(a,c) \subset B \subset (a,c]$ and $(c,b) \subset A \subset [c,b)$. It is also clear that f satisfies (H_{k-1}) in (c,b).

It remains to show that if $a < c$, then $-f$ satisfies (H_{k-1}) in (a,c). Begin by defining a map $\varepsilon \colon (a,c) \to (0,\infty)$ as follows: for $x \in (a,c)$, there exists $\epsilon \in (0,(c-x)/(k-1))$, such that $\Delta_\epsilon^{k-1} f(x) < 0$. By Lemmas 2 and 7, this implies that

$$\Delta_\delta^{k-1} f(y) < 0, \qquad \forall a < y \le x,\ 0 < \delta \le \epsilon.$$

Now define $\varepsilon \colon (a,c) \to (0,\infty)$ by setting

$$\varepsilon(x) := \sup \left\{ \epsilon \in \left(0, \tfrac{c-x}{k-1} \right) : \Delta_\epsilon^{k-1} f(x) < 0 \right\}.$$

By the reasoning just described, ε is a nonincreasing function on (a,c).

With the function ε in hand, we now complete the proof by showing that $-f(x)$ satisfies (H_{k-1}) in (a,c). Let $x \in (a,c)$ and let $h > 0$ be such that $x + (k-1)h < c$. Choose any $y \in (x+(k-1)h,c)$ as well as an integer $n > h/\varepsilon(y)$. It follows that $\Delta_{h/n}^{k-1} f(y) < 0$.

Now recall from Equation (21.2) that

$$\Delta_h^{k-1} f(x) = \sum_{i_1=0}^{n-1} \cdots \sum_{i_{k-1}=0}^{n-1} \Delta_{h/n}^{k-1} f\big(x + [i_1 + \cdots + i_{k-1}]h/n \big).$$

But in each summand, the argument $x + [i_1 + \cdots + i_{k-1}]h/n < y$, so by Lemmas 2 and 7, the previous paragraph implies that each summand is negative. It follows that $\Delta_h^{k-1} f(x) < 0$. This shows that $-f(x)$ satisfies (H_{k-1}) in (a,c), as desired, and concludes the proof. $\square$

Lemma 9   *There are points*

$$a = x_0 < x_1 < \cdots < x_p = b, \qquad \text{with } 1 \le p \le 2^{k-1},$$

such that in each interval $x_j < x < x_{j+1}$, either $f(x)$ or $-f(x)$ satisfies (H_2).

This follows immediately from Lemma 8 by induction on $k \ge 2$.

Lemma 10   *The derivatives $f'_\pm$ both exist and are finite on all of (a,b).*

We remark here that $f'_\pm$ are both needed in what follows, yet Boas and Widder completely avoid discussing f'_- in this lemma or its proof (or in the sequel). For completeness, the proof for f'_- is also now described.

Proof   By Lemmas 9, 4, and 6, the functions $f'_\pm$ exist on all of (a,b), and are finite, possibly except at the points $x_1, \ldots, x_{p-1}$ in Lemma 9. We now show that $f'_\pm$ are finite at each of these points x_j.

First, suppose $f'_+(x_j)$ or $f'_-(x_j)$ equals $+\infty$. Choose $\delta > 0$ small enough, such that

$$x_{j-1} < x_j - (k-2)\delta < x_j < x_j + \delta < x_{j+1}.$$

Now if $f'_+(x_j) = +\infty$, then

$$\Delta_\delta^{k-1} f'_+(x_j - (k-2)\delta) = -\infty$$
$$\implies \lim_{h \to 0^+} \frac{1}{h} \Delta_\delta^{k-1} \Delta_h f(x_j - (k-2)\delta) = -\infty$$
$$\implies \Delta_\delta^{k-1} \Delta_h f(x_j - (k-2)\delta) < 0 \quad \text{for all small positive } h.$$

But this contradicts Lemma 1. Similarly, if $f'_-(x_j) = +\infty$, then

$$\Delta_\delta^{k-1} f'_-(x_j - (k-2)\delta) = -\infty$$
$$\implies \lim_{h \to 0^+} \frac{1}{h} \Delta_\delta^{k-1} \Delta_h f(x_j - (k-2)\delta - h) = -\infty$$
$$\implies \Delta_\delta^{k-1} \Delta_h f(x_j - (k-2)\delta - h) < 0 \quad \text{for all small positive } h,$$

which again contradicts Lemma 1.

The other case is if $f'_+(x_j)$ or $f'_-(x_j)$ equals $-\infty$. The first of these subcases is now treated; the subcase $f'_-(x_j) = -\infty$ is similar. Begin as above by choosing $\delta > 0$, such that

$$x_{j-1} < x_j - (k-1)\delta < x_j < x_{j+1}.$$

Now if $f'_+(x_j) = +\infty$, then a similar computation to above yields

$$\Delta_\delta^{k-1} f'_+(x_j - (k-1)\delta) = -\infty$$
$$\implies \lim_{h \to 0^+} \frac{1}{h} \Delta_\delta^{k-1} \Delta_h f(x_j - (k-1)\delta) = -\infty$$
$$\implies \Delta_\delta^{k-1} \Delta_h f(x_j - (k-1)\delta) < 0 \quad \text{for all small positive } h,$$

which contradicts Lemma 1. $\qquad\square$

The above trick of studying $\Delta_\delta^n g(y - p\delta)$ where $p = k - 1$ or $k - 2$ (and $n = k - 1$, $g = f'_\pm$ so that we deal with the kth order divided differences/derivatives of f) is a powerful one. Boas and Widder now use the same trick to further study the derivative of f, and show its existence, finiteness, and continuity in Lemmas 11 and 13.

Lemma 11 f' exists and is finite on (a,b).

Proof   We fix $x \in (a, b)$, and work with $\delta > 0$ small, such that $a < a + k\delta < x < b - 2\delta < b$. Let $p \in \{0, 1, \ldots, k\}$, then

$$0 \le \frac{1}{\delta} \Delta_\delta^k f(x - p\delta) = \frac{1}{\delta} \sum_{i=0}^{k} \binom{k}{i} (-1)^{k-i} f(x + (i - p)\delta).$$

Subtract from this the identity $0 = \delta^{-1} f(x)(1 - 1)^k = \delta^{-1} f(x) \sum_{i=0}^{k} \binom{k}{i}(-1)^{k-i}$, so that the $i = p$ term cancels, and multiply and divide the remaining terms by $(i - p)$ to obtain

$$0 \le \frac{1}{\delta} \Delta_\delta^k f(x - p\delta) = \sum_{\substack{i=0, \\ i \ne p}}^{i=k} \binom{k}{i} (-1)^{k-i} \frac{f(x + (i - p)\delta) - f(x)}{(i - p)\delta} (i - p).$$

Letting $\delta \to 0^+$, it follows that

$$A_p f_-'(x) + B_p f_+'(x) \ge 0, \qquad \text{where } A_p := \sum_{i=0}^{p-1} \binom{k}{i} (-1)^{k-i} (i - p),$$

$$B_p := \sum_{i=p+1}^{k} \binom{k}{i} (-1)^{k-i} (i - p);$$

$$(21.5)$$

note here that

$$A_p + B_p = \sum_{i=0}^{k} \binom{k}{i} (-1)^{k-i} (i - p)$$

$$= k \sum_{i=1}^{k-1} \binom{k-1}{i-1} (-1)^{k-i} - p \sum_{i=0}^{k} \binom{k}{i} (-1)^{k-i} = 0.$$

Now specialize p to be $k - 1$ and $k - 2$. In the former case $B_p = 1$, so $A_p = -1$, and by (21.5) we obtain $f_+'(x) \ge f_-'(x)$. In the latter case $p = k - 2$ (with $k \ge 3$), we have $B_p = 2 - k < 0$. Thus, $A_p = k - 2 > 0$, and by (21.5) we obtain $f_-'(x) \ge f_+'(x)$. Therefore, $f'(x)$ exists and by Lemma 10 it is finite. □

Lemma 12   *If $a < x < x + (k - 1)h < b$, then $\Delta_h^{k-1} f'(x) \ge 0$.*

Proof $\Delta_h^{k-1} f'(x) = \lim_{\delta \to 0^+} \dfrac{\Delta_\delta \Delta_h^{k-1} f(x)}{\delta}$ and   this   is   nonnegative   by Lemma 1. □

Lemma 13   *f' is continuous on (a, b).*

Remark 21.7   We record here a minor typo in the Boas–Widder paper [44]. Namely, the authors begin the proof of Lemma 13 by claiming that f' is monotonic. However, this is not true as stated: for any $k \geq 3$, the function $f(x) = x^3$ satisfies (H_k) on $I = (-1, 1)$ but f' is not monotone on I. The first paragraph of the following proof addresses this issue, using that f' is *piecewise* monotone on (a, b).

Proof of Lemma 13   By Lemmas 9 and 4, there are finitely many points x_j with $0 \leq j \leq p \leq 2^{k-1}$, such that on each (x_j, x_{j+1}), $f'_{\pm} = f'$ is monotone (where this last equality follows from Lemma 11). Thus, f' is piecewise monotone on (a, b).

Now define the limits

$$f'(x^{\pm}) := \lim_{h \to 0^+} f'(x \pm h), \qquad x \in (a, b).$$

It is clear that $f'(x^{\pm})$ exists on (a, b), including at each $x_j \neq a, b$. Note that $f'(x_j^{\pm}) \in [-\infty, +\infty]$, while $f'(x^{\pm}) \in \mathbb{R}$ for all other points $x \neq x_j$. First, claim that $f'(x^+) = f'(x^-)$ – where this common limit is possibly infinite – and then that $f'(x^+) = f'(x)$, which will rule out the infinitude using Lemma 11, and complete the proof.

For each of the two steps, we proceed as in the proof of Lemma 11. Begin by fixing $x \in (a, b)$, and let $\delta > 0$ be such that $a < x - k\delta < x < x + 2\delta < b$. Let $p \in \{0, 1, \ldots, k\}$, then by Lemma 12,

$$0 \leq \Delta_{\delta}^{k-1} f'\left(x - \left(p - \tfrac{1}{2}\right)\delta\right)$$

$$= \sum_{i=0}^{k-1} \binom{k-1}{i} (-1)^{k-1-i} f'\left(x + \left(i - p + \tfrac{1}{2}\right)\delta\right).$$

Let $\delta \to 0^+$, then,

$$A_p f'(x^-) - A_p f'(x^+) \geq 0,$$

$$\text{where } A_p := \sum_{i=0}^{p-1} \binom{k-1}{i}(-1)^{k-1-i} = -\sum_{i=p}^{k-1}\binom{k-1}{i}(-1)^{k-1-i}.$$

Now specialize p to be $k - 1$ and $k - 2$. In the former case $A_p = -1$, hence $f'(x^-) \leq f'(x^+)$; whereas if $p = k - 2$, then $A_p = k - 2 > 0$, hence $f'(x^-) \geq f'(x^+)$. These inequalities and the trichotomy of the extended real line $[-\infty, +\infty]$ imply that $f'(x^-) = f'(x^+)$.

Using the same $\delta \in ((x - a)/k, (b - x)/2)$ and $p \in \{0, 1, \ldots, k\}$, Lemma 12 also implies

$$0 \leq \Delta_\delta^{k-1} f'(x - p\delta).$$

Taking $\delta \to 0^+$ and using that $f'(x^-) = f'(x^+)$ yields

$$B_p f'(x) - B_p f'(x^+) \geq 0,$$

where $B_p := \binom{k-1}{p}(-1)^{k-1-p} = -\sum_{\substack{i=0, \\ i \neq p}}^{k-1} \binom{k-1}{i}(-1)^{k-1-i}.$

Now specialize p to be $k - 1$ and $k - 2$. In the former case $B_p = 1$, hence $f'(x) \geq f'(x^+)$; whereas if $p = k - 2$, then $B_p = 1 - k < 0$, hence $f'(x) \leq f'(x^+)$. These inequalities imply that $f'(x^+) = f'(x^-)$ equals $f(x)$, and in particular is finite for all $x \in (a,b)$. □

The final lemma simply combines the preceding two:

Lemma 14 f' satisfies the condition (H_{k-1}) in (a,b).

Proof   This follows immediately from Lemmas 12 and 13.   □

Having shown the 14 lemmas above, we conclude with:

Proof of the Boas–Widder Theorem 21.1   The proof is by induction on $k \geq 2$. The induction step is clear: use Lemma 14. We now show the base case of $k = 2$. By Lemma 4, the functions $f'_\pm$ exist and are nondecreasing on (a,b). Moreover, f is continuous by assumption. To prove its convexity, we make use of the following basic result from one-variable calculus:

Proposition 21.8   Let $f : [p,q] \to \mathbb{R}$ be a continuous function whose right-hand derivative f'_+ exists on $[p,q)$ and is Lebesgue integrable. Then,

$$f(y) = f(p) + \int_p^y f'_+(t)\, dt, \qquad \forall y \in [p,q].$$

Proposition 21.8 applies to our function f satisfying (H_2), since f'_+ is nondecreasing by Lemma 4, and hence Lebesgue integrable. Therefore, $f(y) - f(x) = \int_x^y f'_+(t)\, dt$ for $a < x < y < b$. Now repeat the proof of Proposition 21.4 to show that f is convex on (a,b). This completes the base case of $k = 2$ and concludes the proof.   □

22

Menger's Results and Euclidean Distance Geometry

We conclude this part of the text with a brief detour into the same area where we started this part of the text: metric geometry, specifically, that of Euclidean spaces $\mathbb{R}^n$ – and of their closure, Hilbert space ℓ^2. This is a beautiful area of mathematical discovery, which has featured work by several prominent mathematicians, including Birkhoff, Cauchy, Cayley, Gödel, Menger, Schoenberg, and von Neumann, among others. See [156] for a modern exposition of some of the gems of distance geometry (which begins, interestingly, with Heron's formula for the area of a triangle, from two millennia ago).

The main result of this chapter is a 1928 theorem of Menger:

Theorem 22.1 (Menger [169]; see also [203]) *A metric space (X,d) can be isometrically embedded in Hilbert space ℓ^2 if and only if X is separable and every subset of X of size $n + 1$ can be isometrically embedded in $\mathbb{R}^n$ (equivalently, in ℓ^2) for $n \geq 2$.*

This result, together with Schoenberg's theorems 11.10 and 11.14 on Hilbert space embeddings of finite metric spaces X, immediately yields the same characterizations for arbitrary separable X:

Theorem 22.2 (Schoenberg) *Suppose (X,d) is a separable metric space.*

(1) *X embeds isometrically into Hilbert space ℓ^2 if and only if for every integer $n \geq 2$ and $(n + 1)$-tuple of points $Y := (x_0, \ldots, x_n)$ in X, the "alternate Cayley–Menger matrix" $CM'(Y) := \big(d(x_0, x_j)^2 + d(x_0, x_k)^2 - d(x_j, x_k)^2 \big)_{j,k=1}^n$ is positive semidefinite.*

(2) *X embeds isometrically into Hilbert space ℓ^2 if and only if the Gaussian kernel $\exp(-\sigma(\cdot)^2)$ is a positive definite function on X for all $\sigma > 0$ (equivalently, for some sequence σ_m of positive numbers decreasing to 0^+).*

190

The goal in this chapter is to explore some simple, yet beautiful observations in Euclidean distance geometry, which help prove Theorem 22.1, and also provide connections to Cayley–Menger matrices [55, 170] and to n-point homogeneous spaces (see Remark 11.18). We begin with the latter.

22.1 n-Point Homogeneity of Euclidean and Hilbert Spaces

As early as 1944, in his influential work [39] Birkhoff defines a metric space (X, d) to be *n-point homogeneous* if given two equinumerous subsets of X of size at most n, an isometry between them extends to a self-isometry of X. The heart of the present proof of Theorem 22.1 is to show that Euclidean space $\mathbb{R}^k$ is n-point homogeneous for all $k, n \geq 1$:

Theorem 22.3   *Fix an integer $k \geq 1$.*

(1) *The Euclidean space $\mathbb{R}^k$ with the Euclidean metric is n-point homogeneous for all n. More strongly: any isometry between two subsets $M, N \subset \mathbb{R}^k$ is, up to a translation, the restriction of an orthogonal linear transformation of $\mathbb{R}^k$.*
(2) *Hilbert space ℓ^2 is n-point homogeneous for all n.*

The first step in proving Theorem 22.3 is the following observation about Gram matrices:

Lemma 22.4   *Given vectors $y_0, \ldots, y_n \in \ell^2$ for some $n \geq 0$, the Gram matrix $(\langle y_j, y_k \rangle)_{j,k=0}^{n}$ is invertible if and only if the y_j are linearly independent.*

Proof   We prove the contrapositive. If $\sum_{k=0}^{n} c_k y_k = 0$ is a nontrivial linear combination, then applying $\langle y_j, - \rangle$ for all j yields $\mathrm{Gram}((y_k)_k)\mathbf{c} = 0$, where $\mathbf{c} = (c_0, \ldots, c_n)^T \neq 0$. Conversely, if $\mathrm{Gram}((y_k)_k)\mathbf{c} = 0$ and $\mathbf{c} \neq 0$, then

$$0 = \mathbf{c}^T \mathrm{Gram}((y_k)_k)\mathbf{c} = \left\| \sum_{k=0}^{n} c_k y_k \right\|^2,$$

so that some nontrivial linear combination of the y_k vanishes, as desired.   □

This simple lemma leads to striking consequences. We will presently mention two, the first of which involves a classical concept (which already featured in Theorem 11.10):

Definition 22.5   Given a metric space $X = (\{x_0, x_1, \ldots, x_n\}, d)$, the associated Cayley–Menger matrix is

$$CM(X)_{(n+2)\times(n+2)} := \begin{pmatrix} 0 & d_{01}^2 & d_{02}^2 & \cdots & d_{0n}^2 & 1 \\ d_{10}^2 & 0 & d_{12}^2 & \cdots & d_{1n}^2 & 1 \\ d_{20}^2 & d_{21}^2 & 0 & \cdots & d_{2n}^2 & 1 \\ \vdots & \vdots & \vdots & \ddots & \vdots & \vdots \\ d_{n0}^2 & d_{n1}^2 & d_{n2}^2 & \cdots & 0 & 1 \\ 1 & 1 & 1 & \cdots & 1 & 0 \end{pmatrix}, \tag{22.1}$$

where $d_{jk} := d(x_j, x_k)$ for $0 \le j, k \le n$. Similarly, the "alternate form" of the Cayley–Menger matrix here is

$$CM'(X)_{n \times n} := \begin{pmatrix} 2d_{01}^2 & d_{01}^2 + d_{02}^2 - d_{12}^2 & \cdots & d_{01}^2 + d_{0n}^2 - d_{1n}^2 \\ d_{01}^2 + d_{02}^2 - d_{12}^2 & 2d_{02}^2 & \cdots & d_{01}^2 + d_{0n}^2 - d_{1n}^2 \\ \vdots & \vdots & \ddots & \vdots \\ d_{01}^2 + d_{0n}^2 - d_{1n}^2 & d_{02}^2 + d_{0n}^2 - d_{2n}^2 & \cdots & 2d_{0n}^2 \end{pmatrix}.$$
$$\tag{22.2}$$

Recall that the positive semidefiniteness of the second matrix features in Schoenberg's recasting of Menger and Fréchet's results on Hilbert space embeddings of finite metric spaces. We now write down a preliminary observation that relates the two determinants above:

Lemma 22.6   *For all finite metric spaces X with at least two points,*

$$\det CM(X) = (-1)^{|X|} \det CM'(X). \tag{22.3}$$

Proof   Starting with the matrix $CM(X)$, perform elementary row and column operations, leaving the determinant unchanged. First, subtract the first row from all nonextremal rows. Then subtract the first and last columns each from the nonextremal columns. This yields the bordered matrix

$$\begin{pmatrix} 0 & \mathbf{0}_n^T & 1 \\ \mathbf{0}_n & -CM'(X) & \mathbf{0}_n \\ 1 & \mathbf{0}_n^T & 0 \end{pmatrix},$$

whose determinant is $(-1)^{n+1} \det CM'(X)$, as desired. $\square$

We can now state and prove the two consequences of Lemma 22.4 promised above. The first of these is a well-known result, proved in 1841 by Cayley during his undergraduate days [55]. The second is the underlying principle behind the Global Positioning System, or GPS – *trilateration* (also referred to more colloquially as "triangulation"): every point in the plane (or on the surface of a sphere "like" the Earth) is uniquely determined by intersecting three circles that denote distances from three noncollinear points (or four spheres centered at four noncoplanar points).

Proposition 22.7

(1) *(Cayley [55]). Suppose an isometry Ψ sends a finite metric space $(X = \{x_0, \ldots, x_n\}, d)$ into Hilbert space ℓ^2. Then the vectors $\Psi(x_0), \ldots, \Psi(x_n)$ are affine linearly dependent (i.e., lie on an $(n-1)$-dimensional subspace) in ℓ^2, if and only if the Cayley–Menger determinant of X vanishes.*

(2) *Fix vectors $y_0 = 0, y_1, \ldots, y_n \in \ell^2$. Given an arbitrary vector $y \in \ell^2$, the following are equivalent:*

    (a) *y is (uniquely) determined by the tuple of Euclidean distances $(\|y\|, \|y - y_1\|, \ldots, \|y - y_n\|) \in \mathbb{R}^{n+1}$.*

    (b) *y is in the span of $y_1, \ldots, y_n$.*

Proof

(1) Denote $y_j := \Psi(x_j)$ for $0 \le j \le n$. Now compute, as in Equation (11.2) in the proof of Theorem 11.10:

$$d(y_0, y_j)^2 + d(y_0, y_k)^2 - d(y_j, y_k)^2 = \langle y_0 - y_j, y_0 - y_k \rangle,$$

so that $CM'(X) = \text{Gram}((y_0 - y_j)_{j=1}^n)$. Now $CM(X)$ is singular if and only if so is $CM'(X)$. From above and by Lemma 22.4, this happens if and only if the vectors $y_0 - y_j$, $1 \le j \le n$ are linearly dependent. This completes the proof.

(2) For this part, let $V \subset \ell^2$ denote the span of the y_j. First, suppose $y \notin V$. Write $y = y_V \oplus y_{V^\perp}$ as the orthogonal decomposition of y. One verifies that for any unit vector $v \in V^\perp$ (for instance, $v = \pm y_{V^\perp} / \|y_{V^\perp}\|$), both y as well as the vector

$$y_V \oplus \|y_{V^\perp}\| v$$

have the same distances from every vector in V – in particular, from each of $0, y_1, \ldots, y_n$. This shows (the contrapositive of) one implication.

    Conversely, suppose $y \in V$. We show that y is uniquely determined by the distances to the y_j and to 0 – in fact, it suffices to consider the distances to a basis of V. Thus, suppose without loss of generality that the

y_j are linearly independent. Let $y := \sum_{j=1}^n c_j y_j$, and let
$d_0 := \|y\|, d_j := \|y - y_j\|$. We show that the d_j uniquely determine the
c_j, and hence y. Indeed, a straightforward computation yields

$$d_0^2 - d_j^2 = \left\| \sum_{k=1}^n c_k y_k \right\|^2 - \left\| \sum_{k=1}^n c_k y_k - y_j \right\|^2 = -\|y_j\|^2 + 2 \sum_{k=1}^n c_k \langle y_j, y_k \rangle$$

for all $1 \le j \le n$. Rewriting this system of linear equations (in
$\mathbf{c} = (c_1, \ldots, c_n)$) yields

$$\mathrm{Gram}((y_j)_j)\mathbf{c} = \frac{1}{2}(\|y\|^2 + \|y_j\|^2 - \|y - y_j\|^2)_{j=1}^n. \tag{22.4}$$

Hence, $\mathbf{c}$ is unique, by Lemma 22.4.   □

Equipped with these preliminaries, we are now ready to proceed toward
proving Menger's result. We first show:

Proof of Theorem 22.3

(1) First suppose that both M, N contain the origin, and $\Psi \colon M \to N$ sends 0
to itself. This is not really a constraint: if here we can show that $\Psi = T|_M$
for some orthogonal matrix $T \in O_k(\mathbb{R})$, then for a general isometry Ψ
and an arbitrary (base)point $m_\circ \in M$, the isometry

$$\Phi \colon M - m_\circ \to \Psi(M) - \Psi(m_\circ), \qquad v \mapsto \Psi(m_\circ + v) - \Psi(m_\circ)$$

sends 0 to 0, hence equals the restriction to M of some $T \in O_k(\mathbb{R})$. Thus,

$$\Psi(m) = T(m) + (\Psi(m_\circ) - T(m_\circ)), \qquad \forall m \in M.$$

Therefore, we may assume without loss of generality that
$m_\circ := 0 \in M \cap N$ and $\Psi(0) = 0$. In this case, we need to show that Ψ is
the restriction to M of an orthogonal matrix.

Begin by isolating a basis $m_1, \ldots, m_r \in M$ of the $\mathbb{R}$-span of M. We
assume $r > 0$, else M is a singleton, hence so is N, and then the result is
immediate. Let $y_j := \Psi(m_j)$, then

$$2\langle y_j, y_k \rangle$$
$$= \|\Psi(m_j) - \Psi(0)\|^2 + \|\Psi(m_k) - \Psi(0)\|^2 - \|\Psi(m_j) - \Psi(m_k)\|^2$$
$$= \|m_j\|^2 + \|m_k\|^2 - \|m_j - m_k\|^2 = 2\langle m_j, m_k \rangle, \qquad \forall 1 \le j, k \le r. \tag{22.5}$$

Hence, by Lemma 22.4, the y_j are also linearly independent. By the
same reasoning, for any $m \in M \setminus \{m_1, \ldots, m_r\}$ we have that the Gram
matrix of $m_1, \ldots, m_r, m$ is singular, hence so is the Gram matrix of

$y_1, \ldots, y_r, \Psi(m)$. Again, using Lemma 22.4, the image $\Psi(M) = N$ is contained in the $\mathbb{R}$-span of $\{y_j = \Psi(m_j) : 1 \le j \le r\}$.

At this point, we define the linear map

$$T : \mathrm{span}_{\mathbb{R}}(M) \to \mathrm{span}_{\mathbb{R}}(N), \qquad m_j \mapsto y_j \ \forall 1 \le j \le r.$$

The next claim is that $T \equiv \Psi$ on M. Indeed, given $m \in M$, write

$$m = \sum_{j=1}^{r} c_j(m) m_j, \qquad \Psi(m) = \sum_{j=1}^{r} c'_j(m) y_j.$$

We now apply Proposition 22.7 to both m and $\Psi(m)$. Since $\mathrm{Gram}((m_j)_j) = \mathrm{Gram}((y_j)_j)$ by (22.5), the computation in Equation (22.4) for both y_j and m_j reveals that

$$c_j(m) = c'_j(m), \qquad \forall m \in M, \ 1 \le j \le r.$$

In particular,

$$T(m) = \sum_{j=1}^{r} c_j(m) T(m_j) = \sum_{j=1}^{r} c'_j(m) y_j = \Psi(m), \qquad \forall m \in M,$$

which proves the claim.

The final claim is that T preserves lengths on $\mathrm{span}_{\mathbb{R}}(M)$. But this is clear by (22.5): if $v := \sum_j c_j m_j$ with all $c_j \in \mathbb{R}$, then

$$\langle Tv, Tv \rangle = \sum_{j,k=1}^{r} c_j c_k \langle y_j, y_k \rangle = \sum_{j,k=1}^{r} c_j c_k \langle m_j m_k \rangle = \langle v, v \rangle.$$

To conclude the proof of this part, choose orthonormal bases of the ortho-complements in $\mathbb{R}^k$ of $\mathrm{span}_{\mathbb{R}}(M)$ and $\mathrm{span}_{\mathbb{R}}(N)$, and map one basis to another to extend T to an orthogonal linear map on all of $\mathbb{R}^k$.

(2) This is clear by the previous part: given $x_j, y_j \in \ell^2$ with $1 \le j \le n$, such that

$$\|x_j - x_k\| = \|y_j - y_k\|, \qquad \forall 1 \le j, k \le n,$$

choose a finite-dimensional subspace of ℓ^2 which contains all x_j, y_j. Apply the previous part to this subspace; modulo the translation, one has an orthogonal transformation of this subspace, which we augment by the identity map on its orthocomplement. $\qquad \square$

Remark 22.8   The proof of Theorem 22.3 is reminiscent of the well-known "lurking isometry" method – so named by J. Ball – in bounded analytic interpolation. This involves using Hilbert space realizations, and has numerous applications, including to the problems of Pick–Nevanlinna and Carathéodory–Fejer, among others (see, e.g., [1, 2]), and also indirectly in H_∞ methods in control theory (see [11] and the references therein).

Finally, we use Theorem 22.3 to prove Menger's result:

Proof of Theorem 22.1   The "only if" part is immediate, modulo a translation in order to map one of the points to the origin. Conversely, if X is finite, then the result is again easy. Thus, we now assume that X is both infinite and separable. Let $D := \{x_n : n \geq 0\}$ denote a countably infinite dense subset of X and define $D_n := \{x_0, \ldots, x_n\}$ for $n \geq 0$. We are given isometric embeddings $\Psi_n : D_n \hookrightarrow \ell^2$ for each $n \geq 2$, where we assume without loss of generality that $\Psi_n(x_0) = 0 \, \forall n \geq 0$. We now construct an isometric embedding: $D \hookrightarrow \ell^2$, once again sending x_0 to 0.

To do so, fix and start at any integer $n_0 \geq 2$, say. Given $n \geq n_0$, we have $\Psi_n : D_n \hookrightarrow \ell^2$ (sending x_0 to 0). Now

$$\Psi_n \circ \Psi_{n+1}^{-1} : \Psi_{n+1}(D_n) \to \Psi_n(D_n)$$

is an isometry of an $(n + 1)$-point set in ℓ^2, sending 0 to 0. Extend this to an orthogonal linear transformation on ℓ^2 by (the proof of) Theorem 22.3, say T_{n+1}. Thus, we have "increased" $\Psi_n(D_n)$ to an isometric image of D_{n+1}, namely $T_{n+1}(\Psi_{n+1}(D_{n+1}))$, while not changing the images of $x_0, x_1, \ldots, x_n$.

We now repeatedly compose the T_n, to obtain the increasing family of sets

$$S_n := (T_{n_0+1} \circ \cdots \circ T_{n-1} \circ T_n)(\Psi_n(D_n)), \quad n \geq n_0,$$

which satisfy

$$0 \in S_{n_0} = \Psi_{n_0}(D_{n_0}) \subset S_{n_0+1} \subset S_{n_0+2} \subset \cdots.$$

Moreover, each $S_n = \{y_0 = 0, y_1, \ldots, y_n\}$ for $n \geq n_0$, together with an isometry: $D_n \to S_n$, sending $x_j \mapsto y_j$ for $0 \leq j \leq n$. The union of these sets provides the desired isometric embedding $\Phi : D \hookrightarrow \bigcup_n S_n = \lim_{n \to \infty} S_n$.

The final step is to apply the following standard fact from analysis, with $Y = \ell^2$:

Suppose $(X, d), (Y, d')$ are metric spaces, with Y complete. If $D \subset X$ is dense, any isometric embedding $\Phi : D \to Y$ extends uniquely to an isometric embedding $\widetilde{\Phi} : X \to Y$.   □

22.2 Cayley–Menger Determinants, Simplex Volumes and Heron's Formula

It is impossible to discuss Cayley–Menger matrices $CM(X)$ without explaining their true content: their connection to the squared volume of the simplex with vertices the elements of X.

Theorem 22.9   *Suppose $n \geq 1$ and $X = \{x_0, \ldots, x_n\} \subset \mathbb{R}^n$. Then the volume $V_n(X)$ of the $(n+1)$-simplex with vertices x_j satisfies*

$$V_n(X)^2 = \frac{(-1)^{n+1} \det CM(X)}{2^n (n!)^2} = \frac{\det CM'(X)}{2^n (n!)^2}.$$

As a special case, if the points x_j are affine linearly dependent, then the volume of the corresponding simplex is zero, as is the determinant by Cayley's proposition 22.7.

Corollary 22.10   *For all finite subsets X of Euclidean space or Hilbert space, $\det CM'(X) \geq 0$.*

Remarkably, Theorem 22.9 can be proved using only determinants (and a bit of visual geometry). Variants of the following proof can be found in several sources, including online.

Proof of Theorem 22.9   We begin with the "usual" description of the volume of the simplex via determinants. Recall that the n-volume of a simplex in $\mathbb{R}^n$ having $n+1$ vertices is obtained inductively by integrating the area of cross section as one goes from the base (which is a simplex in $\mathbb{R}^{n-1}$ with n vertices) to the apex/remaining vertex along an "altitude" of height h_n. An easy undergraduate calculus exercise reveals that if the base has $(n-1)$-volume V_{n-1}, then

$$V_n = \frac{h_n V_{n-1}}{n}.$$

One can now proceed inductively. Thus, let h_1 denote the length $\|x_0 - x_1\|$, let h_2 denote the "height" of x_2 "above" the segment joining x_0, x_1 (so it can be written as the norm of a suitable orthogonal complement), and so on. Then,

$$V_n(X) = \frac{1}{n!} h_n h_{n-1} \cdots h_1.$$

We now show that this product expression equals (up to sign) a determinant. Write

$$x_j := \left(x_j^{(1)}, \ldots, x_j^{(n)} \right)^T \in \mathbb{R}^n, \qquad 0 \leq j \leq n,$$

and **claim** that the volume equals, up to a sign,

$$V_n(X) = \pm\frac{1}{n!} \det(A), \quad \text{where } A := \left(x_j^{(k)} - x_0^{(k)}\right)_{j,k=1}^n. \tag{22.6}$$

To show the claim, note that working with A essentially amounts to assuming $x_0 = 0$. Choosing a suitable orthonormal basis (i.e., by applying a suitable orthogonal transformation), we may further assume that $x_1, \ldots, x_{n-1} \in \mathbb{R}^{n-1}$ – thus, the final column of A has all entries zero except at most the (n, n) entry. Now the final row of A, which denotes the vector $x_n - x_0$, may be replaced by its orthogonal complement to the span of $\{x_j - x_0 : j < n\}$ without changing the determinant, and so we obtain (up to a sign) the height h_n – and in the nth coordinate since $x_j - x_0 \in \mathbb{R}^{n-1}$ for $j < n$. This scalar can be taken out of the determinant and we are now left with the determinant of an $(n-1) \times (n-1)$ matrix.

Applying the same arguments for $x_j - x_0$ with $j \le n-2$ now, we obtain h_{n-1}, and so on. Proceeding by downward induction (and taking the absolute value), we obtain (22.6).

The remainder of the proof consists of matrix manipulations. The next variant is to observe that (up to a sign), the volume of the simplex on X equals

$$\frac{1}{n!} \det \left(x_j^{(k)} - x_0^{(k)}\right)_{j,k=1}^n = \frac{1}{n!} \det A = \frac{1}{n!} \det \begin{pmatrix} 1 & \mathbf{0}_n^T \\ \mathbf{1}_n & A \end{pmatrix}.$$

Add $x_0^{(k)}$ times the first column to the kth column in the final matrix to get

$$\frac{1}{n!} \begin{pmatrix} 1 & x_0^T \\ \vdots & \vdots \\ 1 & x_n^T \end{pmatrix}.$$ Post-multiplying this matrix by its transpose gives,

$$V_n(X)^2 = \frac{1}{(n!)^2} \det \left(1 + x_j^T x_k\right)_{j,k=0}^n$$

$$= \frac{1}{(n!)^2} \det \begin{pmatrix} 1 & \mathbf{1}_{n+1}^T \\ \mathbf{0}_{n+1} & \left(1 + x_j^T x_k\right)_{j,k=0}^n \end{pmatrix}.$$

Subtracting the first row from every other row gives,

$$V_n(X)^2 = \frac{1}{(n!)^2} \det \begin{pmatrix} 1 & \mathbf{1}_{n+1}^T \\ -\mathbf{1}_{n+1} & \left(\langle x_j, x_k\rangle\right)_{j,k=0}^n \end{pmatrix}.$$

One can expand this determinant along the first row. Now the cofactor of the $(1, 1)$ entry is zero, since it is a $(n+1) \times (n+1)$ Gram matrix of vectors in $\mathbb{R}^n$ (so has rank at most n). Thus, we may replace the $(1, 1)$-entry in the above matrix by 0 and leave the determinant unchanged. First do this, then take out a factor of -1 from the first column and of $-1/2$ from all other columns, and finally, a factor of -2 from the first row, to obtain

$$V_n(X)^2 = \frac{(-1)^{n+1}}{2^n(n!)^2} \det \begin{pmatrix} 0 & \mathbf{1}_{n+1}^T \\ \mathbf{1}_{n+1} & \left(-2\langle x_j, x_k \rangle \right)_{j,k=0}^n \end{pmatrix}.$$

Finally, add $\langle x_j, x_j \rangle$ times the initial row to the row containing $-2\langle x_j, x_k \rangle$ – for each $0 \le j \le n$. Also perform the analogous column operations. This yields precisely the Cayley–Menger matrix $CM(X)$, with the final row and column moved to the initial row and column, respectively. As these permutations leave the determinant unchanged, it follows that

$$V_n(X)^2 = \frac{(-1)^{n+1}}{2^n(n!)^2} \det CM(X).$$

The proof is complete by Equation (22.3).   □

As a special case, this result leads to a well-known formula from two thousand years ago:

Corollary 22.11 (Heron's formula)   *A (Euclidean) triangle with edge lengths a, b, c and semi-perimeter $s = \frac{1}{2}(a + b + c)$ has area $\sqrt{s(s-a)(s-b)(s-c)}$.*

Proof   Explicitly expand the determinant in Theorem 22.9 for $n = 2$, to obtain

$$V_2^2 = \frac{-1}{16} \det \begin{pmatrix} 0 & a^2 & b^2 & 1 \\ a^2 & 0 & c^2 & 1 \\ b^2 & c^2 & 0 & 1 \\ 1 & 1 & 1 & 0 \end{pmatrix}$$

$$= \frac{-1}{16} \left(-(a+b+c)(a+b-c)(b+c-a)(c+a-b) \right),$$

and this is precisely $s(s-a)(s-b)(s-c)$.   □

22.3 Completely Monotone Functions and Distance Transforms

In parallel to the use of absolutely monotone functions earlier in this text – to characterize positivity preservers of kernels on infinite domains (or all finite domains) – we present here a related result by Ressel that features completely monotone functions. This is followed by a sampling of early results in metric geometry, again by Schoenberg, that feature such functions.

Definition 22.12   A function $f : (0, \infty) \to \mathbb{R}$ is *completely monotone* if f is smooth and $(-1)^k f^{(k)}$ is nonnegative on $(0, \infty)$ for all $k \geq 0$. A continuous function $f : [0, \infty) \to \mathbb{R}$ is completely monotone if the restriction of f to $(0, \infty)$ is completely monotone.

For instance, $x^{-\alpha}$ for $\alpha \leq 0$ is completely monotone on $(0, \infty)$.

We start with two results which are easily reformulated in the language of kernels:

(1) In his 1974 paper [191], Ressel characterized the functions that are positive definite in a different sense: given an abelian semigroup $(S, +)$, a function $f : S \to \mathbb{R}$ is said to be *positive semidefinite* if f is bounded and for any finite set of elements $s_1, \dots, s_n \in S$, the matrix $(f(s_j + s_k))_{j,k=1}^{n}$ is positive semidefinite. Ressel then showed for all $p \geq 1$ that the continuous and positive semidefinite functions on the semigroup $[0, \infty)^p$ are precisely Laplace transforms of finite nonnegative Borel measures on $[0, \infty)^p$. In particular, for $p = 1$, this is further equivalent – by a result attributed to Bernstein, Hausdorff, and Widder – to f being completely monotone on $[0, \infty)$.

(2) A related result to this was shown by Schoenberg [204] in 1938. It says that a continuous function $f : [0, \infty) \to \mathbb{R}$ satisfies the property that for all integers $m, n \geq 1$ and vectors $x_1, \dots, x_m \in \mathbb{R}^n$, the matrix $\left(f(\|x_j - x_k\|^2)\right)_{j,k=1}^{m}$ is positive semidefinite, if and only if f is completely monotone – i.e., as mentioned in the previous part, there exists a finite nonnegative measure μ on $[0, \infty)$, such that

$$f(x) = \int_0^\infty \exp(-xt)\, d\mu(t), \qquad \forall x \geq 0.$$

Completely monotone functions also feature in the study of metric "endomorphisms" of Euclidean spaces. For instance, Schoenberg proved (in the aforementioned 1938 paper):

Theorem 22.13   *Given a continuous map $f : [0, \infty) \to [0, \infty)$, the following are equivalent:*

(1) *For all integers $m, n \geq 1$ and vectors $x_1, \dots, x_m \in \mathbb{R}^n$, the matrix $\left(f(\|x_j - x_k\|)\right)_{j,k=1}^{m}$ is Euclidean – i.e., $\{x_j\}$ with the metric $f \circ \| \cdot \|$ isometrically embeds into ℓ^2.*

(2) *$f(0) = 0$ and the function $\frac{d}{dx}\left(f(\sqrt{x})^2\right)$ is completely monotone on $(0, \infty)$.*

This result and paper are part of Schoenberg's program [202, 204, 205, 206, 232] to understand the transforms taking distance matrices from Euclidean space E_n of one dimension n, *isometrically* to those from another, say E_m. Schoenberg denoted this problem by $\{E_n; E_m\}$, with $1 \leq m, n \leq \infty$, where $E_\infty \cong \ell^2$ is Hilbert space. Schoenberg showed:

(1) If $n > m$, then $\{E_n; E_m\}$ is given by only the trivial function $f(t) \equiv 0$. Indeed, first observe by induction on n that the only Euclidean configuration of $n + 1$ points that are equidistant from one another is an "equilateral" $(n + 1)$-simplex Δ in $\mathbb{R}^n$ (or E_n), hence in any higher-dimensional Euclidean space – and this cannot exist in $\mathbb{R}^{n-1}$, hence not in $\mathbb{R}^m$ for $m < n$. If now $f(x_0) \neq 0$ for some $x_0 > 0$, then applying f to the distance matrix between vertices of the rescaled simplex $x_0 \Delta \subset \mathbb{R}^n$, produces $n + 1$ equidistant points in $\mathbb{R}^m$, which is not possible if $1 \leq m < n$.

(2) If $2 \leq n \leq m < \infty$, then $\{E_n; E_m\}$ consists only of the homotheties $f(x) = cx$ for some $c \geq 0$. (With von Neumann in 1941 [232], Schoenberg then extended this to answer the question for $n = 1 \leq m \leq \infty$.) Schoenberg also provided answers for $\{E_2; E_\infty\}$.

(3) The solution to the problem $\{E_\infty; E_\infty\}$ is precisely Theorem 22.13.

As a special case of Theorem 22.13, all powers $\delta \in (0, 1)$ of the Euclidean metric embed into Euclidean space. We provide an alternate proof using the above results on metric geometry.

Corollary 22.14 (Schoenberg [203, 204])   *Hilbert space ℓ^2, with the metric $\|x - y\|^\delta$, embeds isometrically in "usual" ℓ^2 for any $\delta \in (0, 1)$.*

This was shown in 1936 by Blumenthal [41] for four-point subsets of ℓ^2 and $\delta \in (0, 1/2)$. Schoenberg extended this to all finite sets.

Proof   As observed by Schoenberg in [203], it suffices to show the result for $(n + 1)$-element subsets $\{x_0, \ldots, x_n\} \subset \ell^2$, by Menger's theorem 22.1. Now note that for $c > 0$, the function $g(u) := (1 - e^{-cu})/u$ is bounded and continuous on $(0, 1]$, hence admits a continuous extension to $[0, 1]$. Since $u^{-\delta}$ is integrable in $(0, 1]$, so is the product

$$\varphi: (0, 1] \to \mathbb{R}, \qquad u \mapsto u^{-1-\delta}(1 - e^{-cu}).$$

Clearly, $\varphi : [1, \infty) \to \mathbb{R}$ is also integrable, being continuous, nonnegative, and bounded above by $u^{-1-\delta}$. By changing variables, we obtain a normalization constant $c_\delta > 0$, such that

$$t^{2\delta} = c_\delta \int_0^\infty \left(1 - e^{-\lambda^2 t^2}\right) \lambda^{-1-2\delta} \, d\lambda, \qquad \forall t > 0.$$

Set $t := \|x_j - x_k\|$, and let $\mathbf{u} = (u_0, \ldots, u_n)^T \in \mathbb{R}^{n+1}$ with $\sum_j u_j = 0$. Then,

$$\sum_{j,k=0}^n u_j u_k \|x_j - x_k\|^{2\delta} = c_\delta \int_0^\infty \left(\sum_{j,k=0}^n u_j u_k \left(1 - e^{-\lambda^2 \|x_j - x_k\|^2} \right) \right) \lambda^{-1-2\delta} \, d\lambda.$$

But the double-sum inside the integrand equals $-\sum_{j,k} u_j u_k e^{-\lambda^2 \|x_j - x_k\|^2}$, and this is nonpositive by Theorem 11.14. It follows that the matrix $\left(- (\|x_j - x_k\|^\delta)^2 \right)_{j,k=0}^n$ is conditionally positive semidefinite, and so we are done by Theorem 11.10 and Lemma 11.11.   □

23

Exercises

Question 23.1 Given a metric space (X, d), show that the set of positive definite functions on X forms a closed convex cone, which is moreover closed under taking (pointwise) products.

Question 23.2 Suppose (X, d) is a metric space, and $z \in X$ is a fixed basepoint.

(i) Prove that the Kuratowski embedding $\Psi : X \to Fun(X, \mathbb{R})$ (the real-valued functions on X), given by

$$\Psi(x)(y) := d(x, y) - d(z, y), \qquad y \in X$$

is an isometric embedding of X into $C_b(X)$, the space of continuous bounded real-valued functions on X.

(ii) (This was used in the proof of Theorem 22.1.) Suppose $(X, d), (Y, d')$ are metric spaces, with Y complete. If $D \subset X$ is dense, show that any isometric embedding $\Phi : D \to Y$ extends uniquely to an isometric embedding $\widetilde{\Phi} : X \to Y$.

(iii) Let ℓ^∞ denote the Banach space of uniformly bounded sequences of real numbers $(a_n)_{n \geq 1}$, equipped with the supnorm $\| \cdot \|_\infty$. Show that if X is separable, it admits an isometric embedding into $(\ell^\infty, \| \cdot \|_\infty)$.

(Hint: Choose a countable dense set $D := \{x_n : n \in \mathbb{N}\}$ and consider the embedding $\Psi : x \mapsto (d(x, x_n) - d(z, x_n))_{n \geq 1} \in \ell^\infty$. Clearly, Ψ is continuous. By a preceding part, $\Psi|_D$ is an isometric embedding into $C_b(D) \cong \ell^\infty$.)

Question 23.3 The preceding question worked with arbitrary (separable) metric spaces. In this question, we work with finite metric spaces X of size $n + 1$ for some $n \geq 0$. The goal is to isometrically embed X into $(\mathbb{R}^m, \| \cdot \|_\infty)$ for some $m \geq 1$.

(i) (Fréchet [89]; also [205]) The recipe in the preceding question provides an isometric embedding into $\mathbb{R}^{n+1}$ with the supnorm. Fréchet improved this to an embedding into $(\mathbb{R}^n, \|\cdot\|_\infty)$. Indeed, show that this isometric embedding is achieved by the map

$$x_j \mapsto (d(x_1, x_j), \ldots, d(x_n, x_j)), \qquad 0 \leq j \leq n.$$

(ii) (Witsenhausen [239]) One can further improve the bound on the dimension in the preceding part. This part defines an explicit embedding for $n = 3$. Namely, given a metric space $(X = \{x_0, x_1, x_2, x_3\}, d)$, verify that the following map is an isometry: $X \to (\mathbb{R}^2, \|\cdot\|_\infty)$:

$$x_0 \mapsto (0, d(x_1, x_2) - d(x_0, x_2)),$$
$$x_1 \mapsto (d(x_0, x_1), 0),$$
$$x_2 \mapsto (d(x_0, x_1) - d(x_1, x_3) + d(x_2, x_3), d(x_1, x_2)),$$
$$x_3 \mapsto (d(x_0, x_1) - d(x_1, x_3), d(x_1, x_2) - d(x_0, x_2) + d(x_0, x_3)).$$

(iii) To extend the result in the preceding part to higher values of n, we require the *one-dimensional Helly theorem:* given finitely many convex sets (i.e., nonempty intervals) in $\mathbb{R}$, every two of which intersect, prove that there exists a point common to all of them.

(iv) This part and the next are again from [239]. Suppose $(X = \{x_0, \ldots, x_n\}, d)$ is a metric space, and

$$\Psi_n : (\{x_1, \ldots, x_n\}, d) \to (\mathbb{R}^m, \|\cdot\|_\infty), \qquad x_j \mapsto \left(y_j^{(1)}, \ldots, y_j^{(m)}\right)$$

is a given isometric embedding (so $1 \leq j \leq n$). Verify that the map

$$\Psi_{n+1} : x_j \mapsto \left(y_j^{(1)}, \ldots, y_j^{(m)}, d(x_0, x_j)\right)$$

is an isometric embedding: $(\{x_1, \ldots, x_n\}, d) \to (\mathbb{R}^{m+1}, \|\cdot\|_\infty)$.

(v) It remains to embed the point $x_0 \mapsto \Psi_{n+1}(x_0) := \left(y_0^{(1)}, \ldots, y_0^{(m+1)}\right)$ in an isometric fashion. Set $y_0^{(m+1)} := 0$; this will achieve $d\left(\Psi_{n+1}(x_0), \Psi_{n+1}(x_j)\right) = d(x_0, x_j)$ as long as the $y_0^{(k)}$ are each (independently) chosen to satisfy

$$\left|y_0^{(k)} - y_j^{(k)}\right| \leq d(x_0, x_j), \qquad \forall 1 \leq j \leq n,\ 1 \leq k \leq m.$$

Thus, for each $1 \leq k \leq m$, we require $y_0^{(k)}$ to lie in the simultaneous intersection of the intervals

$$\left[y_j^{(k)} - d(x_0, x_j), y_j^{(k)} + d(x_0, x_j)\right], \qquad 1 \leq j \leq n.$$

Show that such a solution exists for each k, using Helly's theorem.

(vi) This completes the proof that every metric space of size $n + 1$ embeds into $(\mathbb{R}^{n-1}, \| \cdot \|_\infty)$ for $n \geq 3$. Show that this fails to hold for $n = 1, 2$.

Question 23.4   Suppose $y_1, \ldots, y_n$ are linearly independent vectors in Hilbert space ℓ^2, and y is in their span. Find a closed-form expression for y purely in terms of the inner products

$$\langle y_j, y_k \rangle, \ 1 \leq j, k \leq n, \qquad \langle y_j, y \rangle, \ 1 \leq j \leq n.$$

(Hint: Imitate the proof of Proposition 22.7.)

Question 23.5   Show the following extension of Menger's theorem 22.1: Fix an increasing (unbounded) sequence of integers $0 < n_1 < n_2 < \cdots$ and a complete, separable metric space (Y, d') that is n_j-point homogeneous for all $j \geq 1$. A metric space (X, d) isometrically embeds into (Y, d') if and only if X is separable and every finite subset of (X, d) embeds isometrically into (Y, d').

Question 23.6   This question concerns the Gaussian kernel on Hilbert space ℓ^2, i.e.,

$$T_{G_\sigma} : \ell^2 \times \ell^2 \to \mathbb{R}, \qquad (x, y) \mapsto \exp(-\sigma \| x - y \|^2).$$

Fix $\delta \in (0, 1)$. Using Corollary 22.14, show that $e^{-\sigma(\cdot)^{2\delta}}$ is a positive definite function on ℓ^2 for each scalar $\sigma > 0$.

Question 23.7   (This is found in the book [30] by Berg et al. in 1984 the authors attribute it to Schoenberg [205]) Let $A' := (a_{jk})_{j,k=0}^n$ be a real symmetric matrix, with $n \geq 1$.

  (i) Show that Lemma 11.11 holds more generally: A' is conditionally positive semidefinite if and only if the matrix

$$A := (a_{jk} - a_{j0} - a_{0k} + a_{00})_{j,k=1}^n$$

or equivalently,

$$\tilde{A} := (a_{jk} - a_{j0} - a_{0k} + a_{00})_{j,k=0}^n$$

is positive semidefinite.
 (ii) The above is equivalent to two additional conditions. Namely, show that the following are equivalent:

  (a) A' is conditionally positive semidefinite.
  (b) The entrywise exponential matrix $\exp[\sigma A']$ is positive semidefinite for all $\sigma > 0$.
  (c) The matrix $\exp[\sigma_m A']$ is positive semidefinite for some decreasing sequence of positive scalars $\sigma_m \downarrow 0^+$.

(Hint: For one implication, check that $\sigma_m^{-1}(\exp[\sigma_m A'] - \mathbf{1}_{n \times n})$ is conditionally positive semidefinite for all $m \geq 1$. To show that (a) implies (b) for $\sigma = 1$, note that $\widetilde{A}$ is positive semidefinite by the preceding part, and

$$\exp[A] = \exp[\widetilde{A}] \circ (e^{-a_{00}} \mathbf{1}_{(n+1) \times (n+1)}) \circ uu^T,$$

where $u := (\exp(a_{j0}))_{j=0}^n \in \mathbb{R}^{n+1}$.)

Question 23.8   Suppose $f : [0, \infty)$ is differentiable at 0^+ and $f'(0^+) < 0$ (e.g., $f(x) = e^{-x}$). Let $(X, d) = \{x_0, \ldots, x_n\}$ be a finite metric space with $n \geq 1$ and let $\sigma_m \downarrow 0^+$ be a decreasing sequence of positive real numbers. If $f[\sigma_m A]$ is conditionally positive semidefinite for all m, where A is the $(n+1) \times (n+1)$ matrix with entries $d(x_j, x_k)^2$, then (X, d) isometrically embeds into $\mathbb{R}^r$, with $r \leq n$.

Question 23.9   Every finite simple connected graph $G = (V, E)$ can be thought of as a metric space, by setting each edge to have unit length and assigning the distance between two nodes to be the length of the shortest path joining them. This question proves that the only graphs that isometrically embed into Hilbert space ℓ^2 are path graphs and complete graphs. (Check that these do embed.)

(i) Show that if $|V| \leq 3$, then G embeds isometrically into Hilbert space ℓ^2.
(ii) Show that if $|V| = 4$, then G embeds isometrically into ℓ^2 if and only if G is either the path graph or the complete graph.
(iii) Show that the only cycle that embeds isometrically into ℓ^2 is $C_3 = K_3$.
(iv) Now suppose G is neither a path nor a cycle. Then G has a node v_0 of degree at least 3. (Why?) Assuming G embeds isometrically into ℓ^2, show that (a) v_0 is simplicial, i.e., its neighbors in G are all adjacent to each other. Now show that (b) G is complete.

Question 23.10 (Khare [143])   Extend Loewner's computations in Proposition 12.4 to the following more general result: Fix real scalars $\epsilon > 0$ and a, and integers

$$n \geq 1, \qquad 0 \leq m_0 < m_1 < \cdots < m_{n-1}.$$

Define $M := m_0 + \cdots + m_{n-1}$. Suppose a function $f : [a, a + \epsilon) \to \mathbb{R}$ is M-times differentiable at a. Now fix vectors $\mathbf{u}, \mathbf{v} \in \mathbb{R}^n$ and define $\Delta : [0, \epsilon') \to \mathbb{R}$ via

$$\Delta(t) := \det f[a\mathbf{1}_{n \times n} + t\mathbf{u}\mathbf{v}^T],$$

for a sufficiently small $\epsilon' \in (0, \epsilon)$. Denoting $\mathbf{u}^{\circ \mathbf{m}} := (\mathbf{u}^{\circ m_0} | \cdots | \mathbf{u}^{\circ m_{n-1}})$, we have

$$\Delta^{(M)}(0^+) = \sum_{\mathbf{m} \vdash M} \binom{M}{m_0, m_1, \ldots, m_{n-1}} \det(\mathbf{u}^{\circ \mathbf{m}}) \det(\mathbf{v}^{\circ \mathbf{m}}) \prod_{k=0}^{n-1} f^{(m_k)}(a).$$

Here, the first factor in the summand on the right-hand side is the multinomial coefficient and the sum is over all integer partitions $\mathbf{m} = (m_{n-1}, \ldots, m_0)$ of M, i.e., $M = m_0 + \cdots + m_{n-1}$ and $m_{n-1} \geq \cdots \geq m_0$.

As a special case, we recover Loewner's computation for $M = \binom{n}{2}$

$$\Delta(0) = \Delta'(0) = \cdots = \Delta^{(N-1)}(0) = 0, \quad N = \binom{n}{2}.$$

This result reveals the emergence of *Schur polynomials* from arbitrary smooth – or even sufficiently differentiable – functions. It also extends certain computations in the next part of this text; and it has an algebraic counterpart that extends Proposition 12.6 – see [143] for details.

Question 23.11   Verify the preliminary result in the use of mollifiers above: the function $\exp(-1/x) \cdot \mathbf{1}(x > 0)$ is smooth on $\mathbb{R}$, and all of its derivatives vanish at the origin.

Question 23.12   Work out the "other case" in the proof of Lemma 15.3 – when the mass of σ is positive on $(-\infty, -1)$.

Question 23.13   This question concerns the Wiener norm $\| \cdot \|_{1,+}$ defined in (15.2).

  (i) Prove that $\| \cdot \|_{1,+}$ is indeed a norm on the space of real polynomials. Inside what Banach space does this space embed isometrically?
 (ii) Prove that $\| \cdot \|_{1,+}$ is stronger than the uniform norm – i.e., when considering the polynomials as continuous functions, evaluated on any *compact* subset of $\mathbb{R}$.
(iii) Extend the definition of the Wiener norm to the space of d-variable real polynomials (for any $d \geq 1$) and prove that the preceding two parts of this question also hold in this setting.

Question 23.14   Work out the first equality in (15.6).

Question 23.15   Show that for $a > 0$, the function $L_a(y) := \log(a + y)$ is real analytic on $(-a, a)$. (This was used at the end of Chapter 16.)

Question 23.16   Let $0 < \rho \leq \infty$. In this part, we have seen two results – for domains $I = (0, \rho)$ and $(-\rho, \rho)$ – by Vasudeva (essentially) and Rudin, respectively. Each of these results asserts that if $f : I \to \mathbb{R}$ is a function that entrywise preserves positivity on $\mathbb{P}_n(I)$ for all $n \geq 1$, then f is a power series on I with nonnegative Maclaurin coefficients. (One can also replace $\mathbb{P}_n(I)$ by the subset of matrices of rank at most 2 or 3 depending on I, or the smaller subset of Toeplitz matrices with entries in $(-\rho, \rho)$ – as considered by Rudin – or of Hankel matrices with entries in either domain I as used in the above sections.)

(i) Let $I = (0, \rho)$. Prove that each such result
   over I for a single, finite $\rho \in (0, \infty)$ implies the same result for every finite or infinite $\rho \in (0, \infty]$. (Hint: You may need to use Lemma 16.3.)
(ii) Repeat the preceding part, now for $I = (-\rho, \rho)$.

Question 23.17   (This problem and the next elaborate on Remark 14.5) So far we have seen that the preservers (powers in a fixed dimension; functions in all dimensions) of $\mathbb{P}_n((0, \infty))$ and of $\text{HTN}_n \cap \mathbb{P}_n((0, \infty))$ coincide. Here is one case in which they slightly differ: when the domain $(0, \infty)$ is replaced by $[0, \infty)$.

We begin by classifying, in this question, the entrywise preservers of HTN_n – or of rank ≤ 2 matrices in HTN_n – for all $n \geq 1$:

(i) Show that the only matrices in HTN_n that contain a zero entry correspond to Hankel moment matrices of Dirac measures supported at the origin.
(ii) As a consequence, show that if $f : [0, \infty) \to \mathbb{R}$ is absolutely monotonic on $(0, \infty)$ (hence a power series with nonnegative Maclaurin coefficients), and $0 \leq f(0) \leq \lim_{x \to 0^+} f(x)$, then the entrywise map $f[-]$ preserves HTN_n for all n.
(iii) Show the converse using the stronger Vasudeva theorem 14.1.
(iv) Show that this answer coincides with the *positivity preservers* on the same test sets.

Question 23.18   In contrast to the previous question, we now show that the entrywise preservers of positivity on $\mathbb{P}_n([0, \infty))$ for all n – or of rank ≤ 2 matrices in $\mathbb{P}_n([0, \infty))$ – are precisely the absolutely monotonic functions on $[0, \infty)$, i.e., they are also continuous at the origin.

(i) Show that all such functions are entrywise preservers.
(ii) Conversely, show (again using Theorem 14.1) that any such function must be absolutely monotonic on $(0, \infty)$. To examine the behavior of f at 0^+, define the Hankel rank-2 matrix $A := \begin{pmatrix} 2 & 1 & 1 \\ 1 & 1 & 0 \\ 1 & 0 & 1 \end{pmatrix}$, and by considering the inequality

$$\lim_{x \to 0^+} \det f[xA] \geq 0,$$

show that f is continuous at 0^+.

Question 23.19   Prove Proposition 21.8, stated and used in Chapter 21 at the end of the proof of the Boas–Widder theorem.

Question 23.20   Fill in the details of the sketched implication in the proof of Theorem 18.3.

PART THREE

ENTRYWISE POLYNOMIALS PRESERVING POSITIVITY IN A FIXED DIMENSION

24

Entrywise Polynomial Preservers and Horn–Loewner-Type Conditions

In Part II of this text, we classified the entrywise functions preserving positivity in all dimensions; these are precisely the power series with nonnegative coefficients. Earlier in Part I, we had classified the entrywise powers preserving positivity in fixed dimension. In this final part of the text, we study polynomials that entrywise preserve positive semidefiniteness in fixed dimension.

Recall from the Schur product theorem 2.10 and its converse, Schoenberg's theorem 11.2 (with Rudin's strengthening), that the only polynomials that entrywise preserve positivity in all dimensions are the ones with all non-negative coefficients. Thus, if one fixes the dimension $N \geq 3$ of the test set of positive matrices, then it is reasonable to expect that there should exist more polynomial preservers – in other words, polynomial preservers with negative coefficients. However, this problem remained completely open until very recently ($\sim$ 2016): *not a single* polynomial preserver was known with a negative coefficient, nor was a nonexistence result proved!

In this final part, we answer this existence question as well as stronger variants of it. Namely, not only do we produce such polynomial preservers, we also fully resolve the more challenging question: *which coefficients of polynomial preservers on $N \times N$ matrices can be negative?* Looking ahead in this part:

- We classify the sign patterns of entrywise polynomial preservers on $\mathbb{P}_N$ for fixed N.
- We extend this to all power series; but also, countable sums of real powers, such as $\sum_{\alpha \in \mathbb{Q}, \ \alpha \geq N-2} c_\alpha x^\alpha$. This case is more subtle than that of polynomial preservers.

- We will also completely classify the sign patterns of polynomials that entrywise preserve totally nonnegative (*TN*) Hankel matrices of a fixed dimension. Recall from the discussions around Theorems 6.11 and 14.1 that this is expected to be very similar to (maybe even the same as) the classification for positivity preservers.

In what follows, we work with $\mathbb{P}_N((0, \rho))$ for $N > 0$ fixed and $0 < \rho < \infty$. Since we work with polynomials and power series, this is equivalent to working over $\mathbb{P}_N([0, \rho))$ by density and continuity. If $\rho = +\infty$, one can prove results that are similar to the ones shown in this part of the text; but for a first look at the proofs and techniques used, we restrict ourselves to $\mathbb{P}_N((0, \rho))$. For full details of the $\rho = +\infty$ case, as well as for the proofs, ramifications, and applications of the results below we refer the reader to the paper by Khare and Tao [146].

24.1 Horn–Loewner-Type Condition: Matrices with Negative Entries

In this chapter and beyond, we work with polynomials or power series

$$f(x) = c_{n_0} x^{n_0} + c_{n_1} x^{n_1} + \cdots, \qquad \text{with } n_0, n_1, \ldots \text{ pairwise distinct,}$$

(24.1)

and $c_{n_j} \in \mathbb{R}$ typically nonzero. Recall the (stronger) Horn–Loewner theorem 12.1, which shows that if $f \in C^{(N-1)}(I)$ for $I = (0, \infty)$, and $f[-]$ preserves positivity on (rank-2 Hankel *TN* matrices in) $\mathbb{P}_N(I)$, then $f, f', \ldots, f^{(N-1)} \geq 0$ on I. In the special case that f is a polynomial or a power series, one can say more, and under weaker assumptions:

Lemma 24.1 (Horn–Loewner-type necessary condition)   *Fix an integer $N > 0$. Let $\rho > 0$ and $f: (0, \rho) \to \mathbb{R}$ be a function of the form (24.1) satisfying*

(1) *f is absolutely convergent on $(0, \rho)$, i.e., $\sum_{j \geq 0} |c_{n_j}| x^{n_j} < \infty$ on $(0, \rho)$.*
(2) *$f[-]$ preserves positivity on rank-1 Hankel TN matrices in $\mathbb{P}_N((0, \rho))$.*

If $c_{n_{j_0}} < 0$ for some $j_0 \geq 0$, then $c_{n_j} > 0$ for at least N values of j for which $n_j < n_{j_0}$.

Remark 24.2   In both (24.1) as well as Lemma 24.1 (and its proof), we have deliberately **not** insisted on the exponents n_j being nonnegative integers.

In fact, one can choose $\{n_j : j \geq 0\}$ to be an arbitrary sequence of pairwise distinct real numbers.

Proof of Lemma 24.1   By the properties of f, the function

$$g(x) := \sum_{j \neq j_0 : c_{n_j} < 0} |c_{n_j}| x^{n_j}$$

entrywise preserves positivity on rank-1 Hankel TN matrices in $\mathbb{P}_N((0, \rho))$. Hence, so does

$$f(x) + g(x) = \sum_{j : c_{n_j} > 0} c_{n_j} x^{n_j} + c_{n_{j_0}} x^{n_{j_0}}.$$

Now suppose the result is false. Then the preceding sum contains at most k terms n_j that lie in $(0, n_{j_0})$ (for some $0 \leq k < N$), and which we label by $n_0, \ldots, n_{k-1}$. Also, set $m := n_{j_0}$. Choose any $u_0 \in (0, 1)$ and define $\mathbf{u} := \left(1, u_0, \ldots, u_0^{N-1}\right)^T \in \mathbb{R}^N$. Then $\mathbf{u}^{\circ n_0}, \ldots, \mathbf{u}^{\circ n_{k-1}}; \mathbf{u}^{\circ m}$ are linearly independent, forming (some of) the columns of a generalized Vandermonde matrix. Hence, there exists $\mathbf{v} \in \mathbb{R}^N$, such that

$$\mathbf{v} \perp \mathbf{u}^{\circ n_0}, \ldots, \mathbf{u}^{\circ n_{k-1}} \quad \text{and} \quad \mathbf{v}^T \mathbf{u}^{\circ m} = 1.$$

For $0 < \epsilon < \rho$, we let $A_\epsilon := \epsilon \mathbf{u}\mathbf{u}^T$, which is a rank-1 Hankel moment matrix in $\mathbb{P}_N((0, \rho))$ (and hence TN). Now compute using the hypotheses

$$0 \leq \mathbf{v}^T (f + g)[A_\epsilon] \mathbf{v} = \mathbf{v}^T \left(\sum_{j : c_{n_j} > 0} c_{n_j} \epsilon^{n_j} \mathbf{u}^{\circ n_j} \left(\mathbf{u}^{\circ n_j}\right)^T + c_m \epsilon^m \mathbf{u}^{\circ m} \left(\mathbf{u}^{\circ m}\right)^T \right) \mathbf{v}$$

$$= c_m \epsilon^m \left(\mathbf{v}^T \mathbf{u}^{\circ m}\right)^2 + \sum_{j : c_{n_j} > 0, \, n_j > n_{j_0}} c_{n_j} \epsilon^{n_j} \left(\mathbf{v}^T \mathbf{u}^{\circ n_j}\right)^2$$

$$= c_m \epsilon^m + o(\epsilon^m).$$

Thus, $0 \leq \lim_{\epsilon \to 0^+} \dfrac{\mathbf{v}^T (f + g)[A_\epsilon] \mathbf{v}}{\epsilon^m} = c_m < 0$, which is a contradiction. Hence, $k \geq N$, proving the claim.   □

By Lemma 24.1, every polynomial that entrywise preserves positivity on $\mathbb{P}_N((0, \rho))$ must have its N nonzero Maclaurin coefficients of "lowest degree" to be positive. The obvious question is if any of the other terms can be negative, e.g., the immediate next coefficient.

We tackle this question in the remainder of this text and show that, in fact, every other coefficient can indeed be negative. For now, we point out that working with positive matrices with other entries cannot provide such a structured answer (in the flavor of Lemma 24.1). As a simple example, consider the family of polynomials

$$p_{k,t}(x) := t(1 + x^2 + \cdots + x^{2k}) - x^{2k+1}, \qquad t > 0,$$

where $k \geq 0$ is an integer. Now claim that $p_{k,t}[-]$ can never preserve positivity on $\mathbb{P}_N((-\rho, \rho))$ for $N \geq 2$. Indeed, if $\mathbf{u} := (1, -1, 0, \ldots, 0)^T$ and $A := (\rho/2)\mathbf{u}\mathbf{u}^T \in \mathbb{P}_N((-\rho, \rho))$, then

$$\mathbf{u}^T p_{k,t}[A]\mathbf{u} = -4(\rho/2)^{2k+1} < 0.$$

Therefore, $p_{k,t}[A]$ is not positive semidefinite for any $k \geq 0$. If one allows complex entries, similar examples with higher-order roots of unity can be constructed, in which such negative results (compared to Lemma 24.1) can be obtained.

24.2 Classification of Sign Patterns for Polynomials

In light of the above discussion, henceforth we restrict ourselves to working with matrices in $\mathbb{P}_N((0, \rho))$ for $0 < \rho < \infty$. By Lemma 24.1, every polynomial preserver on $\mathbb{P}_N((0, \rho))$ must have its N lowest-degree Maclaurin coefficients (which are nonzero) to be positive.

We are interested in understanding if any (or every) other coefficient can be negative. If, say, the next lowest-degree coefficient could be negative, this would achieve two goals:

- It would provide (the first example of) a polynomial preserver in fixed dimension, which has a negative Maclaurin coefficient.
- It would provide (the first example of) a polynomial that preserves positivity on $\mathbb{P}_N((0, \rho))$, but necessarily *not* on $\mathbb{P}_{N+1}((0, \rho))$. In particular, this would show that the Horn–Loewner-type necessary condition in Lemma 24.1 is "best possible." (See Remark 12.3 in the parallel setting of entrywise power preservers for the original Horn condition.)

We show in this part of the text that these goals are indeed achieved:

Theorem 24.3 (Classification of sign patterns, fixed dimension)   *Fix integers $N > 0$ and $0 \leq n_0 < n_1 < \cdots < n_{N-1}$, as well as a sign $\varepsilon_M \in \{-1, 0, 1\}$ for each integer $M > n_{N-1}$. Given reals $\rho, c_{n_0}, c_{n_1}, \ldots, c_{n_{N-1}} > 0$, there exists a power series*

$$f(x) = c_{n_0}x^{n_0} + \cdots + c_{n_{N-1}}x^{n_{N-1}} + \sum_{M > n_{N-1}} c_M x^M,$$

satisfying the following properties:

(1) *f is convergent on $(0, \rho)$.*
(2) *$f[-] \colon \mathbb{P}_N((0, \rho)) \to \mathbb{P}_N$.*
(3) *$\mathrm{sgn}(c_M) = \varepsilon_M$ for each $M > n_{N-1}$.*

This is slightly stronger than classifying the sign patterns, in that the "initial coefficients" are also specified. In fact, this result can be strengthened in two different ways, see (1) Theorem 24.6, in which the set of powers allowed is vastly more general; and (2) Theorem 27.8 and the discussion preceding it, in which the coefficients for $M > n_{N-1}$ are also specified.

Proof   Suppose we can prove the theorem in the special case when exactly one ε_M is negative. Then for each $M > n_{N-1}$, there exists $0 < \delta_M < \frac{1}{M!}$, such that

$$f_M(x) := \sum_{j=0}^{N_1} c_{n_j} x^{n_j} + c_M x^M$$

preserves positivity on $\mathbb{P}_N((0, \rho))$ whenever $|c_M| \leq \delta_M$. Set $c_M := \varepsilon_M \delta_M$ for each $M > n_{N-1}$ and define $f(x) := \sum_{M > n_{N-1}} 2^{n_{N-1}-M} f_M(x)$. If $x \in (0, \rho)$, then we have

$$|f(x)| \leq \sum_{M > n_{N-1}} 2^{n_{N-1}-M} |f_M(x)|$$

$$\leq \sum_{M > n_{N-1}} 2^{n_{N-1}-M} \left(\sum_{j=0}^{N-1} c_{n_j} x^{n_j} + \delta_M x^M \right)$$

$$\leq \sum_{j=0}^{N-1} c_{n_j} x^{n_j} + e^x < \infty.$$

Hence, f converges on $(0, \rho)$. As each $f_M[-]$ preserves positivity and $\mathbb{P}_N$ is a closed convex cone, $f[-]$ also preserves positivity. It therefore remains to show that the result holds when one coefficient is negative. But this follows from Theorem 24.4.   □

Thus, it remains to show the following "qualitative" result:

Theorem 24.4   *Let $N > 0$, $0 \leq n_0 < n_1 < \cdots < n_{N-1} < M$ be integers, and $\rho, c_{n_0}, c_{n_1}, \ldots, c_{n_{N-1}} > 0$ be real. Then the function $f(x) = \sum_{j=0}^{N-1} c_{n_j} x^{n_j} + c_M x^M$ entrywise preserves positivity on $\mathbb{P}_N((0, \rho))$, for some $c_M < 0$.*

We will show this result in Chapters 25 and 26.

24.3 Classification of Sign Patterns for Sums of Real Powers

After proving Theorem 24.4, we further strengthen it by proving a *quantitative* version – see Theorem 27.1 – which gives a sharp lower bound on c_M. For now, we list a special case of that result (without proof, as we show the more general Theorem 27.1). In the following result and beyond, the set $\mathbb{Z}^{\geq 0} \cup [N - 2, \infty)$ comes from Theorem 5.3.

Theorem 24.5   *Theorem 24.4 holds even when the exponents $n_0, n_1, \ldots,$ n_{N-1} and M are real and lie in the set $\mathbb{Z}^{\geq 0} \cup [N - 2, \infty)$.*

With Theorem 24.5 at hand, it is possible to classify the sign patterns of a more general family of preservers, of the form $f(x) = \sum_{j=0}^{\infty} c_{n_j} x^{n_j}$, where $n_j \in \mathbb{Z}^{\geq 0} \cup [N-2, \infty)$ are an arbitrary countable collection of pairwise distinct nonnegative (real) exponents.

Theorem 24.6 (Classification of sign patterns of power series preservers, fixed dimension)   *Let $N \geq 2$ and let $n_0, n_1, \ldots$ be a sequence of pairwise distinct real numbers in $\mathbb{Z}^{\geq 0} \cup [N - 2, \infty)$. For each $j \geq 0$, let $\varepsilon_j \in \{-1, 0, 1\}$ be a sign, such that whenever $\varepsilon_{j_0} = -1$, one has $\varepsilon_j = +1$ for at least N choices of j satisfying: $n_j < n_{j_0}$. Then for every $\rho > 0$, there exists a series with real exponents and real coefficients*

$$f(x) = \sum_{j=0}^{\infty} c_{n_j} x^{n_j},$$

which is convergent on $(0, \rho)$ and entrywise preserves positivity on $\mathbb{P}_N((0, \rho))$, and in which $\operatorname{sgn}(c_{n_j}) = \varepsilon_j$ for all $j \geq 0$.

That the sign patterns necessarily satisfy the given hypotheses follows from Lemma 24.1. In particular, Theorem 24.6 shows that the Horn–Loewner-type necessary condition in Lemma 24.1 remains the best possible in this generality as well.

Remark 24.7   A key difference between the classifications in Theorems 24.3 and 24.6 is that the latter is more flexible, since the sequence $n_0, n_1, \ldots$ can now contain an infinite decreasing subsequence of exponents. This is more general than even the Hahn or Puiseux series, not just power series. For instance, the sum may be over all rational powers in $\mathbb{Z}^{\geq 0} \cup [N - 2, \infty)$.

Proof of Theorem 24.6   Given any set $\{n_j : j \geq 0\}$ of (pairwise distinct) nonnegative powers,

$$\sum_{j \geqslant 0} \frac{x^{n_j}}{j! \lceil n_j \rceil!} < \infty, \qquad \forall x > 0. \tag{24.2}$$

Indeed, if we partition $\mathbb{Z}^{\geqslant 0}$ into the disjoint union of $J_k := \{j \geqslant 0 : n_j \in (k-1, k]\}$, $k \geqslant 0$, then using Tonelli's theorem, we can estimate

$$\sum_{j \geqslant 0} \frac{x^{n_j}}{j! \lceil n_j \rceil!} = \sum_{k \geqslant 0} \frac{1}{k!} \sum_{j \in J_k} \frac{x^{n_j}}{j!} \leqslant e$$

$$+ \sum_{k \geqslant 1} \frac{1}{k!} \sum_{j \in J_k} \frac{x^k + x^{k-1}}{j!} < e + e(e^x + x^{-1}e^x) < \infty.$$

We now turn to the proof. Set $J := \{j : \varepsilon_j = -1\} \subset \mathbb{Z}^{\geqslant 0}$. By the hypotheses, for each $l \in J$ there exist $j_1(l), \ldots, j_N(l)$, such that $\varepsilon_{j_k(l)} = 1$ and $n_{j_k(l)} < n_l$, for $k = 1, \ldots, N$. Define

$$f_l(x) := \sum_{k=1}^{N} \frac{x^{n_{j_k(l)}}}{\lceil n_{j_k(l)} \rceil!} - \delta_l \frac{x^{n_l}}{\lceil n_l \rceil!},$$

where $\delta_l \in (0,1)$ is chosen, such that $f_l[-]$ preserves positivity on $\mathbb{P}_N((0,\rho))$ by Theorem 24.5. Let $J' \subset \mathbb{Z}^{\geqslant 0}$ consist of all $j \geqslant 0$, such that $\varepsilon_j = +1$ but $j \neq j_k(l)$ for any $l \in J, k \in [1, N]$. Finally, define

$$f(x) := \sum_{l \in J} \frac{f_l(x)}{l!} + \sum_{j \in J'} \frac{x^{n_j}}{j! \lceil n_j \rceil!}, \qquad x > 0.$$

Repeating the calculation in (24.2), one can verify that f converges absolutely on $(0, \infty)$ and hence on $(0, \rho)$. By the above hypotheses and the Schur product theorem, it follows that $f[-]$ preserves positivity on $\mathbb{P}_N((0, \rho))$. $\qquad \square$

25

Polynomial Preservers for Rank-1 Matrices, via Schur Polynomials

The goal in this chapter and Chapter 26 is to prove the "qualitative" Theorem 24.4 from Chapter 24. Thus, we work with polynomials of the form

$$f(x) = \sum_{j=0}^{N-1} c_{n_j} x^{n_j} + c_M x^M,$$

where $N > 0, 0 \leq n_0 < n_1 < \cdots < n_{N-1} < M$ are integers, and $\rho, c_{n_0}, c_{n_1}, \ldots, c_{n_{N-1}} > 0$ are real.

25.1 Basic Properties of Schur Polynomials

In this chapter, we begin by defining the key tool required here and beyond: Schur polynomials. We then use these functions – via the Cauchy–Binet formula – to understand when polynomials of the above form entrywise preserve positivity on a generic rank-1 matrix in $\mathbb{P}_N((0, \rho))$.

Definition 25.1 Fix integers $m, N > 0$, and define $\mathbf{n}_{\min} := (0, 1, \ldots, N-1)$. Now suppose $0 \leq n_0' \leq n_1' \leq \cdots \leq n_{N-1}'$ are also integers.

(1) A *column-strict Young tableau*, with shape $\mathbf{n}' := (n_0', n_1', \ldots, n_{N-1}')$ and cell entries $1, 2, \ldots, m$, is a left-aligned two-dimensional rectangular array T of cells, with n_0' cells in the bottom row, n_1' cells in the second lowest row, and so on, such that:

- Each cell in T has integer entry j with $1 \leq j \leq m$.
- Entries weakly decrease in each row, from left to right.
- Entries strictly decrease in each column, from top to bottom.

(2) Given variables $u_1, u_2, \ldots, u_m$ and a column-strict Young tableau T, define its *weight* to be

$$\text{wt}(T) := \prod_{j=1}^{m} u_j^{f_j},$$

where f_j equals the number of cells in T with entry j.

(3) Given an increasing sequence of integers $0 \le n_0 < \cdots < n_{N-1}$, define the tuple $\mathbf{n} := (n_0, n_1, \ldots, n_{N-1})$, and the corresponding *Schur polynomial* over $\mathbf{u} := (u_1, u_2, \ldots, u_m)^T$ to be

$$s_{\mathbf{n}}(\mathbf{u}) := \sum_T \text{wt}(T), \tag{25.1}$$

where T runs over all column-strict Young tableaux of shape $\mathbf{n}' := \mathbf{n} - \mathbf{n}_{\min}$ with cell entries $1, 2, \ldots, m$. (We will also abuse notation slightly and write $s_{\mathbf{n}}(\mathbf{u}) = s_{\mathbf{n}}(u_1, \ldots, u_m)$ on occasion.)

Example 25.2 Suppose $N = m = 3$ and $\mathbf{n} = (0, 2, 4)$. The column-strict Young tableaux with shape $\mathbf{n} - \mathbf{n}_{\min} = (0, 1, 2)$ and cell entries $(1, 2, 3)$ are

3	3
2

3	3
1

3	2
2

3	2
1

3	1
2

3	1
1

2	2
1

2	1
1

As a consequence,

$$s_{(0,2,4)}(u_1, u_2, u_3) = u_3^2 u_2 + u_3^2 u_1 + u_3 u_2^2 + 2 u_3 u_2 u_1 + u_3 u_1^2 + u_2^2 u_1 + u_2 u_1^2$$
$$= (u_1 + u_2)(u_2 + u_3)(u_3 + u_1).$$

Remark 25.3 A notational distinction with the literature is that column-strict Young tableaux traditionally have entries that are *increasing* as one moves down each column, and weakly increasing as one moves across rows. Since we only work with sets of tableaux through the sums of their weights occurring in Schur polynomials, this distinction is unimportant in the text, for the following reason: define an involutive bijection $\iota: j \mapsto m + 1 - j$, where $\{1, \ldots, m\}$ is the alphabet of possible cell entries. Then the column-strict Young tableaux in our notation bijectively correspond under ι – applied to each cell entry – to the "usual" column-strict Young tableaux (in the literature); and as Schur polynomials are symmetric under permuting the variables by ι (see Proposition 25.5), the sums of weights of the two sets of tableaux coincide.

Remark 25.4   Schur polynomials are fundamental objects in type A representation theory (of the general linear group, or the special linear Lie algebra), and are characters of irreducible finite-dimensional polynomial representations (over fields of characteristic zero). The above example 25.2 is a special case, corresponding to the adjoint representation for the Lie algebra of 3×3 traceless matrices. This interpretation will not be used in this text.

Schur polynomials are always homogeneous – and also symmetric, because they can be written as a quotient of two generalized Vandermonde determinants. This is Cauchy's definition; the definition (25.1) using Young tableaux is by Littlewood. One can show that these two definitions are equivalent, among other basic properties:

Proposition 25.5   *Fix integers $m = N > 0$ and $0 \leq n_0 < n_1 < \cdots < n_{N-1}$.*

(1) *(Cauchy's definition) If $\mathbb{F}$ is a field and $\mathbf{u} = (u_1, \ldots, u_N)^T \in \mathbb{F}^N$, then*

$$\det \left(\mathbf{u}^{\circ n_0} \mid \mathbf{u}^{\circ n_1} \mid \cdots \mid \mathbf{u}^{\circ n_{N-1}} \right)_{N \times N} = V(\mathbf{u}) s_{\mathbf{n}}(\mathbf{u}),$$

where for a (column) vector or (row) tuple $\mathbf{u}$, we denote by

$$V(\mathbf{u}) := \prod_{1 \leq j < k \leq N} (u_k - u_j)$$

the Vandermonde determinant as in (12.1). In particular, $s_{\mathbf{n}}(\mathbf{u})$ is symmetric and homogeneous of degree $\sum_{j=0}^{N-1} (n_j - j)$.

(2) *(Principal specialization formula) For any $q \in \mathbb{F}$ that is not a root of unity or else has order $\geq N$, we have*

$$s_{\mathbf{n}}(1, q, q^2, \ldots, q^{N-1}) = \prod_{0 \leq j < k \leq N-1} \frac{q^{n_k} - q^{n_j}}{q^k - q^j}.$$

(3) *(Weyl dimension formula) Specialized to $q = 1$, we have*

$$s_{\mathbf{n}}(1, 1, \ldots, 1) = \frac{V(\mathbf{n})}{V(\mathbf{n}_{\min})} \in \mathbb{N}.$$

In particular, there are $V(\mathbf{n})/V(\mathbf{n}_{\min})$ column-strict tableaux of shape $\mathbf{n} - \mathbf{n}_{\min}$ and cell entries $1, \ldots, N$. Here and below, we will mildly abuse notation and write $V(\mathbf{n})$ for a tuple/row vector $\mathbf{n}$ to denote $V(\mathbf{n}^T)$.

Proof   The first part is proved in Theorem 29.1 below. Using this, we show the second part. Set $\mathbf{u} := (1, q, q^2, \ldots, q^{N-1})^T$ with q as given. Then it is easy to verify that

$$s_{\mathbf{n}}(\mathbf{u}) = \frac{\det(\mathbf{u}^{\circ n_0} \mid \mathbf{u}^{\circ n_1} \mid \cdots \mid \mathbf{u}^{\circ n_{N-1}})}{V(\mathbf{u})} = \frac{V((q^{n_0}, \ldots, q^{n_{N-1}}))}{V((q^0, \ldots, q^{N-1}))}$$

$$= \prod_{0 \le j < k \le N-1} \frac{q^{n_k} - q^{n_j}}{q^k - q^j},$$

as desired.

Finally, to prove the Weyl dimension formula, notice that by the first part, the Schur polynomial has integer coefficients and hence makes sense over $\mathbb{Z}$, and then specializes to $s_{\mathbf{n}}(\mathbf{u})$ over any ground field. Now work over the ground field $\mathbb{Q}$, and let $f_{\mathbf{n}}(T) := s_{\mathbf{n}}(1, T, \ldots, T^{N-1}) \in \mathbb{Z}[T]$ be the corresponding "principally specialized" polynomial. Then,

$$V\big((q^0, \ldots, q^{N-1})\big) f_{\mathbf{n}}(q) = V\big((q^{n_0}, \ldots, q^{n_{N-1}})\big), \qquad \forall q \in \mathbb{Q}.$$

In particular, for every $q \ne 1$, dividing both sides by $(q-1)^{\binom{N}{2}}$, we obtain

$$\prod_{0 \le j < k \le N-1} \big(q^{n_j} + q^{n_j+1} + \cdots + q^{n_k-1}\big) - f_{\mathbf{n}}(q) \times$$

$$\times \prod_{0 \le j < k \le N-1} \big(q^j + q^{j+1} + \cdots + q^{k-1}\big) = 0,$$

for all $q \in \mathbb{Q} \setminus \{1\}$. This means that the left-hand side is (the specialization of) a polynomial with infinitely many roots, and hence the polynomial vanishes identically on $\mathbb{Q}$. Specializing this polynomial to $q = 1$ now yields the Weyl dimension formula

$$\frac{V(\mathbf{n})}{V(\mathbf{n}_{\min})} = \prod_{0 \le j < k \le N-1} \frac{n_k - n_j}{k - j} = f_{\mathbf{n}}(1) = s_{\mathbf{n}}(1, \ldots, 1).$$

The final assertion now follows from Littlewood's definition (25.1) of $s_{\mathbf{n}}(\mathbf{u})$. $\qquad \square$

25.2 Polynomials Preserving Positivity on Rank-1 Positive Matrices

We return to proving Theorem 24.4, and hence Theorem 24.3 on sign patterns. As we have shown, it suffices to prove the theorem for one higher degree (leading) term with a negative coefficient. Before proving the result in full, we tackle the following (simpler) versions. Thus, we are given a real polynomial as above: $f(x) = \sum_{j=0}^{N-1} c_{n_j} x^{n_j} + c_M x^M$, where $c_{n_j} > 0 \; \forall j$.

(1) Does there exist $c_M < 0$, such that $f[-] \colon \mathbb{P}_N((0, \rho)) \to \mathbb{P}_N$?

Here is a reformulation: dividing the expression for $f(x)$ throughout by $|c_M| = 1/t > 0$, define

$$p_t(x) := t \sum_{j=0}^{N-1} c_{n_j} x^{n_j} - x^m, \quad \text{where } c_{n_j} > 0 \, \forall j. \qquad (25.2)$$

Then it is enough to ask for which $t > 0$ (if any) does $p_t[-]$:
$\mathbb{P}_N((0,\rho)) \to \mathbb{P}_N$?

(2) Here are two simplifications: Can we produce such a constant $t > 0$ for only the subset of rank-1 matrices in $\mathbb{P}_N((0,\rho))$? How about for a *single* rank-1 matrix $\mathbf{u}\mathbf{u}^T$?

(3) A further special case: let $\mathbf{u}$ be generic, in that $\mathbf{u} \in (0,\rho)^N$ has distinct coordinates, and p_t is as above. Can one now compute all $t > 0$, such that $p_t[\mathbf{u}\mathbf{u}^T] \in \mathbb{P}_N$? How about all $t > 0$, such that $\det p_t[\mathbf{u}\mathbf{u}^T] \geq 0$?

We begin by answering the last of these questions – the answer crucially uses Schur polynomials. The following result shows that, in fact, $\det p_t[\mathbf{u}\mathbf{u}^T] \geq 0$ implies $p_t[\mathbf{u}\mathbf{u}^T]$ is positive semidefinite!

Proposition 25.6 *With $N \geq 1$ and notation as in (25.2), define the vectors*

$$\mathbf{n} := (n_0, \ldots, n_{N-1}),$$
$$\mathbf{n}_j := (n_0, \ldots, n_{j-1}, \widehat{n_j}, n_{j+1}, \ldots, n_{N-1}, M), \quad 0 \leq j < N, \qquad (25.3)$$

where $0 \leq n_0 < \cdots < n_{N-1} < M$. Now if the n_j and M are integers, and a vector $\mathbf{u} \in (0,\infty)^N$ has pairwise distinct coordinates, then the following are equivalent:

(1) $p_t[\mathbf{u}\mathbf{u}^T]$ *is positive semidefinite.*
(2) $\det p_t[\mathbf{u}\mathbf{u}^T] \geq 0$.
(3) $t \geq \displaystyle\sum_{j=0}^{N-1} \frac{s_{\mathbf{n}_j}(\mathbf{u})^2}{c_{n_j} s_{\mathbf{n}}(\mathbf{u})^2}.$

In particular, at least for "most" rank-1 matrices, it is possible to find polynomial preservers of positivity (on that one matrix), with a negative coefficient.

The proof of Proposition 25.6 uses the following even more widely applicable equivalence between the nonnegativity of the determinant and of the entire spectrum for "special" linear pencils of matrices:

Lemma 25.7 *Fix $\mathbf{w} \in \mathbb{R}^N$ and a positive definite matrix H. Define the linear pencil $P_t := tH - \mathbf{w}\mathbf{w}^T$, for $t > 0$. Then the following are equivalent:*

(1) P_t *is positive semidefinite.*

(2) $\det P_t \geq 0$.

(3) $t \geq \mathbf{w}^T H^{-1} \mathbf{w} = 1 - \dfrac{\det \left(H - \mathbf{w}\mathbf{w}^T \right)}{\det H}$.

This lemma is naturally connected to the theory of *generalized Rayleigh quotients*; see Questions 30.5 and 30.6 in the Exercises.

Proof   We show a cyclic chain of implications. That (1) $\implies$ (2) is immediate.

(2) $\implies$ (3): Using the identity (1.5) from Section 1.4 on Schur complements, we obtain by taking determinants

$$\det \begin{pmatrix} A & B \\ B' & D \end{pmatrix} = \det D \cdot \det(A - BD^{-1}B')$$

whenever A, D are square matrices, with D invertible. Using this, we compute

$$0 \leq \det \left(tH - \mathbf{w}\mathbf{w}^T \right) = \det \begin{pmatrix} tH & \mathbf{w} \\ \mathbf{w}^T & 1 \end{pmatrix}$$

$$= \det \begin{pmatrix} 1 & \mathbf{w}^T \\ \mathbf{w} & tH \end{pmatrix} = \det(tH) \det \left(1 - \mathbf{w}^T (tH)^{-1} \mathbf{w} \right).$$

Since the last quantity is a scalar, and $\det(tH) > 0$ by assumption, it follows from (2) that

$$1 \geq t^{-1} \left(\mathbf{w}^T H^{-1} \mathbf{w} \right) \qquad \implies \qquad t \geq \mathbf{w}^T H^{-1} \mathbf{w}.$$

Now substitute $t = 1$ in the above computation, to obtain

$$\det \left(H - \mathbf{w}\mathbf{w}^T \right) = \det(H) \det \left(1 - \mathbf{w}^T H^{-1} \mathbf{w} \right)$$

$$\implies \quad \frac{\det \left(H - \mathbf{w}\mathbf{w}^T \right)}{\det H} = 1 - \mathbf{w}^T H^{-1} \mathbf{w} \geq 1 - t,$$

which implies (3).

(3) $\implies$ (1): It suffices to show that $\mathbf{x}^T P_t \mathbf{x} \geq 0$ for all nonzero vectors $\mathbf{x} \in \mathbb{R}^N$. Using a change of variables $\mathbf{y} = \sqrt{H}\mathbf{x} \neq \mathbf{0}$, we compute

$$\mathbf{x}^T P_t \mathbf{x} = t\mathbf{y}^T \mathbf{y} - \left(\mathbf{y}^T \sqrt{H}^{-1} \mathbf{w} \right)^2$$

$$= \|\mathbf{y}\|^2 \left(t - \left((\mathbf{y}')^T \sqrt{H}^{-1} \mathbf{w} \right)^2 \right), \qquad \text{where } \mathbf{y}' := \frac{\mathbf{y}}{\|\mathbf{y}\|}$$

$$\geq \|\mathbf{y}\|^2 \left(t - \|\mathbf{y}'\|^2 \left\| \sqrt{H^{-1}} \mathbf{w} \right\|^2 \right) \qquad \text{(using Cauchy–Schwarz)}$$

$$= \|\mathbf{y}\|^2 \left(t - \mathbf{w}^T H^{-1} \mathbf{w} \right) \geq 0 \qquad \text{(by assumption).} \qquad \square$$

We can now answer the last of the above questions on positivity preservers, for generic rank-1 matrices.

Proof of Proposition 25.6   The result is easily shown for $N = 1$, so we now assume $N \geq 2$. We are interested in the following matrix and its determinant:

$$p_t[\mathbf{uu}^T] = t \sum_{j=0}^{N-1} c_{n_j} (\mathbf{u}^{\circ n_j})(\mathbf{u}^{\circ n_j})^T - (\mathbf{u}^{\circ M})(\mathbf{u}^{\circ M})^T.$$

We first work more generally: over any field $\mathbb{F}$, and with matrices $\mathbf{uv}^T$, where $\mathbf{u}, \mathbf{v} \in \mathbb{F}^N$. Thus, we study

$$p_t[\mathbf{uv}^T] = t \sum_{j=0}^{N-1} c_{n_j} \mathbf{u}^{\circ n_j} (\mathbf{v}^{\circ n_j})^T - \mathbf{u}^{\circ M} (\mathbf{v}^{\circ M})^T,$$

where $t, c_{n_j} \in \mathbb{F}$, and $c_{n_j} \neq 0 \; \forall j$. Setting $D = \mathrm{diag}(t c_{n_0}, \ldots, t c_{n_{N-1}}, -1)$, we have the decomposition

$$p_t[\mathbf{uv}^T] = U(\mathbf{u}) D U(\mathbf{v})^T,$$
$$\text{where} \quad U(\mathbf{u})_{N \times (N+1)} := \left(\mathbf{u}^{\circ n_0} \mid \cdots \mid \mathbf{u}^{\circ n_{N-1}} \mid \mathbf{u}^{\circ M} \right).$$

Applying the Cauchy–Binet formula to $A = U(\mathbf{u}), B = DU(\mathbf{v})^T$, as well as Cauchy's definition in Proposition 25.5(1), we obtain the following general determinantal identity, valid over any field:

$$\det p_t[\mathbf{uv}^T] = V(\mathbf{u}) V(\mathbf{v}) t^{N-1} \prod_{j=0}^{N-1} c_{n_j} \cdot \left(s_{\mathbf{n}}(\mathbf{u}) s_{\mathbf{n}}(\mathbf{v}) t - \sum_{j=0}^{N-1} \frac{s_{\mathbf{n}_j}(\mathbf{u}) s_{\mathbf{n}_j}(\mathbf{v})}{c_{n_j}} \right).$$

$$(25.4)$$

Now specialize this identity to $\mathbb{F} = \mathbb{R}$, with $t, c_{n_j} > 0$ and $\mathbf{u} = \mathbf{v} \in (0, \infty)^N$ having distinct coordinates. From this we deduce the following consequences. First, set

$$H := \sum_{j=0}^{N-1} c_{n_j} (\mathbf{uu}^T)^{\circ n_j} = U'(\mathbf{u}) D' U'(\mathbf{u})^T, \qquad \mathbf{w} := \mathbf{u}^{\circ M},$$

where $D' := \mathrm{diag}(c_{n_0}, \ldots, c_{n_{N-1}})$ is a positive definite matrix and $U'(\mathbf{u}) := (\mathbf{u}^{\circ n_0} \mid \cdots \mid \mathbf{u}^{\circ n_{N-1}})$ is a generalized Vandermonde matrix which has determinant $V(\mathbf{u}) s_{\mathbf{n}}(\mathbf{u}) \neq 0$. From this it follows that H is positive definite, so Lemma 25.7 applies. Moreover, $H - \mathbf{ww}^T = p_1[\mathbf{uu}^T]$, so using the above calculation (25.4) and the Cauchy–Binet formula respectively, we have

$$\det\left(H - \mathbf{w}\mathbf{w}^T\right) = V(\mathbf{u})^2 \prod_{j=0}^{N-1} c_{n_j} \cdot s_\mathbf{n}(\mathbf{u})^2 \left(1 - \sum_{j=0}^{N-1} \frac{s_{\mathbf{n}_j}(\mathbf{u})^2}{c_{n_j} s_\mathbf{n}(\mathbf{u})^2}\right),$$

$$\det H = V(\mathbf{u})^2 \prod_{j=0}^{N-1} c_{n_j} \cdot s_\mathbf{n}(\mathbf{u})^2.$$

In particular, from Lemma 25.7(3) we obtain

$$\mathbf{w}^T H^{-1} \mathbf{w} = \sum_{j=0}^{N-1} \frac{s_{\mathbf{n}_j}(\mathbf{u})^2}{c_{n_j} s_\mathbf{n}(\mathbf{u})^2}.$$

Now the proposition follows directly from Lemma 25.7, since $P_t = p_t[\mathbf{u}\mathbf{u}^T]$ for all $t > 0$. $\qquad\square$

26

First-Order Approximation and the Leading Term of Schur Polynomials

In Chapter 25, we computed the exact threshold for the leading term of a polynomial

$$p_t(x) := t \sum_{j=0}^{N-1} c_{n_j} x^{n_j} - x^M, \quad \text{where } c_{n_j} > 0 \ \forall j$$

(and where $0 \le n_0 < \cdots < n_{N-1} < M$ are integers), such that $p_t[\mathbf{u}\mathbf{u}^T] \in \mathbb{P}_N$ for a single vector $\mathbf{u} \in (0, \infty)^N$ with pairwise distinct coordinates. Recall that our (partial) goal is to find a threshold that works for all rank-1 matrices $\mathbf{u}\mathbf{u}^T \in \mathbb{P}_N((0, \rho))$ – i.e., for $\mathbf{u} \in (0, \sqrt{\rho})^N$. Thus, we need to show that the supremum of the threshold over all such $\mathbf{u}$ is bounded:

$$\sup_{\mathbf{u} \in (0, \sqrt{\rho})^N} \sum_{j=0}^{N-1} \frac{s_{\mathbf{n}_j}(\mathbf{u})^2}{c_{n_j} s_{\mathbf{n}}(\mathbf{u})^2} < \infty.$$

Since we only consider vectors $\mathbf{u}$ with positive coordinates, it suffices to bound $s_{\mathbf{n}_j}(\mathbf{u})/s_{\mathbf{n}}(\mathbf{u})$ from above, for each j. In turn, for this it suffices to find lower and upper bounds for every Schur polynomial evaluated at $\mathbf{u} \in (0, \infty)^N$. This is achieved by the following result:

Theorem 26.1 (First-order approximation/Leading term of Schur polynomials) *Say $N \ge 1$ and $0 \le n_0 < \cdots < n_{N-1}$ are integers. Then for all real numbers $0 < u_1 \le u_2 \le \cdots \le u_N$, we have the bounds*

$$1 \times \mathbf{u}^{\mathbf{n} - \mathbf{n}_{\min}} \le s_{\mathbf{n}}(\mathbf{u}) \le \frac{V(\mathbf{n})}{V(\mathbf{n}_{\min})} \times \mathbf{u}^{\mathbf{n} - \mathbf{n}_{\min}},$$

where $\mathbf{u}^{\mathbf{n} - \mathbf{n}_{\min}} = u_1^{n_0 - 0} u_2^{n_1 - 1} \dots u_N^{n_{N-1} - (N-1)}$ and $V(\mathbf{n})$ is as in (12.1). Moreover, the constants 1 and $\frac{V(\mathbf{n})}{V(\mathbf{n}_{\min})}$ cannot be improved.

228

Proof   Recall that $s_{\mathbf{n}}(\mathbf{u})$ is obtained by summing the weights of all column-strict Young tableaux of shape $\mathbf{n} - \mathbf{n}_{\min}$ with cell entries $1, \ldots, N$. Moreover, by the Weyl dimension formula in Proposition 25.5(3), there are precisely $V(\mathbf{n})/V(\mathbf{n}_{\min})$ such tableaux. Now each such tableau can have weight at most $\mathbf{u}^{\mathbf{n}-\mathbf{n}_{\min}}$, as follows: the cells in the top row each have entries at most N; the cells in the next row at most $N - 1$; and so on. The tableau $T_{\max}$ obtained in this fashion has weight precisely $\mathbf{u}^{\mathbf{n}-\mathbf{n}_{\min}}$. Hence, by definition, we have

$$\mathbf{u}^{\mathbf{n}-\mathbf{n}_{\min}} = \mathrm{wt}(T_{\max}) \leq \sum_T \mathrm{wt}(T)$$

$$= s_{\mathbf{n}}(\mathbf{u}) \leq \sum_T \mathrm{wt}(T_{\max}) = \frac{V(\mathbf{n})}{V(\mathbf{n}_{\min})} \mathbf{u}^{\mathbf{n}-\mathbf{n}_{\min}}.$$

This proves the bounds; we claim that both bounds are sharp. If $\mathbf{n} = \mathbf{n}_{\min}$ then all terms in the claimed inequalities are 1, and we are done. Thus, assume henceforth that $\mathbf{n} \neq \mathbf{n}_{\min}$. Let $A > 1$ and define $\mathbf{u}(A) := (A, A^2, \ldots, A^N)$. Then $\mathrm{wt}(T_{\max}) = A^M$ for some $M > 0$. Hence, for every column-strict Young tableau $T \neq T_{\max}$ as above, $\mathrm{wt}(T)$ is at most $\mathrm{wt}(T_{\max})/A$ and at least $1 = \mathrm{wt}(T_{\max})/A^M$. Now summing over all such tableaux T yields

$$s_{\mathbf{n}}(\mathbf{u}(A)) \leq \mathbf{u}(A)^{\mathbf{n}-\mathbf{n}_{\min}} \left(1 + \left(\frac{V(\mathbf{n})}{V(\mathbf{n}_{\min})} - 1\right)\frac{1}{A}\right),$$

$$s_{\mathbf{n}}(\mathbf{u}(A)) \geq \mathbf{u}(A)^{\mathbf{n}-\mathbf{n}_{\min}} \left(1 + \left(\frac{V(\mathbf{n})}{V(\mathbf{n}_{\min})} - 1\right)\frac{1}{A^M}\right).$$

Divide throughout by $\mathbf{u}(A)^{\mathbf{n}-\mathbf{n}_{\min}}$; now taking the limit as $A \to \infty$ yields the sharp lower bound 1 while taking the limit as $A \to 1^+$ yields the sharp upper bound $V(\mathbf{n})/V(\mathbf{n}_{\min})$. □

We now use Theorem 26.1, and Proposition 25.6 in Chapter 25, to find a threshold for $t > 0$ beyond which $p_t[-]$ preserves positivity on all rank-1 matrices in $\mathbb{P}_N((0, \rho))$ – and, in fact, on *all* matrices in $\mathbb{P}_N((0, \rho))$.

Theorem 26.2   *Fix integers $N \geq 1$, $0 \leq n_0 < n_1 < \cdots < n_{N-1} < M$, and scalars $\rho, t, c_{n_0}, \ldots, c_{n_{N-1}} > 0$. The polynomial $p_t(x) := t \sum_{j=0}^{N-1} c_{n_j} x^{n_j} - x^M$ entrywise preserves positivity on $\mathbb{P}_N((0, \rho))$, if $t \geq t_0 := \sum_{j=0}^{N-1} \frac{V(\mathbf{n}_j)^2}{c_{n_j} V(\mathbf{n}_{\min})^2} \rho^{M-n_j}$.*

The following notation is useful here and in the sequel:

Definition 26.3   Given an integer $n \geq 1$, and a totally ordered set X, define $X^{N,\uparrow}$ to be the set of all N-tuples $\mathbf{x} = (x_1, \ldots, x_N) \in X$ with strictly increasing coordinates: $x_1 < \cdots < x_N$. (Karlin calls this the open simplex $\Delta_N(X)$ in his book [136].)

Also recall (from above) that $X^{N,\neq}$ for any set X and an integer $N \geq 1$ denotes the set of N-tuples from X with pairwise distinct entries.

Proof of Theorem 26.2   Given $\mathbf{u} \in (0, \sqrt{\rho})^{N,\neq}$, from Proposition 25.6 it follows that $p_t[\mathbf{uu}^T] \in \mathbb{P}_N$ if and only if $t \geq \sum_{j=0}^{N-1} \frac{s_{\mathbf{n}_j}(\mathbf{u})^2}{c_{n_j} s_{\mathbf{n}}(\mathbf{u})^2}$. Now suppose $\mathbf{u} \in (0, \sqrt{\rho})^{N,\uparrow}$. Then by Theorem 26.1,

$$\sum_{j=0}^{N-1} \frac{s_{\mathbf{n}_j}(\mathbf{u})^2}{c_{n_j} s_{\mathbf{n}}(\mathbf{u})^2} \leq \sum_{j=0}^{N-1} \frac{\mathbf{u}^{2(\mathbf{n}_j - \mathbf{n}_{\min})} V(\mathbf{n}_j)^2 / V(\mathbf{n}_{\min})^2}{c_{n_j} \mathbf{u}^{2(\mathbf{n} - \mathbf{n}_{\min})}}$$

$$= \sum_{j=0}^{N-1} \frac{V(\mathbf{n}_j)^2}{c_{n_j} V(\mathbf{n}_{\min})^2} \mathbf{u}^{2(\mathbf{n}_j - \mathbf{n})},$$

and this is bounded above by t_0, since if $\mathbf{v} := \sqrt{\rho}(1, \dots, 1)^T$ then $\mathbf{u}^{2(\mathbf{n}_j - \mathbf{n})} \leq \mathbf{v}^{2(\mathbf{n}_j - \mathbf{n})} = \rho^{M - n_j}$ for all j. Thus, we conclude that

$$t \geq t_0 \quad \Longrightarrow \quad p_t[\mathbf{uu}^T] \in \mathbb{P}_N \; \forall \mathbf{u} \in (0, \sqrt{\rho})^{N,\uparrow}$$

$$\Longrightarrow \quad p_t[\mathbf{uu}^T] \in \mathbb{P}_N \; \forall \mathbf{u} \in (0, \sqrt{\rho})^{N,\neq}$$

$$\Longrightarrow \quad p_t[\mathbf{uu}^T] \in \mathbb{P}_N \; \forall \mathbf{u} \in (0, \sqrt{\rho})^{N},$$

where the first implication was proved above, the second follows by (the symmetric nature of Schur polynomials and by) relabeling the rows and columns of $\mathbf{uu}^T$ to rearrange the entries of $\mathbf{u}$ in increasing order, and the third implication follows from the continuity of p_t and the density of $(0, \sqrt{\rho})^{N,\neq}$ in $(0, \sqrt{\rho})^{N}$.

This validates the claimed threshold t_0 for all rank-1 matrices. To prove the result on all of $\mathbb{P}_N((0, \rho))$, we use induction on $N \geq 1$, with the base case of $N = 1$ already done since 1×1 matrices have rank 1.

For the induction step, recall the Extension Principle 5.9, which said that: *Suppose $I = (0, \rho)$ or $(-\rho, \rho)$ or its closure, for some $0 < \rho \leq \infty$. If $h \in C^1(I)$ is such that $h[-]$ preserves positivity on rank-1 matrices in $\mathbb{P}_N(I)$ and $h'[-] \colon \mathbb{P}_{N-1}(I) \to \mathbb{P}_{N-1}$, then $h[-] \colon \mathbb{P}_N(I) \to \mathbb{P}_N$.*

We will apply this result to $h(x) = p_{t_0}(x)$, with t_0 as above. By the extension principle, we need to show that $h'[-] \colon \mathbb{P}_{N-1}((0, \rho)) \to \mathbb{P}_{N-1}$. Note that

$$h'(x) = t_0 \sum_{j=0}^{N-1} n_j c_{n_j} x^{n_j - 1} - M x^{M-1} = M g(x) + t_0 n_0 c_{n_0} x^{n_0 - 1},$$

where we define

$$g(x) := \frac{t_0}{M} \sum_{j=1}^{N-1} n_j c_{n_j} x^{n_j-1} - x^{M-1}.$$

We **claim** that the entrywise polynomial map $g[-]: \mathbb{P}_{N-1}((0,\rho)) \to \mathbb{P}_{N-1}$. If this holds, then by the Schur product theorem, the same property is satisfied by $Mg(x) + t_0 n_0 c_{n_0} x^{n_0-1}$ (regardless of whether $n_0 = 0$ or $n_0 > 0$). But this function is precisely h' and the theorem would follow.

It thus remains to prove the claim, and we do so via a series of reductions and simplifications – i.e., "working backward." By the induction hypothesis, Theorem 26.2 holds in dimension $N - 1 \geq 1$, for the polynomials

$$q_t(x) := t \sum_{j=1}^{N-1} n_j c_{n_j} x^{n_j-1} - x^{M-1}.$$

For this family, the threshold is now given by

$$\sum_{j=1}^{N-1} \frac{V(\mathbf{n}'_j)^2}{n_j c_{n_j} V(\mathbf{n}'_{\min})^2} \rho^{M-1-(n_j-1)},$$

where

$$\mathbf{n}'_{\min} := (0, 1, \ldots, N-2),$$

$$\mathbf{n}'_j := \left(n_1, \ldots, n_{j-1}, \widehat{n}_j, n_{j+1}, \ldots, n_{N-1}, M\right) \; \forall j > 0.$$

Thus, the proof is complete if we show that

$$\sum_{j=1}^{N-1} \frac{V(\mathbf{n}'_j)^2}{n_j c_{n_j} V(\mathbf{n}'_{\min})^2} \rho^{M-1-(n_j-1)} \leq \frac{t_0}{M} = \sum_{j=0}^{N-1} \frac{V(\mathbf{n}_j)^2}{M c_{n_j} V(\mathbf{n}_{\min})^2} \rho^{M-n_j}.$$

In turn, comparing just the jth summand for each $j > 0$, it suffices to show that

$$\frac{V(\mathbf{n}'_j)}{\sqrt{n_j} V(\mathbf{n}'_{\min})} \leq \frac{V(\mathbf{n}_j)}{\sqrt{M} V(\mathbf{n}_{\min})}, \qquad \forall j > 0.$$

Dividing the right-hand side by the left-hand side, and cancelling common factors, we obtain the expression

$$\prod_{k=1}^{N-1} \frac{n_k - n_0}{k} \cdot \frac{\sqrt{n_j}}{\sqrt{M}} \cdot \frac{M - n_0}{n_j - n_0}.$$

Since every factor in the product term is at least 1, it remains to show that

$$\frac{M - n_0}{n_j - n_0} \geq \frac{\sqrt{M}}{\sqrt{n_j}}, \qquad \forall j > 0.$$

But this follows from a straightforward calculation

$$(M - n_0)^2 n_j - (n_j - n_0)^2 M = (M - n_j)(Mn_j - n_0^2) > 0,$$

and the proof is complete. □

Finally, we recall our original goal of classifying the sign patterns of positivity preservers in a fixed dimension – see Theorem 24.3. We showed this result holds if one can prove its special case, Theorem 24.4. Now this latter result follows from Theorem 26.2, by setting $c_M := -t_0^{-1}$, where $t_0 = \sum_{j=0}^{N-1} \frac{V(\mathbf{n}_j)^2}{c_{n_j} V(\mathbf{n}_{\min})^2} \rho^{M - n_j}$ as in Theorem 26.2. □

27

Exact Quantitative Bound – Monotonicity
of Schur Ratios

In Chapters 25 and 26, we proved the existence of a negative threshold c_M for polynomials

$$f(x) = \sum_{j=0}^{N-1} c_{n_j} x^{n_j} + c_M x^M$$

to entrywise preserve positivity on $\mathbb{P}_N((0, \rho))$. (Here $N > 0$ and $0 \leq n_0 < \cdots < n_{N-1} < M$ are integers.) We now compute the exact value of this threshold, more generally for *real* powers; this has multiple consequences which are described after stating Theorem 27.1. Thus, our goal is to prove the following quantitative result, for real powers – including negative powers:

Theorem 27.1 *Fix an integer $N > 0$ and real powers $n_0 < \cdots < n_{N-1} < M$. Also fix real scalars $\rho > 0$ and $c_{n_0}, \ldots, c_{n_{N-1}}, c_M$, and define*

$$f(x) := \sum_{j=0}^{N-1} c_{n_j} x^{n_j} + c_M x^M. \tag{27.1}$$

Then the following are equivalent:

(1) *The entrywise map $f[-]$ preserves positivity on all rank-1 matrices in $\mathbb{P}_N((0, \rho))$.*
(2) *The map $f[-]$ preserves positivity on rank-1 totally nonnegative (TN) Hankel matrices in $\mathbb{P}_N((0, \rho))$.*
(3) *Either all $c_{n_j}, c_M \geq 0$; or $c_{n_j} > 0 \; \forall j$ and $c_M \geq -\mathcal{C}^{-1}$, where*

$$\mathcal{C} = \sum_{j=0}^{N-1} \frac{V(\mathbf{n}_j)^2}{c_{n_j} V(\mathbf{n})^2} \rho^{M-n_j}. \tag{27.2}$$

Here $V(\mathbf{u}), \mathbf{n}, \mathbf{n}_j$ are defined as in (12.1) and (25.3).

233

If, moreover, we assume that $n_j \in \mathbb{Z}^{\geqslant 0} \cup [N - 2, \infty)$ for all j, then the above conditions are further equivalent to the "full-rank" version:

(4) *The entrywise map $f[-]$ preserves positivity on $\mathbb{P}_N([0, \rho])$, where we set $0^0 := 1$.*

Theorem 27.1 is a powerful result. It has multiple applications; we now list some of them.

(1) Suppose $M = N$ and $n_j = j$ for $0 \leq j \leq N - 1$. Then the result provides a complete characterization of which polynomials of degree $\leq N$ entrywise preserve positivity on $\mathbb{P}_N((0, \rho))$ – or more generally, on any intermediate set between $\mathbb{P}_N((0, \rho))$ and the rank-1 Hankel *TN* matrices inside it.

(2) In fact, a similar result to the previous characterization is implied, whenever one considers linear combinations of at most $N + 1$ monomial powers.

(3) The result provides information on positivity preservers beyond polynomials, since n_j, M are now allowed to be real, even negative if one works with rank-1 matrices.

(4) In particular, the result implies Theorem 24.5, and hence Theorem 24.6 (see its proof). This latter theorem provides a full classification of the sign patterns of possible "countable sums of real powers" which entrywise preserve positivity on $\mathbb{P}_N((0, \rho))$.

(5) The result also provides information on preservers of total nonnegativity on Hankel matrices in fixed dimension; see Corollary 27.7.

(6) There are further applications, two of which are (i) to the matrix cube problem and to sharp linear matrix inequalities/spectrahedra involving entrywise powers; and (ii) to computing the simultaneous kernels of entrywise powers and a related "Schubert cell-type" stratification of the cone $\mathbb{P}_N(\mathbb{C})$. These are explained in the 2016 paper of Belton et al. [16]; see also the 2021 paper by Khare and Tao [146] (mentioned a few lines above (24.1)).

(7) Theorem 27.1 is proved using a monotonicity phenomenon for ratios of Schur polynomials; see Theorem 27.3. This latter result is also useful in extending a 2011 conjecture by Cuttler–Greene–Skandera (and its proof). In fact, this line of attack ends up *characterizing* majorization and weak majorization – for *real* tuples – using Schur polynomials. See the aforementioned paper by Khare and Tao [146] for more details.

(8) One further application is Theorem 27.8, which finds a threshold for bounding by $\sum_{j=0}^{N-1} c_{n_j} A^{\circ n_j}$, any power series – and more general "Laplace transforms" – applied entrywise to a positive matrix A.

This extends Theorem 27.1, where the power series is simply x^M, because Theorem 27.1 says in particular that $(x^M)[A] = A^{\circ M}$ is dominated by a multiple of $\sum_{j=0}^{N-1} c_{n_j} A^{\circ n_j}$.

(9) As mentioned in the remarks prior to Theorem 24.3, Theorem 27.1 also provides examples of power series preservers on $\mathbb{P}_N((0, \rho))$ with negative coefficients; and of such functions which preserve positivity on $\mathbb{P}_N((0, \rho))$ but not on $\mathbb{P}_{N+1}((0, \rho))$.

27.1 Monotonicity of Ratios of Schur Polynomials

The proof of Theorem 27.1 uses the same ingredients as developed in previous chapters. A summary of what follows is now provided. In the rank-1 case, we use a variant of Proposition 25.6 for an individual matrix; the result does not apply as is, since the powers may now be real. Next, in order to find the sharp threshold for all rank-1 matrices, even for real powers we crucially appeal to the integer power case. Namely, we will first understand the behavior and supremum of the function $s_{\mathbf{n}_j}(\mathbf{u})/s_{\mathbf{n}}(\mathbf{u})$ over $\mathbf{u} \in (0, \sqrt{\rho})^N$ (and for each $0 \le j \le N - 1$). One may hope that these suprema behave well enough that the sharp threshold can be computed for rank-1 matrices; the further hope would be that this threshold bound is tight enough to behave well with respect to the extension principle in Theorem 5.9, and hence to work for all matrices in $\mathbb{P}_N((0, \rho))$. Remarkably, these two hopes are indeed justified, proving the theorem.

We begin with the key result required to be able to take suprema over ratios of Schur polynomials $s_{\mathbf{m}}(\mathbf{u})/s_{\mathbf{n}}(\mathbf{u})$. To motivate the result, here is a special case.

Example 27.2   Suppose $N = 3, \mathbf{n} = (0, 2, 3)$, and $\mathbf{m} = (0, 2, 4)$. As above, we have $\mathbf{u} = (u_1, u_2, u_3)^T$ and $\mathbf{n}_{\min} = (0, 1, 2)$. Now let $f(\mathbf{u}) := \frac{s_{\mathbf{m}}(\mathbf{u})}{s_{\mathbf{n}}(\mathbf{u})}$: $(0, \infty)^N \to (0, \infty)$. This is a rational function, whose numerator sums weights over tableaux of shape $(0, 1, 2)$, and hence by Example 25.2 above, equals $(u_1 + u_2)(u_2 + u_3)(u_3 + u_1)$. The denominator sums weights over tableaux of shape $(0, 1, 1)$; there are only three such tableaux

$$
\begin{array}{|c|} \hline 3 \\ \hline 2 \\ \hline \end{array}
\qquad
\begin{array}{|c|} \hline 3 \\ \hline 1 \\ \hline \end{array}
\qquad
\begin{array}{|c|} \hline 2 \\ \hline 1 \\ \hline \end{array}
$$

and hence,

$$
f(\mathbf{u}) := \frac{(u_1 + u_2)(u_2 + u_3)(u_3 + u_1)}{u_1 u_2 + u_2 u_3 + u_3 u_1}, \qquad u_1, u_2, u_3 > 0.
$$

Notice that the numerator and denominator are both Schur polynomials, hence positive combinations of monomials (this is called "monomial positivity"). In particular, they are both nondecreasing in each coordinate. One can verify that their ratio $f(\mathbf{u})$ is not a polynomial; moreover, it is not a priori clear if $f(\mathbf{u})$ shares the same coordinatewise monotonicity property. However, we claim that this does hold, i.e., $f(\mathbf{u})$ is *nondecreasing in each coordinate* on $\mathbf{u} \in (0,\infty)^N$.

To see why: by symmetry, it suffices to show that f is nondecreasing in u_3. Using the quotient rule of differentiation, we claim that the expression

$$s_{\mathbf{n}}(\mathbf{u})\partial_{u_3} s_{\mathbf{m}}(\mathbf{u}) - s_{\mathbf{m}}(\mathbf{u})\partial_{u_3} s_{\mathbf{n}}(\mathbf{u}) \tag{27.3}$$

is nonnegative on $(0,\infty)^3$. Indeed, computing this expression yields

$$(u_1 + u_2)(u_1 u_3 + 2u_1 u_2 + u_2 u_3)u_3,$$

and this is clearly nonnegative, as desired. More strongly, the expression (27.3) turns out to be monomial positive, which implies nonnegativity.

Here is the punchline: an even stronger phenomenon holds. Namely, when we write the expression (27.3) in the form $\sum_{j\geq 0} p_j(u_1, u_2)u_3^j$, each polynomial p_j is *Schur positive*! This means that it is a nonnegative integer-linear combination of Schur polynomials:

$$p_0(u_1, u_2) = 0,$$

$$p_1(u_1, u_2) = 2u_1 u_2^2 + 2u_1^2 u_2 = 2\;\boxed{\begin{array}{cc}2 & 2 \\ 1 \end{array}} + 2\;\boxed{\begin{array}{cc}2 & 1 \\ 1 \end{array}} = 2s_{(1,3)}(u_1, u_2),$$

$$p_2(u_1, u_2) = (u_1 + u_2)^2 = \boxed{\begin{array}{cc}2 & 2\end{array}} + \boxed{\begin{array}{cc}2 & 1\end{array}} + \boxed{\begin{array}{cc}1 & 1\end{array}} + \boxed{\begin{array}{c}2 \\ 1\end{array}}$$

$$= s_{(0,3)}(u_1, u_2) + s_{(1,2)}(u_1, u_2),$$

modulo a mild abuse of notation. This yields the sought-for nonnegativity, as each $s_{\mathbf{n}}(\mathbf{u})$ is monomial positive by definition.

The remarkable fact is that the phenomena described in the above example also occur for every pair of Schur polynomials $s_{\mathbf{m}}(\mathbf{u}), s_{\mathbf{n}}(\mathbf{u})$ for which $\mathbf{m} \geq \mathbf{n}$ coordinatewise:

Theorem 27.3 (Monotonicity of Schur polynomial ratios) *Suppose* $0 \leqslant n_0 < \cdots < n_{N-1}$ *and* $0 \leqslant m_0 < \cdots < m_{N-1}$ *are integers satisfying:* $n_j \leqslant m_j \; \forall j$. *Then the symmetric function*

$$f: (0,\infty)^N \to \mathbb{R}, \qquad f(\mathbf{u}) := \frac{s_{\mathbf{m}}(\mathbf{u})}{s_{\mathbf{n}}(\mathbf{u})}$$

is nondecreasing in each coordinate.

More strongly, viewing the expression

$$s_{\mathbf{n}}(\mathbf{u}) \cdot \partial_{u_N} s_{\mathbf{m}}(\mathbf{u}) - s_{\mathbf{m}}(\mathbf{u}) \cdot \partial_{u_N} s_{\mathbf{n}}(\mathbf{u})$$

as a polynomial in u_N, the coefficient of each monomial u_N^j is a Schur positive polynomial in $(u_1, u_2, \ldots, u_{N-1})^T$.

Theorem 27.3 is an application of a deep result in representation theory/ symmetric function theory, by Lam et al. in 2007 [154]. The proof of this latter result is beyond the scope of this text, and hence is not pursued further; but its usage means that in the spirit of Chapters 25 and 26, the proof of Theorem 27.3 once again combines analysis with symmetric function theory. Moreover, this 2007 result in [154] arose from the prior work of Skandera [222] in 2004, on determinant inequalities for minors of *TN* matrices.

To proceed further, we introduce the following notation:

Definition 27.4   Given a vector $\mathbf{u} = (u_1, \ldots, u_m)^T \in (0, \infty)^m$ and a real tuple $\mathbf{n} = (n_0, \ldots, n_{N-1})$ for integers $m, N \geq 1$, define

$$\mathbf{u}^{\circ \mathbf{n}} := \left(\mathbf{u}^{\circ n_0} \mid \cdots \mid \mathbf{u}^{\circ n_{N-1}} \right)_{m \times N} = \left(u_j^{n_{k-1}} \right)_{j=1, k=1}^{m, N}.$$

We now extend Theorem 27.3 to arbitrary real powers (instead of nonnegative integer powers). As one can no longer use Schur polynomials, the next result uses generalized Vandermonde determinants instead:

Theorem 27.5   *Fix an integer $N \geq 1$ and real tuples*

$$\mathbf{n} = (n_0 < n_1 < \cdots < n_{N-1}), \qquad \mathbf{m} = (m_0 < m_1 < \cdots < m_{N-1})$$

with $n_j \leq m_j \ \forall j$ and $\mathbf{n} \neq \mathbf{m}$. Then the symmetric function

$$f_{\neq}(\mathbf{u}) := \frac{\det(\mathbf{u}^{\circ \mathbf{m}})}{\det(\mathbf{u}^{\circ \mathbf{n}})}$$

is strictly increasing in each coordinate on $(0, \infty)^{N, \neq}$. (See Definition 26.3.) If $n_0 = m_0 = 0$, then $f_{\neq}(\mathbf{u})$ is strictly increasing in each coordinate on $[0, \infty)^{N, \neq}$.

While we only require $f_{\neq}$ to be nondecreasing in each coordinate, and only on $(0, \infty)^{N, \neq}$, we will show this stronger result.

Proof   The result is immediate for $N = 1$; henceforth suppose $N \geq 2$. For a fixed $t \in \mathbb{R}$, if for each j we multiply the jth row of the matrix $\mathbf{u}^{\circ \mathbf{m}}$ by u_j^t, we obtain a matrix $\mathbf{u}^{\circ \mathbf{m}'}$ where $m'_j = m_j + t \ \forall j$. In particular, if we start with real powers n_j, m_j, then multiplying the numerator and denominator of $f_{\neq}$ by $(u_1 \ldots u_N)^{-n_0}$ reduces the situation to working with the nonnegative real tuples $\mathbf{n}' := (n_j - n_0)_{j=0}^{N-1}$ and $\mathbf{m}' := (m_j - m_0)_{j=0}^{N-1}$. Thus, we suppose henceforth that $n_j, m_j \geq 0 \, \forall j$.

We first show that $f_{\neq}$ is nondecreasing in each coordinate. If n_j, m_j are all integers, then the result is an immediate reformulation of the first part of Theorem 27.3, via Proposition 25.5(1). Next, suppose n_j, m_j are rational. Choose a (large) integer $L > 0$ such that $Ln_j, Lm_j \in \mathbb{Z} \, \forall j$ and define $y_j := u_j^{1/L}$. By the previous subcase, the symmetric function

$$f(\mathbf{y}) := \frac{\det(\mathbf{y}^{\circ L\mathbf{m}})}{\det(\mathbf{y}^{\circ L\mathbf{n}})} = \frac{\det(\mathbf{u}^{\circ \mathbf{m}})}{\det(\mathbf{u}^{\circ \mathbf{n}})}, \qquad \mathbf{y} := (y_1, \ldots, y_N)^T \in (0, \infty)^{N, \neq}$$

is coordinatewise nondecreasing on $(0, \infty)^{N, \neq}$ in the y_j, and hence on $(0, \infty)^{N, \neq}$ in the u_j.

Finally, in the general case, given nonnegative real powers n_j, m_j satisfying the hypotheses, choose sequences

$$0 \leq n_{0,k} < n_{1,k} < \cdots < n_{N-1,k}, \qquad 0 \leq m_{0,k} < m_{1,k} < \cdots < m_{N-1,k}$$

for $k = 1, 2, \ldots$, which further satisfy:

(1) $n_{j,k}, m_{j,k}$ are rational for $0 \leq j \leq N - 1$, $k \geq 1$;
(2) $n_{j,k} \leq m_{j,k} \, \forall j, k$; and
(3) $n_{j,k} \to n_j$ and $m_{j,k} \to m_j$ as $k \to \infty$, for each $j = 0, 1, \ldots, N - 1$.

By the rational case above, for each $k \geq 1$ the symmetric function

$$f_k(\mathbf{u}) := \frac{\det(\mathbf{u}^{\circ \mathbf{m}_k})}{\det(\mathbf{u}^{\circ \mathbf{n}_k})}$$

is coordinatewise nondecreasing, where $\mathbf{m}_k := (m_{0,k}, \ldots, m_{N-1,k})$ and similarly for $\mathbf{n}_k$. But then their limit $\lim_{k \to \infty} f_k(\mathbf{u}) = f_{\neq}(\mathbf{u})$ is also coordinatewise nondecreasing, as claimed.

The next step is to show that $f_{\neq}$ is strictly increasing on $(0, \infty)^{N, \neq}$ in each coordinate, say in u_N by symmetry. Suppose instead that $f_{\neq}$ is constant on the intersection with $\mathbb{R}^{N, \neq}$ of a line segment

$$\{(u_1, \ldots, u_{N-1}, u_N) : u_N \in [x, x']\} \subset (0, \infty)^N,$$

where $u_1, \ldots, u_{N-1} \in (0, \infty)$ are fixed and $0 < x < x'$. Here we may replace $[x, x']$ by a smaller subinterval (still of positive length) that does not contain $u_1, \ldots, u_{N-1}$; thus, without loss of generality the above segment is contained in $(0, \infty)^{N, \neq}$. Now evaluating $f_{\neq}$ as a function of u_N, we obtain a constant function of the form

$$h(u_N) := \frac{\sum_{j=0}^{N-1} u_N^{m_j} g_j}{\sum_{j=0}^{N-1} u_N^{n_j} g_j'}, \qquad u_N \in [x, x'],$$

where g_j, g_j' are generalized Vandermonde determinants, hence all nonzero. Denoting the numerator and denominator by $h_1(u_N), h_2(u_N)$ respectively, if $h(\cdot) \equiv c$ on $[x, x']$ for some $c \in \mathbb{R}$, then $h_1 - ch_2$ has infinitely many zeros on $[x, x']$. Since $\mathbf{m} \neq \mathbf{n}$, this contradicts Descartes' rule of signs (Lemma 4.2).

Finally, we show that $f_{\neq}$ is strictly increasing in each coordinate u_j at $\mathbf{u}$, where one coordinate of $\mathbf{u}$, say u_1, equals zero – and $n_0 = m_0 = 0$. There are two cases: if $j > 1$, then both $\mathbf{u}^{\circ\mathbf{m}}$ and $\mathbf{u}^{\circ\mathbf{n}}$ are matrices with the first column $(1, \ldots, 1)^T$ and first row $\mathbf{e}_1^T = (1, 0, \ldots, 0)$. But then $\det(\mathbf{u}^{\circ\mathbf{m}}) = \det(\mathbf{u}_1^{\circ\mathbf{m}_1})$, where $\mathbf{v}_1$ for a vector $\mathbf{v}$ is the subvector that removes the first coordinate. Hence,

$$f_{\neq}(\mathbf{u}) = \frac{\det(\mathbf{u}_1^{\circ\mathbf{m}_1})}{\det(\mathbf{u}_1^{\circ\mathbf{n}_1})},$$

and since $\mathbf{u}_1 \in (0, \infty)^{N-1, \neq}$, the right-hand side is strictly increasing in u_j for $j > 1$ by the above analysis – now on $(0, \infty)^{N-1, \neq}$.

The other case is if $j = 1$. Then we consider $\mathbf{v} := \mathbf{u} + \mu\mathbf{e}_1$ for some $\mu \in (0, \infty)$. If $c := \min(\mu, \min_{j>1} u_j)$, then $\mathbf{v}(\epsilon) := \mathbf{u} + \epsilon\mathbf{e}_1$ lies in $(0, \infty)^{N, \neq}$ for $\epsilon \in (0, c)$, so by above, $f_{\neq}(\mathbf{v}(\cdot))$ is strictly increasing as a function of $\epsilon \in (0, c)$. Hence, for $\epsilon \in (0, c/4)$, we have from above

$$f_{\neq}(\mathbf{v}) = f_{\neq}(\mathbf{v}(\mu)) > f_{\neq}(\mathbf{v}(c/4)) > f_{\neq}(\mathbf{v}(\epsilon)).$$

Letting $\epsilon \to 0^+$, the proof is complete:

$$f_{\neq}(\mathbf{v}) > f_{\neq}(\mathbf{v}(c/4)) \geq \lim_{\epsilon \to 0^+} f_{\neq}(\mathbf{v}(\epsilon)) = f_{\neq}(\mathbf{v}(0)) = f_{\neq}(\mathbf{u}). \qquad \square$$

27.2 Proof of the Quantitative Bound

Using Theorem 27.5, we can now prove the main result in this chapter.

Proof of Theorem 27.1   We first work only with rank-1 matrices. Clearly, (1) $\implies$ (2), and we show that (2) $\implies$ (3) $\implies$ (1).

If all coefficients $c_{n_j}, c_M \geq 0$, then $f[-]$ preserves positivity on rank-1 matrices. Otherwise, by the Horn–Loewner-type necessary conditions in Lemma 24.1 (now for real powers, possibly negative!), it follows that $c_{n_0}, \ldots, c_{n_{N-1}} > 0 > c_M$. In this case, the discussion that opens Section 25.2 allows us to reformulate the problem using

$$p_t(x) := t \sum_{j=0}^{N-1} c_{n_j} x^{n_j} - x^M, \qquad t > 0,$$

and the goal is to find a *sharp* positive lower bound for t, above which $p_t[-]$ preserves positivity on rank-1 Hankel *TN* matrices $\mathbf{u}\mathbf{u}^T \in \mathbb{P}_N((0, \rho))$.

But now one can play the same game as in Section 25.2. In other words, Lemma 25.7 shows that the "real powers analogue" of Proposition 25.6 holds: $p_t[\mathbf{u}\mathbf{u}^T] \geq 0$ if and only if

$$t \geq \sum_{j=0}^{N-1} \frac{\det(\mathbf{u}^{\circ \mathbf{n}_j})^2}{c_{n_j} \det(\mathbf{u}^{\circ \mathbf{n}})^2},$$

for all generic rank-1 matrices $\mathbf{u}\mathbf{u}^T$, with $\mathbf{u} \in (0, \sqrt{\rho})^{N, \neq}$. By the same reasoning as in the proof of Theorem 26.2 (see Chapter 26), $p_t[-]$ preserves positivity on a given test set of rank-1 matrices $\{\mathbf{u}\mathbf{u}^T : \mathbf{u} \in S \subset (0, \sqrt{\rho})^N\}$, if and only if (by density and continuity,) t exceeds the following supremum:

$$t \geq \sup_{\mathbf{u} \in S \cap (0, \sqrt{\rho})^{N, \neq}} \sum_{j=0}^{N-1} \frac{\det(\mathbf{u}^{\circ \mathbf{n}_j})^2}{c_{n_j} \det(\mathbf{u}^{\circ \mathbf{n}})^2}. \qquad (27.4)$$

This is, of course, subject to $S \cap (0, \sqrt{\rho})^{N, \neq}$ being dense in the set S, which is indeed the case if $\{\mathbf{u}\mathbf{u}^T : \mathbf{u} \in S \cap (0, \sqrt{\rho})^N\}$ equals the set of rank-1 Hankel *TN* matrices as in assertion (2).

Thus, to prove (2) $\implies$ (3) $\implies$ (1) in the theorem, it suffices to prove: (i) the supremum (27.4) is bounded above by the value $\sum_{j=0}^{N-1} \frac{V(\mathbf{n}_j)^2}{c_{n_j} V(\mathbf{n})^2} \rho^{M-n_j}$; and (ii) this value is attained on (a countable set of) rank-1 Hankel *TN* matrices, hence it equals the supremum.

We now prove both of these claims. By Theorem 27.5, each ratio $\frac{\det(\mathbf{u}^{\circ \mathbf{n}_j})}{\det(\mathbf{u}^{\circ \mathbf{n}})}$ is coordinatewise nondecreasing, hence its supremum on $(0, \sqrt{\rho})^{N, \neq}$ is bounded above by (and in fact equals) its limit as $\mathbf{u} \to \sqrt{\rho}(1^-, \dots, 1^-)$. To see why this limit exists, note that every vector $u \in (0, \rho)^N$ is bounded above – coordinatewise – by a vector of the form

$$\mathbf{u}(\epsilon) := \sqrt{\rho}\left(\epsilon, \epsilon^2, \dots, \epsilon^N\right)^T \in (0, \sqrt{\rho})^{N, \neq}, \qquad \epsilon \in (0, 1).$$

In particular, by Theorem 27.5 the limit as $\mathbf{u} \to \sqrt{\rho}(1^-, \dots, 1^-)$ exists and equals the limit by using the rank-1 Hankel *TN* family $\mathbf{u}(\epsilon)\mathbf{u}(\epsilon)^T$, for any sequence of $\epsilon \to 1^-$ – provided this latter limit exists. We show this presently; thus, we work with a countable sequence of $\epsilon \to 0^+$ in place of Lemma 24.1, and another countable sequence of $\epsilon \to 1^-$ in what follows. First observe:

Lemma 27.6 (Principal specialization formula for real powers) *Suppose $q \in (0, \infty)$ and $n_0 < n_1 < \cdots < n_{N-1}$ are real exponents. If $\mathbf{n} := (n_0, \ldots, n_{N-1})$ and $\mathbf{u} := \left(1, q, \ldots, q^{N-1}\right)^T$, then*

$$\det(\mathbf{u}^{\circ \mathbf{n}}) = \prod_{0 \le j < k \le N-1} (q^{n_k} - q^{n_j}) = V(q^{\circ \mathbf{n}}).$$

The proof is exactly the same as of Proposition 25.5(2), since the transpose of $\mathbf{u}^{\circ \mathbf{n}}$ is a *usual* Vandermonde matrix.

We can now complete the proof of Theorem 27.1. The above lemma immediately implies

$$\frac{\det(\mathbf{u}(\epsilon)^{\circ \mathbf{n}_j})}{\det(\mathbf{u}(\epsilon)^{\circ \mathbf{n}})} = \sqrt{\rho}^{M-n_j} \frac{V\left(\epsilon^{\circ \mathbf{n}_j}\right)}{V(\epsilon^{\circ \mathbf{n}})}, \qquad \forall 0 \le j \le N - 1.$$

Dividing the numerator and denominator by $(1 - \epsilon)^{\binom{N}{2}}$ and taking the limit as $\epsilon \to 1^-$ using L'Hôpital's rule, we obtain the expression $\sqrt{\rho}^{M-n_j} \frac{V(\mathbf{n}_j)}{V(\mathbf{n})}$. Since all of these suprema/limits occur as $\epsilon \to 1^-$, we finally have

$$\sup_{\mathbf{u} \in (0, \sqrt{\rho})^{N, \neq}} \sum_{j=0}^{N-1} \frac{\det(\mathbf{u}^{\circ \mathbf{n}_j})^2}{c_{n_j} \det(\mathbf{u}^{\circ \mathbf{n}})^2} = \lim_{\epsilon \to 1^-} \sum_{j=0}^{N-1} \frac{\det(\mathbf{u}(\epsilon)^{\circ \mathbf{n}_j})^2}{c_{n_j} \det(\mathbf{u}(\epsilon)^{\circ \mathbf{n}})^2}$$

$$= \sum_{j=0}^{N-1} \frac{V(\mathbf{n}_j)^2}{V(\mathbf{n})^2} \frac{\rho^{M-n_j}}{c_{n_j}}.$$

This proves the equivalence of assertions (1)–(3) in the theorem, for rank-1 matrices.

Finally, suppose all $n_j \in \mathbb{Z}^{\ge 0} \cup [N - 2, \infty)$. In this case (4) $\implies$ (1) is immediate. Conversely, given that (1) holds, we prove (4) using once again the integration trick of FitzGerald and Horn, as isolated in Theorem 5.9. The proof and calculation are similar to that of Theorem 26.2 above and are left to the interested reader as an exercise. (see Question 30.3). $\qquad \square$

27.3 Applications: Hankel *TN* Preservers, Power Series Preservers

We conclude by discussing some applications of Theorem 27.1. First, the result implies in particular that $A^{\circ M}$ is bounded above by a multiple of $\sum_{j=0}^{N-1} c_{n_j} A^{\circ n_j}$. In particular, the proof of Theorem 24.6 above goes through; thus, we have classified the sign patterns of all entrywise power series preserving positivity on $\mathbb{P}_N((0, \rho))$.

Second, the equivalent conditions in Theorem 27.1 classifying the (entry-wise) polynomial positivity preservers on $\mathbb{P}_N((0, \rho))$ – or on rank-1 matrices – also end up classifying the polynomial preservers of *total nonnegativity* on the corresponding Hankel test sets:

Corollary 27.7   *With notation as in Theorem 27.1, if we restrict to all real powers and only rank-1 matrices, then assertions (1)–(3) in Theorem 27.1 are further equivalent to:*

(1′) $f[-]$ *preserves total nonnegativity on all rank-1 matrices in* HTN_n *with entries in* $(0, \rho)$.

If, moreover, all n_j lie in $\mathbb{Z}^{\geq 0} \cup [N-2, \infty)$, then these conditions are further equivalent to:

(4′) $f[-]$ *preserves total nonnegativity on all matrices in* HTN_n *with entries in* $[0, \rho]$.

Recall here that by Definition 6.10, HTN_n denotes the set of $N \times N$ Hankel totally nonnegative matrices.

Proof   Clearly, (4′) implies (1′), which implies assertion (2) in Theorem 27.1. Conversely, we claim that assertion (1) in Theorem 27.1 implies (1′) via Theorem 3.6. Indeed, if $A \in \mathrm{HTN}_n$ has rank 1 and entries in $(0, \rho)$, then $f[A] \in \mathbb{P}_N$ by Theorem 27.1(1). Similarly, $A^{(1)} \oplus (0)_{1 \times 1} \in \mathbb{P}_N((0, \rho))$ and has rank 1, so $f[A^{(1)}]$ is also positive semidefinite, and hence Theorem 3.6 applies, as desired. The same proof works to show that (4′) follows from Theorem 27.1(4).   □

The third and final application is to bounding $g[A]$, where $g(x)$ is a power series – or more generally, a linear combination of real powers – by a threshold times $\sum_{j=0}^{N-1} c_{n_j} A^{\circ n_j}$. This extends Theorem 27.1 in which $g(x) = x^M$. The idea is that if we fix exponents $0 \leq n_0 < \cdots < n_{N-1}$ and coefficients c_{n_j} for $j = 0, \ldots, N-1$, then

$$A^{\circ M} \leq t_M \sum_{j=0}^{N-1} c_{n_j} A^{\circ n_j}, \qquad \text{where } t_M := \sum_{j=0}^{N-1} \frac{V(\mathbf{n}_j)^2}{c_{n_j} V(\mathbf{n})^2} \rho^{M - n_j}, \qquad (27.5)$$

and this linear matrix inequality holds for all $A \in \mathbb{P}_N((0, \rho))$ – possibly of rank 1 if the n_j are allowed to be arbitrary nonnegative real numbers, else of all ranks if all $n_j \in \mathbb{Z}^{\geq 0} \cup [N - 2, \infty)$. Here the $\leq$ stands for the positive semidefinite ordering, or Loewner ordering – see, e.g., Definition 8.5. Moreover, the constant t_M depends on M through $\mathbf{n}_j$ and $\rho^{M - n_j}$.

If now we consider a power series $g(x) := \sum_{M \geq n_{N-1}+1} c_M x^M$, then by adding several linear matrix inequalities of the form (27.5), it follows that

$$g[A] \leq t_g \sum_{j=0}^{N-1} c_{n_j} A^{\circ n_j}, \qquad \text{where } t_g := \sum_{M \geq n_{N-1}+1} \max(c_M, 0) t_M,$$

and this is a valid linear matrix inequality, as long as the sum t_g is convergent. Thus, we now explore when this sum converges.

Even more generally: notice that a power series is the sum/integral of the power function, over a measure on the *powers* which is supported on the integers. Thus, given any real measure μ supported in $[n_{N-1} + \varepsilon, \infty)$, one can consider its corresponding "Laplace transform"

$$g_\mu(x) := \int_{n_{N-1}+\varepsilon}^\infty x^t \, d\mu(t). \tag{27.6}$$

The final application of Theorem 27.1 explores in this generality, when a finite threshold exists to bound $g_\mu[A]$ by a sum of N lower powers.

Theorem 27.8 *Fix $N \geq 2$ and real exponents $0 \leq n_0 < \cdots < n_{N-1}$ in the set $\mathbb{Z}^{\geq 0} \cup [N-2, \infty)$. Also fix scalars $\rho, c_{n_j} > 0$ for all j.*

Now suppose $\varepsilon, \varepsilon' > 0$ and μ is a real measure supported on $[n_{N-1}+\varepsilon, \infty)$ such that $g_\mu(x)$ – defined as in (27.6) – is absolutely convergent at $\rho(1 + \varepsilon')$. Then there exists a finite constant $t_\mu \in (0, \infty)$, such that the map

$$t_\mu \sum_{j=0}^{N-1} c_{n_j} x^{n_j} - g_\mu(x)$$

entrywise preserves positivity on $\mathbb{P}_N((0, \rho))$. Equivalently,

$$g_\mu[A] \leq t_\mu \sum_{j=0}^{N-1} c_{n_j} A^{\circ n_j}, \qquad \forall A \in \mathbb{P}_N((0, \rho)).$$

Proof If $\mu = \mu_+ - \mu_-$ denotes the decomposition of μ into its positive and negative parts, then notice (e.g., by the FitzGerald–Horn Theorem 5.3) that

$$\int_\mathbb{R} A^{\circ M} \, d\mu_-(M) \in \mathbb{P}_N, \qquad \forall A \in \mathbb{P}_N((0, \rho)).$$

Hence, it suffices to show that

$$t_\mu := \int_{n_{N-1}+\varepsilon}^\infty t_M \, d\mu_+(M) = \int_{n_{N-1}+\varepsilon}^\infty \sum_{j=0}^{N-1} \frac{V(\mathbf{n}_j)^2}{c_{n_j} V(\mathbf{n})^2} \rho^{M-n_j} \, d\mu_+(M) < \infty,$$

$$\tag{27.7}$$

since this would imply

$$t_\mu \sum_{j=0}^{N-1} c_{n_j} A^{\circ n_j} - g_\mu[A]$$

$$= \int_{n_{N-1}+\varepsilon}^\infty \left(t_M \sum_{j=0}^{N-1} c_{n_j} A^{\circ n_j} - A^{\circ M} \right) d\mu_+(M) + \int_{n_{N-1}+\varepsilon}^\infty A^{\circ M} \, d\mu_-(M),$$

and both integrands and integrals are positive semidefinite.

In turn, isolating the terms in (27.7) that depend on M, it suffices to show for each j that

$$\int_{n_{N-1}+\varepsilon}^\infty \prod_{k=0, k\neq j}^{N-1} (M - n_k)^2 \rho^M \, d\mu_+(M) < \infty.$$

By linearity, it suffices to examine the finiteness of the integrals

$$\int_{n_{N-1}+\varepsilon}^\infty M^k \rho^M \, d\mu_+(M), \qquad k \geq 0.$$

But by assumption, $\int_{n_{N-1}+\varepsilon}^\infty \rho^M (1+\varepsilon')^M \, d\mu_+(M)$ is finite; and moreover, for any fixed $k \geq 0$ there is a threshold M_k beyond which $(1+\varepsilon')^M \geq M^k$. $\left(\text{Indeed, this happens when } \frac{\log M}{M} \leq \frac{\log(1+\varepsilon')}{k}. \right)$ Therefore,

$$\int_{n_{N-1}+\varepsilon'}^\infty M^k \rho^M \, d\mu_+(M) \leq \int_{n_{N-1}+\varepsilon'}^{M_k} M^k \rho^M \, d\mu_+(M)$$

$$+ \int_{M_k}^\infty \rho^M (1+\varepsilon')^M \, d\mu_+(M)$$

$$< \infty,$$

which concludes the proof.   □

28

Polynomial Preservers on Matrices with Real or Complex Entries

Having discussed in detail the case of matrices with entries in $(0, \rho)$, we conclude this part of the text with a brief study of entrywise polynomials preserving positivity in fixed dimension – but now on matrices with possibly negative or even complex entries. The first observation is that noninteger powers can no longer be applied, so we restrict ourselves to polynomials. Second, as discussed following the proof of Lemma 24.1, it is not possible to obtain structured results along the same lines as above, for all matrices in $\mathbb{P}_N((-\rho, \rho))$, for every polynomial of the form

$$t\left(c_{n_0} x^{n_0} + \cdots + c_{n_{N-1}} x^{n_{N-1}}\right) - x^M$$

acting entrywise.

The way one now proceeds is as follows. Akin to Chapters 26 and 27, the analysis begins by bounding from above the ratio $s_{\mathbf{n}_j}(\mathbf{u})^2 / s_{\mathbf{n}}(\mathbf{u})^2$ on the domain – in this case, on $[-\rho, \rho]^N$. Since the numerator and denominator both vanish at the origin, a sufficient condition to proceed would be that the zero locus of the denominator $s_{\mathbf{n}}(\cdot)$ is contained in the zero locus of $s_{\mathbf{n}_j}(\cdot)$ for every j. Since the choice of $M > n_{N-1}$ is arbitrary, we therefore try to seek the best possible solution: namely, that $s_{\mathbf{n}}(\cdot)$ *does not vanish on* $\mathbb{R}^N \setminus \{\mathbf{0}\}$. And indeed, it is possible to completely characterize all such tuples $\mathbf{n}$:

Theorem 28.1   *Fix integers $N \geq 2$ and $0 \leq n_0 < \cdots < n_{N-1}$. The following are equivalent:*

(1) *The Schur polynomial $s_{\mathbf{n}}(\cdot) \colon \mathbb{R}^N \to \mathbb{R}$ is positive except possibly at the origin.*
(2) *The Schur polynomial $s_{\mathbf{n}}(\cdot) \colon \mathbb{R}^N \to \mathbb{R}$ is nonvanishing except possibly at the origin.*

(3) *The Schur polynomial $s_{\mathbf{n}}(\cdot)$ does not vanish at the two vectors $\mathbf{e}_1$ and $\mathbf{e}_1 - \mathbf{e}_2$.*

(4) *The tuple $\mathbf{n}$ satisfies: $n_0 = 0, \ldots, n_{N-2} = N - 2$, and $n_{N-1} - (N - 1) = 2r \geq 0$ is an even integer.*

Using Littlewood's definition (25.1), it is easy to see that such a polynomial is precisely the complete homogeneous symmetric polynomial (of even degree $k = 2r$)

$$h_k(u_1, u_2, \ldots) := \sum_{1 \leq j_1 \leq j_2 \leq \cdots \leq j_k} u_{j_1} u_{j_2} \ldots u_{j_k}, \qquad \forall u_j \in \mathbb{R}$$

for $k \geq 0$, where we set $h_0(u_1, u_2, \ldots) \equiv 1$.

In this chapter, we will prove Theorem 28.1 and apply it to study entrywise polynomial preservers of positivity over $\mathbb{P}_N((-\rho, \rho))$. We then study such preservers of $\mathbb{P}_N(D(0, \rho))$.

28.1 Complete Homogeneous Symmetric Polynomials Are Positive

The major part of Theorem 28.1 is to show that the polynomials h_{2r} do not vanish outside the origin. This is a result by Hunter in 1977 [129]. More strongly, Hunter showed that these polynomials are always positive, with a strict lower bound:

Theorem 28.2 (Hunter, 1977 [129])   *Fix integers $r, N \geq 1$. Then we have*

$$h_{2r}(\mathbf{u}) \geq \frac{\|\mathbf{u}\|^{2r}}{2^r r!}, \qquad \mathbf{u} \in \mathbb{R}^N \tag{28.1}$$

with equality if and only if (a) $\min(r, N) = 1$ and (b) $\sum_{j=1}^{N} u_j = 0$.

The proof uses two observations, also made by Hunter in the same work.

Lemma 28.3 (Hunter, 1977 [129])   *Given integers $k, N \geq 1$, and $\mathbf{u} = (u_1, \ldots, u_N)^T \in \mathbb{R}^N$,*

$$\frac{h_k(\mathbf{u}, \xi) - h_k(\mathbf{u}, \eta)}{\xi - \eta} = h_{k-1}(\mathbf{u}, \xi, \eta)$$

for all real $\xi \neq \eta$; and moreover,

$$\frac{\partial h_k}{\partial u_j}(\mathbf{u}) = h_{k-1}(\mathbf{u}, u_j).$$

Proof   Recall from the definition that

$$h_k(\mathbf{u}) = \sum_{s=0}^{k} h_{k-s}(u_1, \ldots, u_{N-1}) u_N^s. \tag{28.2}$$

With (28.2) at hand, the first assertion follows immediately (in fact over any ground field)

$$\frac{h_k(\mathbf{u},\xi) - h_k(\mathbf{u},\eta)}{\xi - \eta} = \sum_{s=0}^{k} h_{k-s}(\mathbf{u}) \frac{\xi^s - \eta^s}{\xi - \eta} = \sum_{s=1}^{k} h_{k-s}(\mathbf{u}) \sum_{t=0}^{s-1} \xi^t \eta^{s-1-t}$$

$$= h_{k-1}(\mathbf{u},\xi,\eta).$$

We isolate the final equality here:

$$h_{k-1}(\mathbf{u},\xi,\eta) = \sum_{s=1}^{k} h_{k-s}(\mathbf{u}) \sum_{t=0}^{s-1} \xi^t \eta^{s-1-t}, \tag{28.3}$$

noting that it holds at all $\xi, \eta \in \mathbb{R}$. Next, we show the second assertion. Since h_k is a symmetric polynomial, it suffices to work with $j = N$. Now compute using (28.2) and (28.3)

$$\frac{\partial h_k}{\partial u_j}(\mathbf{u}) = \sum_{s=1}^{k} h_{k-s}(u_1, \ldots, u_{N-1})\left(s u_N^{s-1}\right)$$

$$= \sum_{s=1}^{k} h_{k-s}(u_1, \ldots, u_{N-1}) \sum_{t=0}^{s-1} u_N^t u_N^{s-1-t} = h_{k-1}(\mathbf{u}, u_N). \qquad \square$$

With Lemma 28.3 at hand, we proceed.

Proof of Theorem 28.2   If $N = 1 \le r$, it is easy to see that (28.1) holds if and only if $u_1 = 0$. Similarly, if $r = 1$, then

$$h_2(\mathbf{u}) = \frac{1}{2}\left[\|\mathbf{u}\|^2 + \left(\sum_{j=1}^{N} u_j\right)^2\right] \ge \frac{\|\mathbf{u}\|^2}{2},$$

with equality if and only if $\sum_{j=1}^{N} u_j = 0$, as desired.

Henceforth we suppose that $r, N \ge 2$, and claim by induction on r that (28.1) holds, with a strict inequality. To show the claim, note that since $h_{2r}(\mathbf{u})$ is homogeneous in $\mathbf{u}$ of total degree $2r$, it suffices to show (28.1) on the unit sphere

$$h_{2r}(\mathbf{u}) > \frac{1}{2^r r!}, \qquad \mathbf{u} \in S^{N-1}.$$

We are thus interested in optimizing (in fact minimizing) the smooth function $h_{2r}(\mathbf{u})$, subject to the constraint $\sum_{j=1}^{N} u_j^2 = 1$. This problem is amenable to the use of Lagrange multipliers, and we obtain that at any extreme point $\mathbf{y} \in S^{N-1}$, there exists $\lambda \in \mathbb{R}$ satisfying

$$\frac{\partial h_{2r}}{\partial u_j}(\mathbf{y}) + 2\lambda y_j = 0, \qquad j = 1, \ldots, N.$$

Multiply this equation by y_j and sum over all j; since h_{2r} is homogeneous of total degree $2r$, Euler's equation yields

$$2r h_{2r}(\mathbf{y}) + 2\lambda \|\mathbf{y}\|^2 = 0 \qquad \Longrightarrow \qquad \lambda = -r h_{2r}(\mathbf{y}).$$

With this at hand, compute using Lemma 28.3

$$h_{2r-1}(\mathbf{y}, y_j) = \frac{\partial h_{2r}}{\partial u_j}(\mathbf{y}) = -2\lambda y_j = 2r h_{2r}(\mathbf{y}) y_j, \qquad j = 1, \ldots, N. \quad (28.4)$$

We now show that at all points $\mathbf{y} \in S^{N-1}$ satisfying (28.4), one has (28.1) with a strict inequality. As one of these points is the global minimum, this would prove the result.

There are two cases. First, the vectors $\mathbf{y}_{\pm} := \frac{\pm 1}{\sqrt{N}} \mathbf{1}_{N \times 1} \in S^{N-1}$ satisfy (28.4); it may help here to observe that the number of terms/monomials in $h_k(u_1, \ldots, u_N)$ is $\binom{N+k-1}{k}$. This observation also implies that at these points $\mathbf{y}_{\pm}$, we have

$$h_{2r}(\mathbf{y}_{\pm}) = \binom{N + 2r - 1}{2r} \frac{1}{N^{2r}} = \frac{(N + 2r - 1)(N + 2r - 2) \cdots N}{N^{2r}} \frac{1}{(2r)!}$$

$$> \frac{1}{(2r)!} > \frac{1}{2^r r!},$$

and this yields (28.1). Otherwise, $\mathbf{y} \neq \mathbf{y}_{\pm}$ has at least two unequal coordinates, say $y_j \neq y_k$, and satisfies (28.4), hence:

$$h_{2r-1}(\mathbf{y}, y_j) - h_{2r-1}(\mathbf{y}, y_k) = 2r h_{2r}(\mathbf{y})(y_j - y_k).$$

Rewriting this and using Lemma 28.3,

$$h_{2r}(\mathbf{y}) = \frac{1}{2r} \frac{h_{2r-1}(\mathbf{y}, y_j) - h_{2r-1}(\mathbf{y}, y_k)}{y_j - y_k} = \frac{1}{2r} h_{2r-2}(\mathbf{y}, y_j, y_k)$$

$$\geq \frac{1}{2r} \frac{\|\mathbf{y}\|^2 + |y_j|^2 + |y_k|^2}{2^{r-1}(r-1)!},$$

where the final inequality follows from the induction hypothesis. Now since $y_j \neq y_k$, the final numerator is strictly greater than 1, and this yields (28.1). □

Theorem 28.2 allows us to prove the existence of polynomials with negative coefficients that entrywise preserve positivity in a fixed dimension. This is discussed presently; we first show for completeness that the polynomials h_{2r} are the only ones that vanish only at the origin.

Proof of Theorem 28.1 That (4) $\implies$ (1) follows directly from Theorem 28.2, and that (1) $\implies$ (2) $\implies$ (3) is immediate. Now suppose (3) holds. Using Littlewood's definition (25.1), if a tableau T of shape $\mathbf{n} - \mathbf{n}_{\min}$ has two nonempty rows, then in any semistandard filling of T, one is forced to use at least two different variables. Now evaluating the weight of T at $\mathbf{e}_1$ yields zero. This argument shows that (3) implies $\mathbf{n} - \mathbf{n}_{\min}$ has at most one row, so by (25.1), $s_{\mathbf{n}}(\mathbf{u}) = h_k(\mathbf{u})$ for some $k \geq 0$. Now $h_k(\mathbf{e}_1 - \mathbf{e}_2)$ is easily evaluated to be a geometric series (consisting of $k+1$ alternating entries 1 and -1). This vanishes if k is odd, so (3) implies k is even, proving (4). $\square$

28.2 Application: Entrywise Polynomials Preserving Positivity

With the above results at hand, we now prove:

Theorem 28.4 *Fix integers $N \geq 1$, $k, r \geq 0$, and $M \geq N + 2r$, as well as positive constants $\rho, c_0, \ldots, c_{N-1}$. There exists a positive constant $t_0 > 0$ such that the polynomial*

$$p_t(x) := t x^k \left(c_0 + c_1 x + \cdots + c_{N-2} x^{N-2} + c_{N-1} x^{N-1+2r} \right) - x^{k+M}$$

entrywise preserves positivity on $\mathbb{P}_N([-\rho, \rho])$ whenever $t \geq t_0$.

Proof The result for $k = 0$ implies that for arbitrary $k \geq 0$, by the Schur product theorem. Thus, we henceforth assume $k = 0$. We now prove the result by induction on $N \geq 1$ with the $N = 1$ case left to the reader as an exercise.

For the induction step, notice that the proof of Proposition 25.6 goes through for $\mathbf{u} \in \mathbb{R}^{N,\neq}$ as long as $s_{\mathbf{n}}(\mathbf{u}) \neq 0$. This is indeed the case if $\mathbf{n} = (0, 1, \ldots, N-2, N-1+2r)$, by Theorem 28.2. Thus, to produce a threshold t_1 as in the theorem, which works for all rank-1 matrices, it suffices to show (by the discussion prior to Theorem 26.1, and using the density of $(-\sqrt{\rho}, \sqrt{\rho})^{N,\neq}$ in $[-\sqrt{\rho}, \sqrt{\rho}]^N$) that

$$\sup_{\mathbf{u} \in (-\sqrt{\rho}, \sqrt{\rho})^{N,\neq}} \sum_{j=0}^{N-1} \frac{s_{\mathbf{n}_j}(\mathbf{u})^2}{h_{2r}(\mathbf{u})^2} \frac{\rho^{M-n_j}}{c_{n_j}} < \infty.$$

In turn, using Theorem 28.2, it suffices to show

$$\sup_{\mathbf{u}\in(-\sqrt{\rho},\,\sqrt{\rho})^{N,\neq}} \frac{s_{\mathbf{n}_j}(\mathbf{u})^2}{\|\mathbf{u}\|^{4r}} < \infty, \qquad j = 0, 1, \ldots, N-1.$$

Now since the polynomial $s_{\mathbf{n}_j}$ is homogeneous of total degree $2r + M - n_j$,

$$\frac{s_{\mathbf{n}_j}(\mathbf{u})^2}{\|\mathbf{u}\|^{4r}} = s_{\mathbf{n}_j}(\mathbf{u}/\|\mathbf{u}\|)^2 \|\mathbf{u}\|^{2(M-n_j)} \leq K_{\mathbf{n}_j}^2 (N\rho)^{M-n_j}$$

for $\mathbf{u} \in (-\sqrt{\rho},\,\sqrt{\rho})^{N,\neq}$, where $K_{\mathbf{n}_j}$ is the maximum of the Schur polynomial $s_{\mathbf{n}_j}(\cdot)$ on the unit sphere S^{N-1}.

This shows the existence of a threshold t_1 that proves the theorem for all rank-1 matrices in $\mathbb{P}_N([-\rho,\rho])$. We will prove the result for all matrices in $\mathbb{P}_N([-\rho,\rho])$ by applying Theorem 5.9; for this, we first note that

$$M^{-1}p_t'(x) = t\left(\sum_{j=1}^{N-2} \frac{jc_j}{M}x^{j-1} + \frac{(N-1+2r)c_{N-1}}{M}x^{N-2+2r}\right) - x^{M-1}$$

is again of the same form as in the theorem. Hence, by the induction hypothesis, there exists a threshold t_2 such that $p_t'[-]$ preserves positivity on $\mathbb{P}_{N-1}([-\rho,\rho])$ for $t \geq t_2$. The induction step is now complete by taking $t_0 := \max(t_1, t_2)$. $\qquad\square$

A natural question that remains, in parallel to the study of polynomial positivity preservers of matrices in $\mathbb{P}_N([0,\rho])$, is as follows:

Question 28.5   Given the data as in the preceding theorem, find the sharp constant t_0.

A first step toward this goal is the related question in rank 1, which can essentially be rephrased as follows:

Question 28.6   Given integers $r \geq 0$, $N \geq 1$, and

$$m_0 \geq 0, \quad m_1 \geq 1, \quad \ldots, \quad m_{N-2} \geq N-2, \quad m_{N-1} \geq N-1+2r,$$

maximize the ratio $\dfrac{s_{\mathbf{m}}(\mathbf{u})^2}{h_{2r}(\mathbf{u})^2}$ over the punctured unit cube $[-1,1]^N \setminus \{\mathbf{0}\}$.

Remark 28.7   Notice by homogeneity that this ratio of squares increases as one travels radially from the origin. Thus, the maximization on the punctured solid cube is equivalent to the same question on the boundary of this cube.

28.3 Matrices with Complex Entries

The final topic along this theme is to explore matrices with complex entries, say in the open disk $D(0, \rho)$ (or its closure) for some $0 < \rho < \infty$. In this case, the set of admissible "initial sequences of powers" $0 \leq n_0 < \cdots < n_{N-1}$ turns out to be far more limited – and (the same) tight threshold bound is available in all such cases:

Theorem 28.8   *Fix integers $M \geq N \geq 2$ and $k \geq 0$, and let $n_j = j + k$ for $0 \leq j \leq N - 1$ – i.e., N consecutive integers. Also fix real scalars $\rho > 0$, $c_0, \ldots, c_{N-1}$, and define*

$$f(z) := z^k (c_0 + c_1 z + \cdots + c_{N-1} z^{N-1}) + c_M z^{k+M}, \qquad z \in \mathbb{C}.$$

Then the following are equivalent:

(1) *The entrywise map $f[-]$ preserves positivity on $\mathbb{P}_N(D(0, \rho))$.*
(2) *The map $f[-]$ preserves positivity on rank-1 totally nonnegative (TN) Hankel matrices in $\mathbb{P}_N((0, \rho))$.*
(3) *Either all $c_j, c_M \geq 0$; or $c_j > 0$ for all $j < N$ and $c_M \geq -\mathcal{C}^{-1}$, where*

$$\mathcal{C} = \sum_{j=0}^{N-1} \frac{V(\mathbf{n}_j)^2}{c_j V(\mathbf{n}_{\min})^2} \rho^{M-j},$$

where $\mathbf{n}_{\min} := (0, 1, \ldots, N - 1)$ and $\mathbf{n}_j := (0, 1, \ldots, j - 1, j + 1, \ldots, N - 1, M)$ for $0 \leq j \leq N - 1$.

A chronological remark: this result was the first instance of entrywise polynomial positivity preservers with negative coefficients to be discovered, in 2016. The more refined and challenging sharp bound for arbitrary polynomials (or tuples of real powers $\mathbf{n}$) operating on $\mathbb{P}_N((0, \rho))$, as well as the existence of a tight threshold for the leading term of a polynomial preserver operating on $\mathbb{P}_N((-\rho, \rho))$, were worked out later – though in this text, we have already proved those results.

Remark 28.9   After proving Theorem 28.8, we will also show that if the initial sequence $\mathbf{n}$ of nonnegative integer powers is nonconsecutive (i.e., not of the form in Theorem 28.8), then such a "structured" result does not hold for infinitely many powers $M > n_{N-1}$.

Proof of Theorem 28.8   Clearly, (1) $\implies$ (2). Next, notice that the constant $\mathcal{C}$ in (3) remains unchanged under a simultaneous shift of all exponents by the same amount k. Thus, (2) $\implies$ (3) by Theorem 27.1 (and Lemma 24.1).

It remains to show that (3) $\implies$ (1). Since the $k \geq 0$ case follows from the $k = 0$ case of (1) by the Schur product theorem, we assume henceforth that $k = 0$. Now we proceed as in Chapters 26 and 27, by first showing the result for rank-1 matrices, and then using an analogue of the Extension principle 5.9 to extend to all ranks via induction on N. The first step here involves extending Lemma 25.7 to complex matrices:

Lemma 28.10   Fix $\mathbf{w} \in \mathbb{C}^N$ and a positive definite (Hermitian) matrix $H \in \mathbb{C}^{N \times N}$. Define the linear pencil $P_t := tH - \mathbf{w}\mathbf{w}^*$, for $t > 0$. Then the following are equivalent:

(1) P_t is positive semidefinite.
(2) $\det P_t \geq 0$.
(3) $t \geq \mathbf{w}^* H^{-1} \mathbf{w} = 1 - \dfrac{\det(H - \mathbf{w}\mathbf{w}^*)}{\det H}$.

The proof is virtually identical to that of Lemma 25.7 and is hence omitted.

Next, using that $s_{\mathbf{n}}(\mathbf{u}^*) = \overline{s_{\mathbf{n}}(\mathbf{u})}$ for all integer tuples $\mathbf{n}$ and all vectors $\mathbf{u} \in \mathbb{C}^N$ (and $s_{\mathbf{n}_{\min}}(\mathbf{u}) \equiv 1$), we apply Lemma 28.10 to computing the sharp threshold bound for a single "generic" rank-1 complex matrix, parallel to how Proposition 25.6 is an adaptation of Lemma 25.7:

Proposition 28.11   With the given positive scalars c_j, and integers $M \geq N \geq 2$ and $n_j = j - 1$, define

$$p_t(z) := t \sum_{j=0}^{N-1} c_j z^j - z^M, \qquad t \in (0, \infty), \; z \in \mathbb{C}.$$

Then the following are equivalent for $\mathbf{u} \in \mathbb{C}^{N, \neq}$:

(1) $p_t[\mathbf{u}\mathbf{u}^*]$ is positive semidefinite.
(2) $\det p_t[\mathbf{u}\mathbf{u}^*] \geq 0$.
(3) $t \geq \displaystyle\sum_{j=0}^{N-1} \dfrac{|s_{\mathbf{n}_j}(\mathbf{u})|^2}{c_j}$.

Once again, the proof is omitted.

We continue to repeat the approach for $\mathbb{P}_N((0, \rho))$ in previous chapters. By the discussion prior to Theorem 26.1, and using the density of $D(0, \sqrt{\rho})^{N, \neq}$ in $\overline{D(0, \sqrt{\rho})}^N$, we next compute

$$\sup_{\mathbf{u} \in D(0, \sqrt{\rho})^{N, \neq}} |s_{\mathbf{n}_j}(\mathbf{u})^2|, \qquad 0 \leq j \leq N - 1.$$

Use Littlewood's definition (25.1) of $s_\mathbf{n}(\cdot)$, and the triangle inequality, to conclude that

$$|s_\mathbf{n}(\mathbf{u})| = \left| \sum_T \prod_{j=1}^N u_j^{f_j(T)} \right| \leq \sum_T \prod_{j=1}^N |u_j|^{f_j(T)} = s_\mathbf{n}(|\mathbf{u}|),$$

where $|\mathbf{u}| := (|u_1|, \ldots, |u_N|)$. Thus, equality is indeed attained here if one works with a vector $\mathbf{u} \in (0, \sqrt{\rho})^{N,\neq}$. For this reason, and since $s_\mathbf{n}(\mathbf{u})$ is coordinatewise nondecreasing on $(0, \infty)^N$,

$$\sup_{\mathbf{u} \in D(0, \sqrt{\rho})^{N,\neq}} |s_{\mathbf{n}_j}(\mathbf{u})^2| = s_{\mathbf{n}_j}(\sqrt{\rho}(1, \ldots, 1))^2 = \frac{V(\mathbf{n}_j)^2}{V(\mathbf{n}_{\min})^2} \rho^{M-j}, \qquad \forall j.$$

Akin to $\mathbb{P}_N((0, \rho))$, we conclude that $p_t[\mathbf{uu}^*] \in \mathbb{P}_N$ for all $\mathbf{u} \in D(0, \sqrt{\rho})^N$, if and only if

$$t \geq C = \sum_{j=0}^{N-1} \frac{V(\mathbf{n}_j)^2}{c_j V(\mathbf{n}_{\min})^2} \rho^{M-j}.$$

The final step is to prove the result for all matrices in $\mathbb{P}_N(D(0, \rho))$, not just those of rank 1. For this we work by induction on $N \geq 1$, with the base case following from above. For the induction step, we will apply the extension principle (Theorem 5.9); to do so, we first extend that result as follows, with essentially the same proof.

Lemma 28.12   *Theorem 5.9 holds if $h(z)$ is a polynomial and $I = D(0, \rho)$ or its closure.*

To apply this result, first note that

$$M^{-1} p_t'(z) = t \sum_{j=1}^{N-1} M^{-1} j c_j z^{j-1} - z^{M-1},$$

and so it suffices to show that this preserves positivity on $\mathbb{P}_{N-1}(D(0, \rho))$ if $t \geq C$. By the induction hypothesis, it suffices to show that $C \geq C'$, where C' is the constant obtained from $M^{-1} p_t'$

$$C' = \sum_{j=1}^{N-1} \frac{M V(\mathbf{n}_j')^2}{j c_j V(\mathbf{n}_{\min}')^2} \rho^{M-j},$$

where $\mathbf{n}_j' := (0, 1, \ldots, j-2, j, \ldots, N-2, M-1)$, and $\mathbf{n}'_{\min} := (0, 1, \ldots, N-2)$. Thus, to show that $\mathcal{C} \geq \mathcal{C}'$, it suffices to show that

$$\frac{V(\mathbf{n}_j)^2}{V(\mathbf{n}_{\min})^2} \geq \frac{M V(\mathbf{n}'_j)^2}{j V(\mathbf{n}'_{\min})^2}$$

for $j = 1, \ldots, N-1$. This is not hard to show; e.g., for "most" cases of j, a straightforward computation yields

$$\left(\frac{V(\mathbf{n}_j)/V(\mathbf{n}'_j)}{V(\mathbf{n}_{\min})/V(\mathbf{n}'_{\min})} \right)^2 = \left(\frac{(N-1)! \, M/j}{(N-1)!} \right)^2 = \frac{M^2}{j^2} > \frac{M}{j}. \qquad \square$$

As promised above, we conclude by showing that for every other tuple of "initial powers," i.e., nonconsecutive powers $\mathbf{n}$, one cannot always have a positivity preserver with a negative coefficient – even on generic one-parameter families of rank-1 matrices.

Theorem 28.13 *Fix integers $N \geq 2$ and $0 \leq n_0 < \cdots < n_{N-1}$, where the n_j are not all consecutive. Also fix $N-1$ distinct numbers $u_1, \ldots, u_{N-1} > 0$, and set*

$$\mathbf{u}(z) := (u_1, \ldots, u_{N-1}, z)^T \in \mathbb{C}^N, \qquad z \in \mathbb{C}.$$

Then there exists $z_0 \in \mathbb{C}$ and infinitely many integers $M > n_{N-1}$, such that for all choices of (a) scalar $\epsilon > 0$ and (b) coefficients $c_{n_0}, \ldots, c_{n_{N-1}} > 0 > c' \in \mathbb{R}$, the polynomial

$$f(z) := c_{n_0} z^{n_0} + \cdots + c_{n_{N-1}} z^{n_{N-1}} + c' z^M$$

does not preserve positivity on the rank-1 matrix $\epsilon \mathbf{u}(z_0)\mathbf{u}(z_0)^*$ when applied entrywise.

Note that if instead all $c_{n_j}, c_M \geq 0$, then $f[-]$ preserves positivity by the Schur product theorem; while if some $c_{n_j} < 0$, then the FitzGerald–Horn argument from Theorem 5.3 can be adapted to show that $f[\epsilon \mathbf{u}(u_N)\mathbf{u}(u_N)^*] \notin \mathbb{P}_N$ for all sufficiently small $\epsilon > 0$, where $u_N \in \mathbb{C}$ is such that the nonzero polynomial $s_{\mathbf{n}}(u_1, \ldots, u_{N-1}, u_N) \neq 0$.

Proof   Since the n_j are not all consecutive, the tableau-shape corresponding to $\mathbf{n} - \mathbf{n}_{\min}$ has at least one row with two cells. It follows by Littlewood's definition (25.1) that $s_{\mathbf{n}}(\mathbf{u})$ has at least two monomials. Now consider $s_{\mathbf{n}}(\mathbf{u}(z))$ as a function only of z, say $g(z)$. Then $g(z)$ is a polynomial that is not a constant multiple of a monomial, so it has a nonzero complex root $z_0 \in \mathbb{C}^\times$. Notice that z_0 is also not in $(0, \infty)$ because the Schur polynomial evaluated at $(u_1, \ldots, u_{N-1}, u_N)$ is positive for every $u_N \in (0, \infty)$. Thus, $z_0 \in \mathbb{C} \setminus [0, \infty)$.

By choice of z_0 and Cauchy's definition of $s_{\mathbf{n}}(\mathbf{u}(z_0))$ (see Proposition 25.5),

$$\mathbf{u}(z_0)^{\circ \mathbf{n}} = [\mathbf{u}(z_0)^{\circ n_0} | \cdots | \mathbf{u}(z_0)^{\circ n_{N-1}}]$$

is a singular matrix. That said, this matrix has rank $N - 1$ by the properties of generalized Vandermonde determinants (see Theorem 4.1); in fact, every subset of $N - 1$ columns here is linearly independent. Let V_0 denote the span of these columns; then the orthocomplement $V_0^\perp \subset \mathbb{C}^N$ is one-dimensional, i.e., there exists unique $\mathbf{v} \in \mathbb{C}^N$ up to rescaling, such that $\mathbf{v}^* \mathbf{u}(z_0)^{\circ n_j} = 0 \ \forall j$.

Now given any N consecutive integers $l + 1, \ldots, l + N$ with $l \geq n_{N-1}$, we claim there exists an integer $M \in [l + 1, l + N]$ such that $\mathbf{v}^* \mathbf{u}(z_0)^{\circ M} \neq 0$. Indeed, the *usual* Vandermonde matrix

$$\left[\mathbf{u}(z_0)^{\circ(l+1)} | \cdots | \mathbf{u}(z_0)^{\circ(l+N)} \right]$$

is nonsingular (since no coordinate in $\mathbf{u}(z_0)$ is zero), so at least one column $\mathbf{u}(z_0)^{\circ M} \notin V_0$. In particular, $\mathbf{v}^* \mathbf{u}(z_0)^{\circ M} \neq 0$, proving the claim.

Finally, choose arbitrary $\epsilon, c_{n_j} > 0 > c'$ as in the theorem. We then assert that $f[\epsilon \mathbf{u}(z_0) \mathbf{u}(z_0)^*]$, where f is defined using this value of M, is not positive semidefinite. Indeed

$$\mathbf{v}^* f[\epsilon \mathbf{u}(z_0) \mathbf{u}(z_0)^*] \mathbf{v} = \sum_{j=0}^{N-1} c_{n_j} \epsilon^{n_j} |\mathbf{v}^* \mathbf{u}(z_0)^{\circ n_j}|^2 + c' \epsilon^M |\mathbf{v}^* \mathbf{u}(z_0)^{\circ M}|^2$$

$$= c' \epsilon^M |\mathbf{v}^* \mathbf{u}(z_0)^{\circ M}|^2,$$

and this is negative, proving that $f[\epsilon \mathbf{u}(z_0) \mathbf{u}(z_0)^*] \notin \mathbb{P}_N$.   □

29

Cauchy and Littlewood's Definitions
of Schur Polynomials

For completeness, in this chapter we show the equivalence of *four* definitions
of Schur polynomials, two of which are named identities. To proceed, first
recall two other families of symmetric polynomials: the *elementary symmetric
polynomials* are simply

$$e_1(u_1, u_2, \ldots) := u_1 + u_2 + \cdots,$$

$$e_2(u_1, u_2, \ldots) := u_1 u_2 + u_1 u_3 + u_2 u_3 + \cdots,$$

and in general,

$$e_k(u_1, u_2, \ldots) := \sum_{1 \le j_1 < j_2 < \cdots < j_k} u_{j_1} u_{j_2} \ldots u_{j_k}.$$

These symmetric functions crucially feature while decomposing polynomials
into linear factors.

We also recall the *complete homogeneous symmetric polynomials*

$$h_k(u_1, u_2, \ldots) := \sum_{1 \le j_1 \le j_2 \le \cdots \le j_k} u_{j_1} u_{j_2} \ldots u_{j_k}.$$

By convention, we set $e_0 = h_0 = 1$, and $e_k = h_k = 0$ for $k < 0$. Now we
have:

Theorem 29.1 *Fix an integer $N \ge 1$ and any unital commutative ground
ring. Given a partition of N – i.e., an N-tuple of nonincreasing nonnegative
integers $\lambda = (\lambda_1 \ge \cdots \ge \lambda_N)$ with $\sum_j \lambda_j = N$ – the following four
definitions give the same expression $s_{\overline{\lambda + \delta}}(u_1, u_2, \ldots, u_N)$, where $\delta := (N - 1,
N - 2, \ldots, 0)$ and $\overline{\lambda + \delta} = (\lambda_N, \lambda_{N-1} + 1, \ldots, \lambda_1 + N - 1)$ in our convention.*

(1) *(Littlewood's definition) The sum of weights over all column-strict Young tableaux of shape λ with cell entries $u_1, \ldots, u_N$.*

(2) *(Cauchy's definition, aka the type A Weyl character formula) The ratio of the (generalized) Vandermonde determinants $a_{\lambda+\delta}/a_\delta$, where*
$$a_\lambda := \det(u_j^{\lambda_k + N - k}).$$

(3) *(The Jacobi–Trudi identity) The determinant $\det(h_{\lambda_j - j + k})_{j,k=1}^N$.*

(4) *(The dual Jacobi–Trudi identity, or von Nägelsbach–Kostka identity) The determinant $\det(e_{\lambda'_j - j + k})$, where λ' is the dual partition, meaning*
$$\lambda'_k := \#\{j : \lambda_j \geq k\}.$$

From this result, we deduce the equivalence of these definitions of the Schur polynomial for fewer numbers of variables $u_1, \ldots, u_n$, where $n \leq N$.

Corollary 29.2   *Suppose $1 \leq r < N$ and $\lambda_{r+1} = \cdots = \lambda_N = 0$. Then the four definitions in Theorem 29.1 agree for the smaller set of variables $u_1, \ldots, u_r$.*

Proof   Using fewer numbers of variables in definitions (3) and (4) amounts to specializing the remaining variables $u_{r+1}, \ldots, u_N$ to zero. The same holds for definition (1) since weights involving the extra variables $u_{r+1}, \ldots, u_N$ now get set to zero. It follows that definitions (1), (3), and (4) agree for fewer numbers of variables.

We will show that Cauchy's definition (2) in Theorem 29.1 has the same property. In this case the definitions are different: Given $u_1, \ldots, u_r$ for $1 \leq r \leq N$, the corresponding ratio of alternating polynomials would only involve $\lambda_1 \geq \cdots \geq \lambda_r$, and would equal $\det\left(u_j^{\lambda_k + r - k}\right)_{j,k=1}^r / \det(u_j^{r-k})_{j,k=1}^r$. Now claim that this equals the ratio in (2), by downward induction on $r \leq N$. Note that it suffices to show the claim for $r = N - 1$. But here, if we set $u_N := 0$, then both generalized Vandermonde matrices have last column $(0, \ldots, 0, 1)^T$. In particular, we may expand along their last columns. Now canceling the common factors of $u_1 \ldots u_{N-1}$ from each of the previous columns reduces to the case of $r = N - 1$, and the proof is completed by similarly continuing inductively.   □

The remainder of this chapter is devoted to proving Theorem 29.1. We will show that (4) $\Longleftrightarrow$ (1) $\Longleftrightarrow$ (3) $\Longleftrightarrow$ (2), and over the ground ring $\mathbb{Z}$, which then carries over to arbitrary ground rings. To do so, we use an idea due to Karlin and Macgregor (1959, [137, 138]), Lindström (1973, [157]), and Gessel and Viennot (1985, [98]), which interprets determinants in terms of tuples of weighted lattice paths. The approach below is taken from the work of Bressoud and Wei (1993, [53]).

Proposition 29.3   *The definitions (1) and (3) are equivalent.*

Proof   The proof is divided into steps, for ease of exposition.

Step 1   In this step we define the formalism of lattice paths and their weights. Define points in the plane

$$P_k := (N - k + 1, N), \quad Q_k := (N - k + 1 + \lambda_k, 1), \quad k = 1, 2, \dots, N,$$

and consider (ordered) N-tuples $\mathbf{p}$ of (directed) lattice paths satisfying the following properties:

(1) The kth path starts at some P_j and ends at Q_k, for each k.
(2) No two paths start at the same point P_j.
(3) From P_j, and at each point (a, b), a path can go either east or south.
    Weight each east step at height (a, b) by u_{N+1-b}.

Notice that one can assign a unique permutation $\sigma = \sigma_{\mathbf{p}} \in S_N$ to each tuple of paths $\mathbf{p}$, so that paths go from $P_{\sigma(k)}$ to Q_k for each k.

We now assign a weight to each tuple $\mathbf{p}$, defined to be $(-1)^{\sigma_{\mathbf{p}}}$ times the product of the weights at all east steps in $\mathbf{p}$. For instance, if $\lambda = (3, 1, 1, 0, 0)$ partitions $N = 5$, then here is a typical tuple of paths:

- For $k = 4, 5$, P_k and Q_k are each connected by vertical straight lines (i.e., four south steps each).
- P_2 and Q_3 are connected by a vertical straight line (i.e., four south steps).
- The steps from P_3 to Q_2 are $SESESS$.
- The steps from P_1 to Q_1 are $SEESSES$.

This tuple $\mathbf{p}$ corresponds to the permutation $\sigma_{\mathbf{p}} = (13245)$, and has weight $-u_2^3 u_3 u_4$.

Step 2   The next goal is to examine the *generating function* of the tuples, i.e., $\sum_{\mathbf{p}} \text{wt}(\mathbf{p})$. Note that given σ, among all tuples $\mathbf{p}$ with $\sigma_{\mathbf{p}} = \sigma$, the kth path contributes a monomial of total degree $\lambda_k - k + \sigma(k)$, which can be any monomial in $u_1, \dots, u_N$ of this total degree. It follows that the generating function equals

$$\sum_{\mathbf{p}} \text{wt}(\mathbf{p}) = \sum_{\sigma \in S_N} (-1)^{\sigma} \prod_{k=1}^{N} h_{\lambda_k - k + \sigma(k)} = \det(h_{\lambda_k - k + j})_{j,k=1}^{N}.$$

Step 3   We next rewrite the above generating function to obtain $\sum_T \text{wt}(T)$ (the sum of weights over all column-strict Young tableaux of shape λ with cell entries $u_1, \ldots, u_N$), which is precisely $s_{\overline{\lambda+\delta}}(u_1, \ldots, u_N)$ by definition. To do so, we will pair off the tuples $\mathbf{p}$ of *intersecting* paths into pairs, whose weights cancel one another.

Suppose $\mathbf{p}$ consists of intersecting paths. Define the *final intersection point* of $\mathbf{p}$ to be the lattice point with maximum x-coordinate where at least two paths intersect, and if there are more than one such points, then the one with minimal y-coordinate. Now claim that exactly two paths in $\mathbf{p}$ intersect at this point. Indeed, if three paths intersect at any point, then all of them have to go either east or south at the next step. By the pigeonhole principle, there are at least two paths that proceed in the same direction. It follows that a point common to three paths in $\mathbf{p}$ cannot be the final intersection point, as desired.

Define the *tail* of $\mathbf{p}$ to be the two paths to the east and south of the final intersection point in $\mathbf{p}$. Given an intersecting tuple of paths $\mathbf{p}$, there exists a unique other tuple $\mathbf{p}'$ with the same final intersection point between the same two paths, but with the tails swapped. It is easy to see that the paths $\mathbf{p}$ and $\mathbf{p}'$ satisfy have opposite signs (for their permutations $\sigma_{\mathbf{p}}, \sigma_{\mathbf{p}'}$), but the same monomials in their weights. Therefore, $\text{wt}(\mathbf{p}) = -\text{wt}(\mathbf{p}')$, and the intersecting paths pair off, as desired.

Step 4 From Step 3, the generating function $\sum_{\mathbf{p}} \text{wt}(\mathbf{p})$ equals the sum over only tuples of nonintersecting paths. Each of these tuples necessarily has $\sigma_{\mathbf{p}} = \text{id}$, so all signs are positive. In such a tuple, the monomial weight for the kth path naturally corresponds to a weakly increasing sequence of λ_k integers in $[1, N]$. That the paths do not intersect corresponds to the entries in the kth sequence being strictly smaller than the corresponding entries in the $(k+1)$st sequence. This yields a natural weight-preserving bijection from the tuples of nonintersecting paths to the "column-strict" Young tableaux of shape λ with cell entries $1, \ldots, N$. (Notice that these tableaux are in direct bijection to the column-strict Young tableaux studied earlier in this part, by switching the cell entries $j \longleftrightarrow N + 1 - j$.) This concludes the proof. □

Proposition 29.4 *The definitions (1) and (4) are equivalent.*

Proof The proof is a variant of that of Proposition 29.3. Now we consider all tuples of paths such that the kth path goes from $P_{\sigma(k)}$, to the point

$$Q'_k := (N - k + 1 + \lambda'_k, 1),$$

and, moreover, each of these paths has at most one east step at each fixed height – i.e., no two east steps are consecutive.

Once again, in summing to obtain the generating function, given a permutation $\sigma = \sigma_\mathbf{p}$, the kth path in $\mathbf{p}$ contributes a monomial of total degree $\lambda'_k - k + \sigma(k)$, but now runs over all monomials with individual variables of degree at most 1 – i.e., all monomials in $e_{\lambda_k - k + \sigma(k)}$. It follows that

$$\sum_\mathbf{p} \mathrm{wt}(\mathbf{p}) = \sum_{\sigma \in S_N} (-1)^\sigma \prod_{k=1}^N e_{\lambda'_k - k + \sigma(k)} = \det(e_{\lambda'_k - k + j})_{j,k=1}^N.$$

On the other side, we once again pair off tuples – this time, leaving the ones that do not *overlap*. In other words, paths in tuples may intersect at a point, but do not share an east/south line segment. Now given a tuple containing two overlapping paths, define the *final overlap segment* similarly as in Proposition 29.3; as in the previous proof, notice that exactly two paths overlap on this segment. Then for every tuple of paths $\mathbf{p}$ that overlaps, there exists a unique other tuple $\mathbf{p}'$ with the same final overlap segment between the same two paths, but with the (new version of) tails swapped. It is easy to see that $\mathbf{p}$ and $\mathbf{p}'$ have the same monomials as weights, but with opposite signs, so they pair off and cancel weights.

This leaves us with tuples of nonoverlapping paths, all of which again corresponding to $\sigma_\mathbf{p} = \mathrm{id}$. In such a tuple, from the kth path we obtain a strictly increasing sequence of λ'_k integers in $[1, N]$. That the paths do not overlap corresponds to the entries in the kth sequence being at most as large as the corresponding entries in the $(k + 1)$st sequence. This gives a bijection to the *conjugates* of column-strict Young tableaux of shape λ, and hence we once again have $\sum_\mathbf{p} \mathrm{wt}(\mathbf{p}) = \sum_T \mathrm{wt}(T)$ in this setting.   □

Corollary 29.5   *Schur polynomials are symmetric and homogeneous.*

Proof   This follows because definition (4) is symmetric and homogeneous in the variables u_j.   □

Finally, we show:

Proposition 29.6   *The definitions (2) and (3) are equivalent.*

Proof   Once again, this proof is split into steps, for ease of exposition. In the proof below, we use the above results and assume that the definitions (1), (3), and (4) are all equivalent. Thus, our goal is to show that

$$\det\left(u_j^{N-k}\right)_{j,k=1}^N \cdot \det\left(h_{\lambda_j - j + k}\right)_{j,k=1}^N = \det\left(u_j^{\lambda_k + N - k}\right)_{j,k=1}^N.$$

Step 1   We explain the formalism, which is a refinement of the one in the proof of Proposition 29.3. Thus, we return to the setting of paths between $P_k = (N - k + 1, N)$ and $Q_k = (N - k + 1 + \lambda_k, 1)$ for $k = 1, \ldots, N$, but now *equipped also with a permutation* $\tau \in S_N$. The weight of an east step now depends on its height: at height $N + 1 - b$ an east step has weight $u_{\tau(b)}$ instead of u_b. Now consider tuples of paths over all τ; let us write their weights as $\mathrm{wt}_\tau(\mathbf{p})$ for notational clarity. In what follows, we also use $\mathbf{p}$ or $(\mathbf{p}, \tau)$ depending on the need to specify and work with $\tau \in S_N$.

For each fixed $\tau \in S_N$, notice first that the generating function $\sum_{\mathbf{p}} \mathrm{wt}_\tau(\mathbf{p})$ of the τ-permuted paths is independent of τ, by Corollary 29.5.

Now we define a new weight for these τ-permuted paths $\mathbf{p}$. Namely, given $\mathbf{p} = (\mathbf{p}, \tau)$, recall there exists a unique permutation $\sigma_{\mathbf{p}} \in S_N$; now define

$$\mathrm{wt}'_\tau(\mathbf{p}) := (-1)^\tau \mu(\tau) \cdot \mathrm{wt}_\tau(\mathbf{p}), \quad \text{where} \quad \mu(\tau) := u_{\tau(1)}^{N-1} u_{\tau(2)}^{N-2} \cdots u_{\tau(N-1)}.$$

The new generating function is

$$\sum_{\tau \in S_N} \sum_{\mathbf{p}} \mathrm{wt}'_\tau(\mathbf{p}) = \sum_{\tau \in S_N} (-1)^\tau \mu(\tau) \sum_{\mathbf{p}} \mathrm{wt}_\tau(\mathbf{p})$$
$$= \det\left(h_{\lambda_k - k + j}\right)_{j,k=1}^N \cdot \det\left(u_j^{N-k}\right)_{j,k=1}^N,$$

where the final equality follows from the above propositions, given that the inner sum is independent of τ from above.

Step 2   Say that a tuple $\mathbf{p} = (P_{\sigma_{\mathbf{p}}(k)} \to Q_k)_k$ is *high enough* if for every $1 \leq k \leq N$, the kth path has no east steps below height $N + 1 - k$. Now **claim** that (summing over all $\tau \in S_N$,) the τ-tuples that are not high enough once again pair up, with canceling weights.

Modulo the claim, we prove the theorem. The first reduction is that for a fixed τ, we may further restrict to the τ-tuples that are high enough *and* are nonintersecting (as in the proof of Proposition 29.3). Indeed, defining the final intersection point and the tail of $\mathbf{p}$ as in that proof, it follows that switching tails in tuples $\mathbf{p}$ of intersecting paths changes neither the monomial part of the weight, nor the high-enough property; and it induces the opposite sign to that of $\mathbf{p}$.

Thus, the generating function of all τ-tuples (over all τ) equals that of all nonintersecting, high-enough τ-tuples (also summed over all $\tau \in S_N$). But each such tuple corresponds to $\sigma_{\mathbf{p}} = \mathrm{id}$, and in it, all east steps in the first path must occur in the topmost row/height/y-coordinate of N. Hence, all east steps in the second path must occur in the next highest row, and so on. It follows that the nonintersecting, high-enough τ-tuples $\mathbf{p} = (\mathbf{p}, \tau)$ are in bijection with $\tau \in S_N$; moreover, each such tuple has weight $(-1)^\tau \mu(\tau) u_{\tau(1)}^{\lambda_1} u_{\tau(2)}^{\lambda_2} \cdots u_{\tau(N)}^{\lambda_N}$. Thus, the above generating function is shown to equal

$$\det \left(u_j^{\lambda_k + N - k} \right)_{j,k=1}^N,$$

and the proof is complete.

Step 3   It thus remains to show the claim in Step 2 above. Given parameters

$$\sigma \in S_N, \quad k \in [1, N], \quad j \in [1, N - k],$$

let $NH_{\sigma,k,j}$ denote the τ-tuples of paths $\mathbf{p} = (\mathbf{p}, \tau)$ (with τ running over S_N), which satisfy the following properties:

(1) $\mathbf{p}$ is not high (NH) enough.
(2) In $\mathbf{p}$, the kth path has an east step at most by height $N - k$, but the paths labeled $1, \ldots, k - 1$ are all high enough.
(3) Moreover, j is the height of the lowest east step in the kth path; thus, $j \in [1, N - k]$.
(4) The permutation associated to the start and end points of the paths in the tuple is $\sigma_\mathbf{p} = \sigma \in S_N$.

Note that the set NH of tuples of paths that are not high enough can be partitioned as

$$NH = \bigsqcup_{\sigma \in S_N, \ k \in [1,N], \ j \in [1,N-k]} NH_{\sigma,k,j}.$$

We now construct an involution of sets $\iota \colon NH \to NH$ which permutes each subset $NH_{\sigma,k,j}$, and such that $\mathbf{p}$ and $\iota(\mathbf{p})$ have the same monomial attached to them but different τ and τ', leading to canceling signs $(-1)^\tau \neq (-1)^{\tau'}$.

Thus, suppose $\mathbf{p}$ is a τ-tuple in $NH_{\sigma,k,j}$. Now define $\tau' := \tau \circ (N - j, N + 1 - j)$; in other words,

$$\tau'(i) := \begin{cases} \tau(i + 1), & \text{if } i = N - j; \\ \tau(i - 1), & \text{if } i = N - j + 1; \\ \tau(i), & \text{otherwise.} \end{cases}$$

In particular,

$$(-1)^{\tau'} = -(-1)^\tau \quad \text{and} \quad \mu(\tau') = \mu(\tau) u_{\tau(N+1-j)} u_{\tau(N-j)}^{-1}.$$

With τ' in hand, we can define the tuple $\iota(\mathbf{p}) = (\iota(\mathbf{p}), \tau') \in NH_{\sigma,k,j}$. First, change the weight of each east step at height $N + 1 - b$, from $u_{\tau(b)}$ to $u_{\tau'(b)}$.

Next, we keep unchanged the paths labeled $1, \ldots, k - 1$, and in the remaining paths we do not change the source and target nodes either (since σ is fixed). Notice that weights change at only two heights j and $j + 1$; hence the first $k - 1$ paths do not see any weights change.

The changes in the (other) paths are now described. In the kth path, change only the numbers n_l of east steps at height $l = j, j + 1$, via: $(n_j, n_{j+1}) \mapsto (n_{j+1} + 1, n_j - 1)$. Note, the product of weights of all east steps in this path changes by a multiplicative factor of $u_{\tau(N+1-j)}^{-1} u_{\tau(N-j)}$ – which cancels the above change from $\mu(\tau)$ to $\mu(\tau')$. Finally, in the mth path for each $m > k$, if n_l again denotes the number of east steps at height l, then we swap $n_j \longleftrightarrow n_{j+1}$ steps in the mth path. This leaves unchanged the weight of those paths, and hence of the tuple $\mathbf{p}$ overall.

It is now straightforward to verify that the map ι is an involution that preserves each of the sets $NH_{\sigma,k,j}$. Since $\mathrm{wt}(\iota(\mathbf{p})) = -\mathrm{wt}(\mathbf{p})$ for all $\mathbf{p} \in NH$, the claim in Step 2 is true, and the proof of the theorem is complete. $\square$

30

Exercises

Question 30.1 (Khare and Tao [146])   The goal in this question and the next is to prove the existence of entrywise polynomials preserving positivity on $\mathbb{P}_N((0, \infty))$ – i.e., where the domain is unbounded. We begin with a Horn–Loewner-type necessary condition:

Suppose $f(x) = \sum_{n \geq 0} c_n x^n$ is a convergent power series on $(0, \infty)$ with real coefficients c_n. Suppose $f[-]$ preserves positivity on rank-1 Hankel matrices in $\mathbb{P}_N((0, \infty))$. Prove that if some coefficient $c_{n_0} < 0$, then we necessarily have $c_n > 0$ for at least N values of $n < n_0$ and at least N values of $n > n_0$.

(Hint: One side follows from Lemma 24.1, and if the other fails, then it fails for f a polynomial – say of degree $d > 0$. Now consider $x^d f(1/x)$.)

Question 30.2 (Khare and Tao [146])   We now construct some "building blocks" – polynomials with negative coefficients – which entrywise preserve positivity on $\mathbb{P}_N((0, \infty))$ (and suitable infinite linear combinations of which can yield power series preservers). These building blocks are polynomials with precisely one negative coefficient and $2N$ positive ones. Thus, suppose

$$0 \leq n_0 < n_1 < \cdots < n_{N-1} < M < n_N < \cdots < n_{2N-1}$$

are integers, and $c_{n_j} > 0$ for $0 \leq j \leq 2N - 1$ are real scalars. Define

$$p_t(x) := t \sum_{j=0}^{2N-1} c_{n_j} x^{n_j} - x^M, \qquad t > 0.$$

(i) Using Theorem 26.1 (together with Lemma 25.7 and the Cauchy–Binet formula), prove that there exists $t > 0$, such that the entrywise map $p_t[-]$ preserves positivity on rank-1 matrices in $\mathbb{P}_N((0, \infty))$.

(ii) Using the extension principle (Theorem 5.9), show that there exists $t > 0$, such that $p_t[-]$ preserves positivity on all of $\mathbb{P}_N([0, \infty))$.

(iii) Now show, using Theorem 3.6, that the same scalar t as in the preceding part is such that $p_t[-]$ preserves the set HTN_n of totally nonnegative (TN) Hankel matrices.

Question 30.3 Recall that Theorem 27.1 provided a sharp bound for the "leading term" coefficient of a sum of real powers, to entrywise preserve positivity on $\mathbb{P}_N((0,\rho))$. The final part of that proof involved using the extension principle (Theorem 5.9) to show that the sharp bound that worked for rank-1 matrices also worked for all matrices. This was left as an exercise for the reader – check that it does work.

Question 30.4 Using the Cauchy–Binet formula, extend Proposition 25.6 to work with polynomials of the form

$$p_t(x) = \sum_{j=0}^{K-1} c_{n_j} x^{n_j} - x^M,$$

where $0 \le N \le K, 0 \le n_0 < n_1 < \cdots < n_{K-1} < M$ are integers, and all c_{n_j} are positive real scalars. The sharp bound of the threshold (for a single matrix $\mathbf{u}\mathbf{u}^T$, with $\mathbf{u}$ having pairwise distinct, positive coordinates) should be expressed in terms of Schur polynomials.

Question 30.5 (Generalized Rayleigh quotients) The existence of the sharp threshold bounds in Lemma 25.7 – and hence in the results in this part of the text – can be recast as an optimization problem involving generalized Rayleigh quotients. We begin in this problem with a "general" result.

Given an integer $N \ge 1$ and matrices $C, D \in \mathbb{P}_N(\mathbb{R})$, show that the following are equivalent:

(i) If $v^T C v = 0$ for some $v \in \mathbb{R}^N$, then $v^T D v = 0$.
(ii) $\ker C \subset \ker D$.
(iii) There exists $t > 0$, such that $t \cdot v^T C v \ge v^T D v$ for all $v \in \mathbb{R}^N$.

Now suppose $D \ne 0$, and let t_0 denote the smallest/"best" such constant t. Then t_0 equals the spectral radius (i.e., the largest eigenvalue) of a positive semidefinite matrix, and is also computable by maximizing a generalized Rayleigh quotient

$$t_0 = \sup_{v \notin \ker D} \frac{v^T D v}{v^T C v} = \varrho\left(C^{\dagger/2} D C^{\dagger/2}\right) = \varrho\left(X^T C^{\dagger} X\right),$$

where $C^{\dagger}$ is the Moore–Penrose inverse of C, $C^{\dagger/2}$ is its positive semidefinite square root, and X is any matrix, such that $D = XX^T$.

Question 30.6 (See [16, 145, 146])   Applying the previous question to entrywise polynomials preserving positivity on matrices in $\mathbb{P}_N$, we now deduce the sharp threshold bound for a single rank-1 matrix $\mathbf{u}\mathbf{u}^T$, where $\mathbf{u} \in (0, \infty)^N$ has pairwise distinct coordinates.

Fix an integer $N \geq 2$, real exponents $n_0 < \cdots < n_{N-1} < M$, and scalars $c_{n_j} > 0$ for all j. Now define

$$h(x) := \sum_{j=0}^{N-1} c_{n_j} x^{n_j}, \qquad x \in (0, \infty).$$

(i) Then $t \cdot h[A] \geq A^{\circ M}$ if and only if $t \geq \varrho(h[A]^{\dagger/2} A^{\circ M} h[A]^{\dagger/2})$.

(ii) If $A = \mathbf{u}\mathbf{u}^T$ has rank 1, with $\mathbf{u} \in (0, \infty)^N$, then show that
$t \cdot h[A] \geq A^{\circ M}$ if and only if $t \geq (\mathbf{u}^{\circ M})^T h[\mathbf{u}\mathbf{u}^T]^{\dagger} \mathbf{u}^{\circ M}$.

(iii) Suppose, moreover, that $\mathbf{u} \in (0, \infty)$ has distinct coordinates. Then show that $h[\mathbf{u}\mathbf{u}^T]$ is invertible, and the sharp threshold bound equals

$$\varrho(h[A]^{\dagger/2} A^{\circ M} h[A]^{\dagger/2}) = (\mathbf{u}^{\circ M})^T h[\mathbf{u}\mathbf{u}^T]^{-1} \mathbf{u}^{\circ M} = \sum_{j=0}^{N-1} \frac{(\det \mathbf{u}^{\circ \mathbf{n}_j})^2}{c_{n_j} (\det \mathbf{u}^{\circ \mathbf{n}})^2},$$

where $\mathbf{n}, \mathbf{n}_j$ were defined in (25.3). When the powers n_j, M are nonnegative integers, one can cancel $V(\mathbf{u})^2$ from the numerator and denominator in each summand, to obtain the desired threshold in terms of Schur polynomials.

(iv) The final equality in the preceding part is a purely algebraic fact that holds over any field. Namely, fix an integer $N \geq 1$ and a field $\mathbb{F}$ with at least $N + 1$ elements. Let $\mathbf{u}, \mathbf{v} \in (\mathbb{F}^\times)^{N, \neq}$, i.e., they each have pairwise distinct – and nonzero – coordinates. Also fix integers $n_0 < \cdots < n_{N-1} < M$ and nonzero scalars $c_{n_j} \in \mathbb{F}^\times$.

If the matrices $\mathbf{u}^{\circ \mathbf{n}}, \mathbf{v}^{\circ \mathbf{n}}$ are invertible, then show that $h[\mathbf{u}\mathbf{v}^T]$ is also invertible for h as above, and (using Cramer's rule) that

$$(\mathbf{v}^{\circ M})^T h[\mathbf{u}\mathbf{v}^T]^{-1} \mathbf{u}^{\circ M} = \sum_{j=0}^{N-1} \frac{\det \mathbf{u}^{\circ \mathbf{n}_j} \cdot \det \mathbf{v}^{\circ \mathbf{n}_j}}{c_{n_j} \cdot \det \mathbf{u}^{\circ \mathbf{n}} \cdot \det \mathbf{v}^{\circ \mathbf{n}}}.$$

In particular, if $n_0 \geq 0$, then the final expression can be written in terms of Schur polynomials as

$$= \sum_{j=0}^{N-1} \frac{s_{\mathbf{n}_j}(\mathbf{u}) s_{\mathbf{n}_j}(\mathbf{v})}{c_{n_j} s_{\mathbf{n}}(\mathbf{u}) s_{\mathbf{n}}(\mathbf{v})}.$$

Question 30.7 (Schur powers and generalized Vandermonde determinant ratios; see [19, 23] for this question and the next)   This question explores multiple settings in which (higher) Schur powers of a given matrix are "linear" combinations of a given set of (lower) Schur powers.

Fix a field $\mathbb{F}$ and a (possibly rectangular) matrix $A \in \mathbb{F}^{m \times N}$. Denote the rows of A by $\mathbf{a}_1^T, \ldots, \mathbf{a}_m^T$.

(i) Show for all integers $M \geq N$ that

$$A^{\circ M} = \sum_{j=0}^{N-1} (-1)^{N-j+1} D_{M,j}(A) A^{\circ j},$$

where $D_{M,j}(A)$ is a diagonal matrix with (k,k)-entry $s_{\mathbf{n}_j}(\mathbf{a}_k^T)$, with

$$\mathbf{n}_j := (0, 1, \ldots, j-1, j+1, \ldots, N-1, M), \qquad \forall 0 \leq j \leq N-1.$$

(ii) Now suppose $\mathbb{F} = \mathbb{R}$, and $A \in (0, \infty)^{m \times N}$ is such that each row of A has pairwise distinct (and positive) entries. Given real exponents $n_0 < \cdots < n_{N-1}$, show for all $M \in \mathbb{R}$ that

$$A^{\circ M} = \sum_{j=0}^{N-1} (-1)^{N-j+1} D'_{M,j}(A) A^{\circ n_j},$$

where $D'_{M,j}(A)$ is a diagonal matrix with (k,k)-entry $\dfrac{\det \mathbf{a}_k^{\circ \mathbf{n}_j}}{\det \mathbf{a}_k^{\circ \mathbf{n}}}$, with $\mathbf{n}, \mathbf{n}_j$ defined in (25.3).

Question 30.8   This is a continuation of the previous question. Thus, $\mathbb{F}$ is an arbitrary field and $A \in \mathbb{F}^{m \times N}$ is a matrix.

(i) Show that the simultaneous kernel of $A^{\circ 0} := \mathbf{1}_{m \times N}, A = A^{\circ 1}, \ldots, A^{\circ (N-1)}$ coincides with the simultaneous kernel of all entrywise integer powers $\{A^{\circ M} : M \geq 0\}$.

(ii) Now suppose $\mathbb{F} = \mathbb{R}$, and $A \in (0, \infty)^{m \times N}$ is such that each row of A has pairwise distinct (and positive) entries. Given real exponents $n_0 < \cdots < n_{N-1}$, show that the simultaneous kernel of the matrices $\{A^{\circ n_j} : 0 \leq j \leq N-1\}$ coincides with the simultaneous kernel of the entrywise real powers $\{A^{\circ M} : M \in \mathbb{R}\}$.

It is natural to try and compute this simultaneous kernel more explicitly. The next two questions explain how to do so when $\mathbb{F} = \mathbb{R}$ and A is in a distinguished class of symmetric matrices.

Question 30.9 (Belton et al. [16, 19])   Fix a real symmetric matrix $A_{N \times N}$.

(i) Show that there exists a unique coarsest partition $\pi_{\min}(A) = \{I_1, \ldots, I_n\}$ of $[N] = \{1, \ldots, N\}$, such that for each $1 \le j \le n$ all entries of the diagonal block $A_{I_j \times I_j}$ of A are equal.

(ii) For this partition, let $m_j := |I_j|$ and $\mathbf{m} := (m_1, \ldots, m_n)$. Then every off-diagonal block $A_{I_i \times I_j}$ (for $i \ne j$) also has all entries equal, so that $A = \Sigma_{\mathbf{m}}^{\uparrow}(C)$ for some $C \in \mathbb{R}^{n \times n}$. (The decompression operator $\Sigma_{\mathbf{m}}^{\uparrow}$ was defined in Question 10.16.)

(iii) Show that C has the same rank as A, and is r-PMP (for some $0 \le r \le n$; see Question 10.28) if and only if A is.

This result has been generalized to only require each diagonal block to have all its entries lie in a single G-orbit, where G is a fixed multiplicative subgroup of $\mathbb{C}^{\times}$. This yields a unique coarsest partition $\pi_{\min}(A)$ of $[n]$. If one further requires each diagonal block to have rank 1, then this yields a unique coarsest partition $\pi^G(A)$; moreover, $\pi_{\min}(A) = \pi^G(A)$ if G is a subgroup of the circle group $S^1 \subset \mathbb{C}^{\times}$. This also leads to the study of a family of Schubert cell-type stratifications of $\mathbb{P}_N$. See [19, 23] for details.

Question 30.10 (Simultaneous kernels; Belton et al. [19])   Suppose a matrix $A \in \mathbb{R}^{N \times N}$ is 3-PMP (i.e., has all 1×1, 2×2, 3×3 principal minors nonnegative). Let $\pi = \pi_{\min}(A) = \{I_1, \ldots, I_n\}$ be the partition defined in the previous question (i.e., for $G = \{1\}$). Show that the following spaces are equal:

(i) The simultaneous kernel of the Schur powers
$$A^{\circ 0} := \mathbf{1}_{N \times N}, A, \ldots, A^{\circ (N-1)}.$$

(ii) The simultaneous kernel of the Schur powers $A^{\circ M}$ for all $M \ge 0$.

(iii) The kernel of $J_\pi(A)$, the block diagonal matrix $\oplus_{j=1}^n \mathbf{1}_{I_j \times I_j}$.

Verify that this equivalence does not hold if A is not 3-PMP. For example, let $T_N \in \mathbb{R}^{N \times N}$ be the 2-PMP Jacobi/Toeplitz tridiagonal matrix, whose entries are $a_{jk} := \mathbf{1}_{|j-k| \le 1}$. (So $T_N = T_N^{\circ 2} = \cdots$.) For this matrix, show that $\pi_{\min}(T_N) = \{\{1\}, \ldots, \{N\}\}$, so that $\ker J_\pi = \{0\}$. However, for $N \equiv 2$ mod 3 with $N \ge 5$, verify that T_N is not 3-PMP and that $(1, -1, 0, 1, -1, 0, \ldots, 1, -1)^T$ lies in $\ker \mathbf{1}_{N \times N}$ and in $\ker T_N$.

Question 30.11   The equivalence in the previous problem shows that the simultaneous kernels of Hadamard powers of a single 3-PMP matrix (e.g., any positive semidefinite matrix) belong to only a *finite* set of subspaces of $\mathbb{R}^N$, of the form $\ker J_\pi$ for some partition π of $[N]$. This problem contrasts that situation with the case of the holomorphic functional calculus.

(i) Given a matrix $A \in \mathbb{R}^{N \times N}$, the simultaneous kernel of the "usual" powers of a matrix $A_{N \times N}$ is far easier to compute. Namely, show that $\bigcap_{n \geq n_0} \ker(A^n) = \ker A^{n_0}$ for any integer $n_0 \geq 0$.

(ii) Now fix an integer $n_0 \geq 1$. Constraining A now to be a positive semidefinite matrix, or even a projection operator, show that the simultaneous kernel in the previous part can equal any subspace $U \subset \mathbb{R}^N$.

Question 30.12   Verify that the proof of Proposition 25.6 goes through for $\mathbf{u} \in \mathbb{R}^{N,\neq}$ as long as $s_{\mathbf{n}}(\mathbf{u}) \neq 0$.

Question 30.13   Similarly, write down proofs of Lemma 28.10 and Proposition 28.11.

Question 30.14   Verify that the vectors $\mathbf{y} = \frac{\pm 1}{\sqrt{n}} \mathbf{1}_{N \times 1}$ indeed satisfy (28.4).

Question 30.15   We remind the reader of the open questions 28.5 and 28.6.

Question 30.16 (Belton et al. [18])   A natural question to ask is if – and when – the matrix inequality in Theorem 28.8 is an *equality*. Certainly, the negative constant must equal the threshold value, but even then, is equality attained? We explore this question here.

(i) First verify that if $N = 1$, then $g[A] = 0$ if and only if $A = (\rho)$.

For the remainder of this question, fix integers $M \geq N \geq 2$ and real scalars $c_0, \ldots, c_{N-1}, \rho > 0$. Let

$$g(z) = \sum_{j=0}^{N-1} c_j z^j - C z^M, \qquad \text{where } C = \sum_{j=0}^{N-1} \frac{V(\mathbf{n}_j)^2}{c_j V(\mathbf{n}_{\min})^2} \rho^{M-j},$$

with notation as in Theorem 28.8. Notice by the continuity of g that $g[-]$ preserves positivity on $\mathbb{P}_N(\overline{D}(0, \rho))$.

(ii) Suppose $A = \mathbf{u}\mathbf{u}^*$, with $\mathbf{u} \in \overline{D}(0, \sqrt{\rho})^{N,\neq}$ having pairwise distinct coordinates. Show that $\det g[\mathbf{u}\mathbf{u}^*] > 0$, so $g[\mathbf{u}\mathbf{u}^*]$ is not only nonzero, but, in fact, positive definite.

(iii) Now suppose $A \in \mathbb{P}_N(D(0, \rho))$ has a row $\mathbf{v}$ with distinct entries. Let $\mathbf{u} := v_0^{-1/2} \mathbf{v}$, where v_0 is the diagonal entry in $\mathbf{v}$, and show that $g[A] \geq g[\mathbf{u}\mathbf{u}^*]$, so $g[A]$ is also positive definite.

(iv) Finally, suppose $A \in \mathbb{P}_N(\overline{D}(0, \rho))$. Choose $\mathbf{u} \in \overline{D}(0, \sqrt{\rho})^{N,\neq}$, such that $\mathbf{u}\mathbf{u}^*$ and A have at least one diagonal entry in common. Conclude from an earlier part that $g[A] \neq 0_{N \times N}$.

Part I: Bibliographic Notes and References

Most of the material in the first four chapters of this text is standard, and can be found in other textbooks on matrix theory; see, e.g., Bapat and Raghavan [12], Bhatia [35, 36], Fallat and Johnson [80], Gantmacher [92], Hiai and Petz [120], Horn and Johnson [127, 128], Karlin [136], Pinkus [183], Zhan [242], and Zhang [244].

We now provide references for the rest of the material in Part I. The matrix factorization in (1.5) involving Schur complements was observed by Schur in [217]. Theorems 1.28 and 1.32, on the positivity of a block-matrix in terms of Schur complements, were shown by Albert [8]. Remark 2.7 on applications of Schur products to other areas is taken from discussions in the books [127, 128]. More broadly, a discussion of the legacy of Schur's contributions in analysis can be found in the comprehensive survey [73].

The Schur product theorem 2.10 was shown by Schur [216] (the proof involving Kronecker products is by Marcus and Khan [166]), and its nonzero lower bound, Theorem 2.12, is by Khare [142] (following a prior bound by Vybíral [233]). Remark 2.13 is by Vybíral [233]; and the previous nonzero lower bounds on the Schur product in (2.5) are from [83] and [190]. For more on the Hamburger and Stieltjes moment problems (see Remarks 1.20 and 3.9, respectively), see the monographs [7, 199, 219].

The notion of *TN* and *TP* matrices and kernels was introduced by Schoenberg in [200], where he showed that *TN* matrices satisfy the variation diminishing property. (Schoenberg then proved in [201] the Budan–Fourier theorem using *TN* matrices.) The characterization in Theorem 3.4 of this property is from Motzkin's thesis [174]. (Instances of total nonnegativity and of variation diminution had appeared in earlier works, e.g., by Fekete [82], Hurwitz [130], Laguerre [153], and others.) Theorem 3.6, relating positivity and total nonnegativity for a Hankel matrix, appears first in [183] for *TP*

Hankel matrices, then in detail in [81] for the TN, TP_p, and TN_p variants. (Neither of these works uses contiguous minors, which have the advantage of only needing to work with Hankel submatrices.) The lemmas used in the proof above are given in [92, 93], and the result of Fekete and its extension by Schoenberg are in [82] and [213], respectively. Corollary 3.8 – on the total nonnegativity of moment matrices of measures on $[0, \infty)$ – is the easy half of the Stieltjes moment problem, and was also proved differently, by Heiligers [113].

Theorem 4.1 on the total positivity of generalized Vandermonde matrices is found in [92]. The "weak" Descartes' rule of signs (Lemma 4.2) was first shown by Descartes in 1637 [71] for polynomials; the proof given in this text via Rolle's theorem is by Laguerre in 1883 [153] and holds equally well for the extension to real powers. See also Jameson's article [133] for a historical account of Descartes' rule of signs. The Basic Composition formula (4.5) can be found in the book by Pólya and Szegő [185] (see also Karlin [136]); the subsequent observations on the total positivity of "most" Hankel moment matrices are taken from [136].

Entrywise calculus was initiated by Schur in 1911, when he defined the map $f[A]$ (although he did not use this notation), in the same paper [216] where he proved the Schur product theorem. Schur also proved the first result involving entrywise maps; see also p. cxii of the survey [73].

Theorem 5.3 and Lemma 5.5, which help classify the Loewner positive powers, are by FitzGerald and Horn [84]. The use of the rank-2 Hankel matrices in the proof, as well as the powers preserving positive definiteness in Corollary 5.8, are by Fallat et al. [81]. The "individual" matrices encoding Loewner positive powers were constructed in Theorem 5.7 by Jain [131]; the extension principle in Theorem 5.9 is by Khare and Tao [146]. Also note the related papers by Bhatia and Elsner [37], Hiai [119], and Guillot et al. [104], which study "two-sided" powers: $\mathbb{R} \to \mathbb{R}$, and which of these are Loewner positive on $\mathbb{P}_n(\mathbb{R})$ for fixed $n \geq 1$.

Section 6.1 on the continuity of bounded midconvex functions is taken from the book of Roberts and Varberg [194]; the first main Theorem 6.2 there closely resembles a result by Ostrowski [180], while the second Theorem 6.4 was proved independently by Blumberg [40] and Sierpińsky [220]. Theorem 6.7, classifying the Loewner positive maps on $\mathbb{P}_2((0, \infty))$ and $\mathbb{P}_2([0, \infty))$, is essentially by Vasudeva [230]; see also [25, 108] for the versions that led to the present formulation. The short argument for midconvexity implying continuity, at the end of that proof, is due to Hiai [119]. Theorem 6.11, which classifies the powers preserving Hankel TN $n \times n$ matrices, is taken from [25].

Chapter 7 on the entrywise functions preserving positivity on $\mathbb{P}_G$ for G a noncomplete graph (specifically, a tree) follows Guillot et al. [105]. Chapter 8 classifying the Loewner positive powers on $\mathbb{P}_G$ for G a chordal graph – and hence, computing the critical exponent of G for Loewner positivity – is due to Guillot et al. [106] (see also the summary in [107]). The two exceptions are Theorem 8.7 by FitzGerald and Horn [84] and Theorem 8.8 by Guillot et al. [104], which classify the Loewner monotone and superadditive powers on $\mathbb{P}_n((0, \infty))$, respectively. Also see [134] for a survey of critical exponents in the matrix theory literature.

Theorem 9.1 and Corollary 9.6, about individual matrices encoding the sets of Loewner positive and monotone powers respectively, are by Jain [132]. The arguments proving these results are taken from [132] (some of these are variants of results in her earlier works) and from Khare [144] – specifically, the homotopy argument in Proposition 9.5, which differs from Jain's similar assertion in [132] and avoids SSR (strictly sign regular) matrices. Finally, the classification of the Loewner convex powers on $\mathbb{P}_n$ (i.e., the equivalence (1) $\Longleftrightarrow$ (3) in Theorem 9.8) was shown by Hiai [119] via the intermediate Proposition 9.9; see also Guillot et al. [104] for a rank-constrained version of Theorem 9.8. The further equivalence to Theorem 9.8(2), which obtains individual matrix-encoders of the Loewner convex powers, is taken from [144].

Part II: Bibliographic Notes and References

Lemma 11.1 is taken from the well-known monograph [185] by Pólya and Szegő. Theorems 11.2, 11.3, 11.4, 11.5, 11.6 and 11.7, classifying the dimension-free entrywise positivity preservers on various domains, are due to Schoenberg [206], Rudin [197], Vasudeva [230], Herz [118], and Christensen and Ressel [57, 58], respectively. The theorems of Schoenberg and Rudin and Vasudeva were recently shown by Belton et al. in [25] using significantly smaller test sets than all positive semidefinite matrices of all sizes; proving these results is the main focus of this part. These results turn out to be further useful in fully classifying the total positivity preservers on bi-infinite domains; see [22].

Loewner's theorem 11.8 on operator/matrix monotone functions is from [158]; see also Donoghue's book [72] and the recent monograph by Simon [221], which contains a dozen different proofs.

The results in distance geometry are but a sampling from the numerous works of Schoenberg. Theorem 11.10, relating Euclidean embedding of a metric and the conditional negativity of the corresponding squared-distance matrix, is from [202], following then-recent works by Menger [169, 170, 171] and Fréchet [90]. Schoenberg's theorem 11.14 (respectively, Proposition 11.15), characterizing Hilbert space (respectively, the Hilbert sphere) in terms of positive definiteness of the Gaussian family (respectively, the cosine), is from [205] (respectively, from [202]). We point out that Schoenberg proved these results more generally for separable (not just finite) metric spaces, as discussed in Chapter 22. See also Schoenberg's paper [204] and another with von Neumann [232] (and its related work [151] by Kolmogorov). For the works of Bochner in this context, we restrict ourselves to mentioning [45, 46]. Theorems 11.16 and 11.20 by Schoenberg on positive definite functions on spheres are from [206]. Also, see the survey [226] of positive definite functions by James Drewry Stewart (who is perhaps somewhat better known for his series of calculus textbooks).

While Schoenberg's motivations in arriving at his theorem lay in metric geometry, as described above, Rudin's motivations were from Fourier analysis. More precisely, Rudin was studying functions operating on spaces of Fourier transforms of L^1 functions on groups G, or of measures on G. Here, G is a locally compact abelian group equipped with its Haar measure; Rudin worked with the torus $G = \mathbb{T}$, while Kahane and Katznelson worked with its dual group $\mathbb{Z}$. These authors together with Helson proved [114] a remarkable result in a converse direction to Wiener–Levy theory in 1959. That same year, Rudin showed Schoenberg's theorem without the continuity hypothesis, i.e., Theorem 11.3. For more details on this part, on the metric geometric motivations of Schoenberg, and other topics, the reader is referred to the detailed recent twin surveys of Belton et al. [20, 21].

The Horn–Loewner Theorem 12.1 (in a special case) originally appeared in Horn's paper [126], where he attributes it to Loewner. The theorem has since been extended by Khare (jointly) in various ways; see, e.g., [25, 108]; a common, overarching generalization of these and other variants has been achieved in [143]. Horn–Loewner's determinant calculation in Proposition 12.4 was also extended to Proposition 12.6 by Khare [143] (see Question 23.10). The second, direct proof of Theorem 12.1 is essentially due to Vasudeva [230].

Mollifiers were introduced by Friedrichs [91], following the famous paper of Sobolev [224], and their basic properties can be found in standard textbooks in analysis, as can Cauchy's mean value theorem for divided differences. The remainder of the proof of the stronger Horn–Loewner theorem is from [126] and the Boas–Widder theorem 13.8(2) is from [44].

Bernstein's theorem 14.3 is from his well-known memoir on absolutely monotone functions [34]; see also Widder's textbook [238]. Boas theorem 14.8 on the analyticity of smooth functions with SSR (strictly sign regular) derivatives is from [43]. Hamburger's theorem 14.12 is a folklore result, found in standard reference books – see, e.g., [7, 199, 219]. The remainder of Chapter 14 is from Belton et al. [25]. Chapter 15 is taken from the same paper, with the exception of Theorem 15.5 which is new, as are "Proofs 1 and 2" of the existence of a positivity certificate/limiting s.o.s. representation for $(1 \pm t)(1 - t^2)^n$. The former proof cites a result by Berg et al. [29], and the latter, direct proof, is new.

Chapters 16 and 17 are again from [25], save for the standard Identity Theorem 16.4, and Proposition 16.2 on the closure of real analytic functions under composition; these can be found in, e.g., [152]. The complex analysis basics, including Montel and Morera's theorems, can be found in standard textbooks; we cite [63]. The multivariate Schoenberg–Rudin theorem was proved by FitzGerald et al. in [85], and subsequently, under significantly weaker hypotheses in [25].

Theorems 19.1 and 19.2 understanding and relating the Loewner positive, monotone, and convex maps, were originally proved without rank constraints on the test sets, by Hiai in [119]. Lemma 19.3 is partly taken from [119] and partly from Rockafellar's book – see [195, Theorems 24.1 and 25.3].

Chapter 20, classifying the dimension-free preservers of positivity when not acting on diagonal blocks, is from the recent work of Vishwakarma [231], with the exception of the textbook Proposition 20.4, and Theorem 20.1 by Guillot–Rajaratnam [110].

Chapter 21 on the Boas–Widder theorem is from [44] except for the initial observations. Boas and Widder mention Popoviciu [186] who had proved the same result previously, using unequally spaced difference operators. The very last "calculus" Proposition 21.8 can be found in standard textbooks.

Coming to Chapter 22: Theorem 22.1 is due to Menger [169]. Theorem 22.2 comes from various works of Schoenberg on metric geometry (cited two pages above). Theorem 22.3 was already known to experts at the time; we cite here Birkhoff's famous paper [39]. The first part of Proposition 22.7 was shown by Cayley [55], and features Cayley–Menger determinants. The proof of Theorem 22.9 can be found in numerous sources, including online. The results in Section 22.3 are taken from the sources mentioned in it (e.g., the proof of Corollary 22.14).

Part III: Bibliographic Notes and References

Most of the material in this part is taken from two papers: one by Belton et al. [16] (see also [17] and [18] for summaries), and the other by Khare and Tao [146] (see also its summary [145]). We list the remaining references. For preliminaries on Schur polynomials, the standard reference is Macdonald's monograph [164]. Theorem 27.3 on the coordinatewise monotonicity of Schur polynomial ratios is proved using a deep result of Lam et al. [154], following previous work by Skandera [222]. There are other ways to show this result, e.g., using Chebyshev blossoming as shown by Ait-Haddou in joint works [5, 6], or by Dodgson condensation (see [146]). Theorem 27.5 is taken in part from [146] (the coordinatewise nondecreasing property on $(0, \infty)_{\neq}^N$). We also remark that Equation (25.4), like Proposition 12.4, was recently extended to arbitrary polynomials and (formal) power series Khare in [143].

Theorem 28.2 and its proof are due to Hunter [129]. Chapter 29 follows Bressoud and Wei [53], relying on the earlier works of Karlin and McGregor [137, 138], Lindström [157], and Gessel and Viennot [98].

References

[1] Jim Agler and John Edward McCarthy. What can Hilbert spaces tell us about bounded functions in the bidisk? In *A Glimpse at Hilbert Space Operators* (S. Axler, P. Rosenthal, D. Sarason, Eds.), pp. 81–97. Operator Theory: Advances and Applications, Vol. 207, Basel: Birkhäuser, 2010.

[2] Jim Agler, John Edward McCarthy, and Nicholas John Young. *Operator Analysis: Hilbert space methods in complex analysis*. Cambridge Tracts in Mathematics, Vol. 219, Cambridge: Cambridge University Press, 2020.

[3] Michael Aissen, Albert Edrei, Isaac Jacob Schoenberg, and Anne M. Whitney. On the generating functions of totally positive sequences. *Proc. Natl. Acad. Sci. USA*, 37(5): 303–307, 1951.

[4] Michael Aissen, Isaac Jacob Schoenberg, and Anne M. Whitney. On the generating functions of totally positive sequences I. *J. d'Analyse Math.*, 2: 93–103, 1952.

[5] Rachid Ait-Haddou and Marie-Laurence Mazure. The fundamental blossoming inequality in Chebyshev spaces – I: Applications to Schur functions. *Found. Comput. Math.*, 18(1): 135–158, 2018.

[6] Rachid Ait-Haddou, Yusuke Sakane, and Taishin Nomura. Chebyshev blossoming in Müntz spaces: Toward shaping with Young diagrams. *J. Comput. Appl. Math.*, 247: 172–208, 2013.

[7] Naum Ilyich Akhiezer. *The classical moment problem and some related questions in analysis*. Translated by N. Kemmer. New York: Hafner Publishing, 1965.

[8] Arthur Albert. Conditions for positive and nonnegative definiteness in terms of pseudoinverses. *SIAM J. Appl. Math.*, 17(2): 434–440, 1969.

[9] Tsuyoshi Ando. Totally positive matrices. *Linear Algebra Appl.*, 90: 165–219, 1987.

[10] Zhi Dong Bai and Li-Xin Zhang. Semicircle law for Hadamard products. *SIAM J. Matrix Anal. Appl.*, 29(2): 473–495, 2007.

[11] Mihály Bakonyi and Hugo J. Woerdeman. *Matrix completions, moments, and sums of Hermitian squares*. Princeton Series in Applied Mathematics, Princeton, NJ: Princeton University Press, 2011.

[12] Ravindra B. Bapat and T.E.S. Raghavan. *Nonnegative matrices and applications*. Encyclopedia of Mathematics and Its Applications, Vol. 64, Cambridge: Cambridge University Press, 1997.

[13] Victor Simões Barbosa and Valdir Antonio Menegatto. Strict positive definiteness on products of compact two-point homogeneous spaces. *Integral Trans. Spec. Funct.*, 28(1): 56–73, 2017.

[14] Richard Keith Beatson and Wolfgang zu Castell. Dimension hopping and families of strictly positive definite zonal basis functions on spheres. *J. Approx. Theory*, 221: 22–37, 2017.

[15] Richard Keith Beatson, Wolfgang zu Castell, and Yuan Xu. A Pólya criterion for (strict) positive-definiteness on the sphere. *IMA J. Numer. Anal.*, 34(2): 550–568, 2014.

[16] Alexander Belton, Dominique Guillot, Apoorva Khare, and Mihai Putinar. Matrix positivity preservers in fixed dimension. I. *Adv. Math.*, 298: 325–368, 2016.

[17] Alexander Belton, Dominique Guillot, Apoorva Khare, and Mihai Putinar. Matrix positivity preservers in fixed dimension. *Comptes Rendus Math.*, 354(2): 143–148, 2016.

[18] Alexander Belton, Dominique Guillot, Apoorva Khare, and Mihai Putinar. Schur polynomials and positivity preservers (Extended abstract). In *FPSAC 2016 Proceedings* (Vol. BC), pp. 155–166. Nancy, France: Discrete Mathematics and Theoretical Computer Science (DMTCS), 2016.

[19] Alexander Belton, Dominique Guillot, Apoorva Khare, and Mihai Putinar. Simultaneous kernels of matrix Hadamard powers. *Linear Algebra Appl.*, 576: 142–157, 2019.

[20] Alexander Belton, Dominique Guillot, Apoorva Khare, and Mihai Putinar. A panorama of positivity. Part I: Dimension free. In *Analysis of Operators on Function Spaces* (The Serguei Shimorin Memorial Volume; A. Aleman, H. Hedenmalm, D. Khavinson, M. Putinar, Eds.), Parts 1 and 2 (unified), pp. 117–165. Trends in Mathematics, Basel: Birkhäuser, 2019. arXiv:math.CA/1812.05482

[21] Alexander Belton, Dominique Guillot, Apoorva Khare, and Mihai Putinar. A panorama of positivity. Part II: Fixed dimension. In *Complex Analysis and Spectral Theory*, proceedings of the CRM Workshop held at Laval University, QC, May 21–25, 2018 (G. Dales, D. Khavinson, J. Mashreghi, Eds.), Parts 1 and 2 (unified), pp. 109–150. CRM Proceedings – AMS Contemporary Mathematics, Vol. 743, Providence, RI: American Mathematical Society, 2020. arXiv: math.CA/1812.05482

[22] Alexander Belton, Dominique Guillot, Apoorva Khare, and Mihai Putinar. Totally positive kernels, Pólya frequency functions, and their transforms. *J. d'Analyse Math.*, in press; arXiv:math.FA/2006.16213

[23] Alexander Belton, Dominique Guillot, Apoorva Khare, and Mihai Putinar. Matrix compression along isogenic blocks. *arXiv*, 2020. arXiv:math.RA/2010.14429

[24] Alexander Belton, Dominique Guillot, Apoorva Khare, and Mihai Putinar. Hirschman–Widder densities. *arXiv*, 2020. arXiv:math.CA/2101.02129

[25] Alexander Belton, Dominique Guillot, Apoorva Khare, and Mihai Putinar. Moment-sequence transforms. *J. Eur. Math. Soc.*, published online, DOI: 10.4171/jems/1145

[26] Arkady Berenstein, Sergey Fomin, and Andrei Zelevinsky. Parametrizations of canonical bases and totally positive matrices. *Adv. Math.*, 122: 49–149, 1996.

[27] Arkady Berenstein and Andrei Zelevinsky. Total positivity in Schubert varieties. *Comment. Math. Helv.*, 72(1): 128–166, 1997.

[28] Arkady Berenstein and Andrei Zelevinsky. Tensor product multiplicities, canonical bases and totally positive varieties. *Invent. Math.*, 143: 77–128, 2001.

[29] Christian Berg, Jens Peter Reus Christensen, and Paul Ressel. Positive definite functions on abelian semigroups. *Math. Ann.*, 223(3): 253–274, 1976.

[30] Christian Berg, Jens Peter Reus Christensen, and Paul Ressel. *Harmonic Analysis on Semigroups: Theory of Positive Definite and Related Functions*. Graduate Texts in Mathematics, Vol. 100, New York: Springer, 1984.

[31] Christian Berg, Ana Paula Peron, and Emilio Porcu. Schoenberg's theorem for real and complex Hilbert spheres revisited. *J. Approx. Theory*, 228: 58–78, 2018.

[32] Christian Berg and Emilio Porcu. From Schoenberg coefficients to Schoenberg functions. *Constr. Approx.*, 45: 217–241, 2017.

[33] Serge Bernstein. *Reçons sur les Propriétés Extrémales et la Meilleure Approximation des Fonctions Analytiques d'une Variable Réelle*. Borel series of monographs, Paris: Gauthier-Villars, 1926.

[34] Serge Bernstein. Sur les fonctions absolument monotones. *Acta Math.*, 52(1): 1–66, 1929.

[35] Rajendra Bhatia. *Matrix analysis*. Graduate Texts in Mathematics, Vol. 169, New York: Springer, 1997.

[36] Rajendra Bhatia. *Positive definite matrices*. Princeton Series in Applied Mathematics, Princeton, NJ: Princeton University Press, 2015.

[37] Rajendra Bhatia and Ludwig Elsner. Positivity preserving Hadamard matrix functions. *Positivity*, 11(4): 583–588, 2007.

[38] Peter John Bickel and Elizaveta Levina. Covariance regularization by thresholding. *Ann. Statist.*, 36(6): 2577–2604, 2008.

[39] Garrett Birkhoff. Metric foundations of geometry. I. *Trans. Amer. Math. Soc.*, 55: 465–492, 1944.

[40] Henry Blumberg. On convex functions. *Trans. Amer. Math. Soc.*, 20: 40–44, 1919.

[41] Leonard Mascot Blumenthal. New theorems and methods in determinant theory. *Duke Math. J.*, 2(2): 396–404, 1936.

[42] Ralph Philip Boas Jr. The Stieltjes moment problem for functions of bounded variation. *Bull. Amer. Math. Soc.*, 45: 399–404, 1939.

[43] Ralph Philip Boas Jr. Functions with positive derivatives. *Duke Math. J.*, 8(1): 163–172, 1941.

[44] Ralph Philip Boas Jr. and David Vernon Widder. Functions with positive differences. *Duke Math. J.*, 7: 496–503, 1940.

[45] Salomon Bochner. Monotone funktionen, Stieltjessche integrale und harmonische analyse. *Math. Ann.*, 108(1): 378–410, 1933.

[46] Salomon Bochner. Hilbert distances and positive definite functions. *Ann. Math. (2)*, 42: 647–656, 1941.

[47] Rafaela Neves Bonfim, Jean Carlo Guella, and Valdir Antonio Menegatto. Strictly positive definite functions on compact two-point homogeneous spaces:

The product alternative. *Symm., Integr. Geom.: Meth. Appl. (SIGMA)*, 14(112): 14 pp., 2018.

[48] Carl de Boor. On calculating with *B*-splines. *J. Approx. Theory*, 6(1): 50–62, 1972.

[49] James Borger. Witt vectors, semirings, and total positivity. In *Absolute Arithmetic and* $\mathbb{F}_1$-*Geometry* (K. Thas, Ed.), pp. 273–331, Eur. Math. Soc., 2016.

[50] James Borger and Darij Grinberg. Boolean Witt vectors and an integral Edrei–Thoma theorem. *Selecta Math. (NS)*, 22: 595–629, 2016.

[51] Francesco Brenti. *Unimodal, log-concave, and Pólya frequency sequences in combinatorics*. Mem. Amer. Math. Soc., Vol. 413, Providence, RI: American Mathematical Society, 1989.

[52] Francesco Brenti. Combinatorics and total positivity. *J. Combin. Theory Ser. A*, 71(2): 175–218, 1995.

[53] David Marius Bressoud and Shi-Yuan Wei. Combinatorial equivalence of definitions of the Schur function. *AMS Contemp. Math.*, 143: 59–64, (A Tribute to Emil Grosswald: Number Theory and Related Analysis), 1993.

[54] Victor Matveevich Buchstaber and Aleksei Antonovich Glutsyuk. Total positivity, Grassmannian and modified Bessel functions. *AMS Contemp. Math.*, 733: 97–108 (Functional Analysis and Geometry: Selim Grigorievich Krein Centennial), 2019. http://dx.doi.org/10.1090/conm/733

[55] Arthur Cayley. On a theorem in the geometry of position. *Cambridge Math. J.*, II: 267–271, 1841.

[56] Debao Chen, Valdir Antonio Menegatto, and Xingping Sun. A necessary and sufficient condition for strictly positive definite functions on spheres. *Proc. Amer. Math. Soc.*, 131(9): 2733–2740, 2003.

[57] Jens Peter Reus Christensen and Paul Ressel. Functions operating on positive definite matrices and a theorem of Schoenberg. *Trans. Amer. Math. Soc.*, 243: 89–95, 1978.

[58] Jens Peter Reus Christensen and Paul Ressel. Positive definite kernels on the complex Hilbert sphere. *Math. Z.*, 180(2): 193–201, 1982.

[59] Henry Cohn and Matthew de Courcy-Ireland. The Gaussian core model in high dimensions. *Duke Math. J.*, 167(13): 2417–2455, 2018.

[60] Henry Cohn and Abhinav Kumar. Universally optimal distribution of points on spheres. *J. Amer. Math. Soc.*, 20(1): 99–148, 2007.

[61] Henry Cohn and Jeechul Woo. Three-point bounds for energy minimization. *J. Amer. Math. Soc.*, 25(4): 929–958, 2012.

[62] Henry Cohn and Yufei Zhao. Sphere packing bounds via spherical codes. *Duke Math. J.*, 163(10): 1965–2002, 2014.

[63] John Bligh Conway. *Functions of one complex variable*. Graduate Texts in Mathematics, Vol. 11, New York: Springer, 1978.

[64] Trevor F. Cox and Michael A.A. Cox. Multidimensional scaling. In *Handbook of Data Visualization* (C.-h. Chen, W. Härdle, and A. Unwin, Eds.), pp. 315–347, Berlin-Heidelberg: Springer, 2008. http://dx.doi.org/10.1007/978-3-540-33037-0

[65] Colin Walker Cryer. The LU-factorization of totally positive matrices. *Linear Algebra Appl.*, 7(1): 83–92, 1973.

[66] Haskell Brooks Curry and Isaac Jacob Schoenberg. On spline distributions and their limits: The Pólya distribution functions. *Bull. Amer. Math. Soc.*, 53(11): 1114, 1947.

[67] Haskell Brooks Curry and Isaac Jacob Schoenberg. On Pólya frequency functions IV: The fundamental spline functions and their limits. *J. d'Analyse Math.*, 17: 71–107, 1966.

[68] Colin Walker Cryer. Some properties of totally positive matrices. *Linear Algebra Appl.*, 15(1): 1–25, 1976.

[69] Yuri L'vovich Dalestkii and Selim Grigorievich Krein. Integration and differentiation of functions of Hermitian operators and applications to the theory of perturbations (Russian). *Voronoež. Gos. Univ. Trudy Sem. Funkcional. Anal.*, Vol. 1, pp. 81–105, 1956; AMS Translations: Ser. 2, Vol. 47, pp. 1–30, 1965.

[70] Aleksey Aleksandrovich Davydov. Totally positive sequences and R-matrix quadratic algebras. *J. Math. Sci.*, 100(1): 1871–1876, 2000.

[71] Réne Descartes. Le Géométrie. Appendix to *Discours de la méthode*, 1637.

[72] William Francis Donoghue Jr. *Monotone matrix functions and analytic continuation*. Grundlehren der mathematischen Wissenschaften, Vol. 207, Berlin: Springer-Verlag, 1974.

[73] Harry Dym and Victor Emmanuilovich Katsnelson. Contributions of Issai Schur to analysis. In *Studies in Memory of Issai Schur* (A. Joseph, A. Melnikov, and R. Rentschler, Eds.), pp. xci–clxxxiii, Progress in Mathematics, Vol. 210, Basel: Birkhäuser, 2003.

[74] Albert Edrei. On the generating functions of totally positive sequences II. *J. d'Analyse Math.*, 2: 104–109, 1952.

[75] Albert Edrei. Proof of a conjecture of Schoenberg on the generating function of a totally positive sequence. *Canad. J. Math.*, 5: 86–94, 1953.

[76] Albert Edrei. On the generation function of a doubly infinite, totally positive sequence. *Trans. Amer. Math. Soc.*, 74: 367–383, 1953.

[77] Bradley Efron. Increasing properties of Pólya frequency functions. *Ann. Math. Statist.*, 36(1): 272–279, 1965.

[78] Noureddine El Karoui. Operator norm consistent estimation of large-dimensional sparse covariance matrices. *Ann. Statist.*, 36(6): 2717–2756, 2008.

[79] Jan Emonds and Hartmut Führ. Strictly positive definite functions on compact abelian groups. *Proc. Amer. Math. Soc.*, 139(3): 1105–1113, 2011.

[80] Shaun Michael Fallat and Charles Royal Johnson. *Totally nonnegative matrices*. Princeton Series in Applied Mathematics, Princeton, NJ: Princeton University Press, 2011.

[81] Shaun Michael Fallat, Charles Royal Johnson, and Alan David Sokal. Total positivity of sums, Hadamard products and Hadamard powers: Results and counterexamples. *Linear Algebra Appl.*, 520: 242–259, 2017; Corrigendum, *Linear Algebra Appl.*, 613: 393–396, 2021.

[82] Mihály Fekete and Georg Pólya. Über ein Problem von Laguerre. *Rend. Circ. Mat. Palermo*, 34: 89–120, 1912.

[83] Miroslav Fiedler and Thomas Markham. An observation on the Hadamard product of Hermitian matrices. *Linear Algebra Appl.*, 215: 179–182, 1995.

[84] Carl Hanson FitzGerald and Roger Alan Horn. On fractional Hadamard powers of positive definite matrices. *J. Math. Anal. Appl.*, 61(3): 633–642, 1977.

[85] Carl Hanson FitzGerald, Charles Anthony Micchelli, and Allan Pinkus. Functions that preserve families of positive semidefinite matrices. *Linear Algebra Appl.*, 221: 83–102, 1995.

[86] Sergey Fomin and Andrei Zelevinsky. Double Bruhat cells and total positivity. *J. Amer. Math. Soc.*, 12: 335–380, 1999.

[87] Sergey Fomin and Andrei Zelevinsky. Total positivity: Tests and parametrizations. *Math. Intelligencer*, 22(1): 23–33, 2000.

[88] Sergey Fomin and Andrei Zelevinsky. Cluster algebras. I. Foundations. *J. Amer. Math. Soc.*, 15(2): 497–529, 2002.

[89] Maurice René Fréchet. Les dimensions d'un ensemble abstrait. *Math. Ann.*, 68: 145–168, 1910.

[90] Maurice René Fréchet. Sur la définition axiomatique d'une classe d'espaces vectoriels distanciés applicables vectoriellement sur l'espace de Hilbert. *Ann. of Math. (2)*, 36(3): 705–718, 1935.

[91] Kurt Otto Friedrichs. The identity of weak and strong extensions of differential operators. *Trans. Amer. Math. Soc.*, 55: 132–151, 1944.

[92] Feliks Ruvimovich Gantmacher. *The theory of matrices. Vols. 1, 2.* Translated by K.A. Hirsch. New York: Chelsea Publishing Co., 1959.

[93] Feliks Ruvimovich Gantmacher and Mark Grigor'evich Krein. Sur les matrices complètement non négatives et oscillatoires. *Compositio Math.*, 4: 445–476, 1937.

[94] Feliks Ruvimovich Gantmacher and Mark Grigor'evich Krein. *Oscillation matrices and kernels and small vibrations of mechanical systems.* Translated by A. Eremenko. New York: Chelsea Publishing Co., 2002.

[95] Jürgen Garloff. Intervals of almost totally positive matrices. *Linear Algebra Appl.*, 363: 103–108, 2003.

[96] Jürgen Garloff and David G. Wagner. Hadamard products of stable polynomials are stable. *J. Math. Anal. Appl.*, 202(3): 797–809, 1996.

[97] Mariano Gasca and Charles A. Micchelli (Eds.) *Total positivity and its applications.* Proceedings of Jaca Meeting (1994), Mathematics and its Applications, Vol. 359, Dordrecht: Kluwer, 1996.

[98] Ira Gessel and Xavier Viennot. Binomial determinants, paths, and hook length formulae. *Adv. Math.*, 58(3): 300–321, 1985.

[99] Tillmann Gneiting. Strictly and non-strictly positive definite functions on spheres. *Bernoulli*, 19(4): 1327–1349, 2013.

[100] Kenneth Ralph Goodearl, Stephane Launois, and Tom H. Lenagan. Totally nonnegative cells and matrix Poisson varieties. *Adv. in Math.*, 226(1): 779–826, 2011.

[101] Karlheinz Gröchenig, José Luis Romero, and Joachim Stöckler. Sampling theorems for shift-invariant spaces, Gabor frames, and totally positive functions. *Invent. Math.*, 211: 1119–1148, 2018.

[102] Karlheinz Gröchenig and Joachim Stöckler. Gabor frames and totally positive functions. *Duke Math. J.*, 162(6): 1003–1031, 2013.

[103] Jean Carlo Guella, Valdir Antonio Menegatto, and Ana Paula Peron. Strictly positive definite kernels on a product of circles. *Positivity*, 21: 329–342, 2017.

[104] Dominique Guillot, Apoorva Khare, and Bala Rajaratnam. Complete characterization of Hadamard powers preserving Loewner positivity, monotonicity, and convexity. *J. Math. Anal. Appl.*, 425(1): 489–507, 2015.

[105] Dominique Guillot, Apoorva Khare, and Bala Rajaratnam. Preserving positivity for matrices with sparsity constraints. *Trans. Amer. Math. Soc.*, 368(12): 8929–8953, 2016.

[106] Dominique Guillot, Apoorva Khare, and Bala Rajaratnam. Critical exponents of graphs. *J. Combin. Theory Ser. A*, 139: 30–58, 2016.

[107] Dominique Guillot, Apoorva Khare, and Bala Rajaratnam. The critical exponent: a novel graph invariant. *Sem. Lothar. Combin.*, 78B, Article 62, 12 pp., 2017.

[108] Dominique Guillot, Apoorva Khare, and Bala Rajaratnam. Preserving positivity for rank-constrained matrices. *Trans. Amer. Math. Soc.*, 369(9): 6105–6145, 2017.

[109] Dominique Guillot and Bala Rajaratnam. Retaining positive definiteness in thresholded matrices. *Linear Algebra Appl.*, 436(11): 4143–4160, 2012.

[110] Dominique Guillot and Bala Rajaratnam. Functions preserving positive definiteness for sparse matrices. *Trans. Amer. Math. Soc.*, 367(1): 627–649, 2015.

[111] Jacques Salomon Hadamard. Théorème sur les séries entières. *Acta Math.*, 22: 55–63, 1899.

[112] Hans Ludwig Hamburger. Über eine Erweiterung des Stieltjesschen Momentenproblems. *Math. Ann.*, Part I, Vol. 81, pp. 235–319, 1920; Part II, Vol. 82, pp. 120–164, 1921; Part III, Vol. 82, pp. 168–187, 1921.

[113] Berthold Heiligers. Totally nonnegative moment matrices. *Linear Algebra Appl.*, 199 (suppl. 1): 213–227, 1994.

[114] Henry Helson, Jean-Pierre Kahane, Yitzhak Katznelson, and Walter Rudin. The functions which operate on Fourier transforms. *Acta Math.*, 102(1): 135–157, 1959.

[115] Alfred Hero and Bala Rajaratnam. Large-scale correlation screening. *J. Amer. Statist. Assoc.*, 106(496): 1540–1552, 2011.

[116] Alfred Hero and Bala Rajaratnam. Hub discovery in partial correlation graphs. *IEEE Trans. Inform. Theory*, 58(9): 6064–6078, 2012.

[117] Daniel Hershkowitz, Michael Neumann, and Hans Schneider. Hermitian positive semidefinite matrices whose entries are 0 or 1 in modulus. *Linear Multilinear Algebra*, 46(4): 259–264, 1999.

[118] Carl S. Herz. Fonctions opérant sur les fonctions définies-positives. *Ann. Inst. Fourier (Grenoble)*, 13: 161–180, 1963.

[119] Fumio Hiai. Monotonicity for entrywise functions of matrices. *Linear Algebra Appl.*, 431(8): 1125–1146, 2009.

[120] Fumio Hiai and Denes Petz. *Introduction to matrix analysis and applications.* Universitext, Cham: Springer, viii+332 pp., 2014.

[121] David Hilbert. Über die Darstellung definiter Formen als Summe von Formenquadraten. *Math. Ann.*, 32: 342–350, 1888.

[122] David Hilbert. Über ternäre definite Formen. *Acta Math.*, 17: 169–197, 1893.

[123] Isidore Isaac Hirschman Jr. and David Vernon Widder. The inversion of a general class of convolution transforms. *Trans. Amer. Math. Soc.*, 66(1): 135–201, 1949.

[124] Isidore Isaac Hirschman Jr. and David Vernon Widder. *The convolution transform.* Princeton Legacy Library, Princeton, NJ: Princeton University Press, 1955.

[125] Phùng Hô Hai. Poincaré series of quantum spaces associated to Hecke operators. *Acta Math. Vietnam.*, 24(2): 235–246, 1999.

[126] Roger Alan Horn. The theory of infinitely divisible matrices and kernels. *Trans. Amer. Math. Soc.*, 136: 269–286, 1969.

[127] Roger Alan Horn and Charles Royal Johnson. *Matrix analysis.* Cambridge: Cambridge University Press, 1985.

[128] Roger Alan Horn and Charles Royal Johnson. *Topics in matrix analysis.* Cambridge: Cambridge University Press, 1991.

[129] David Boss Hunter. The positive-definiteness of the complete symmetric functions of even order. *Math. Proc. Camb. Phil. Soc.*, 82(2): 255–258, 1977.

[130] Adolf Hurwitz. Über die Bedingungen, unter welchen eine Gleichung nur Wurzeln mit negativen reellen Theilen besitzt. *Math. Ann.*, 46: 273–284, 1895.

[131] Tanvi Jain. Hadamard powers of some positive matrices. *Linear Algebra Appl.*, 528: 147–158, 2017.

[132] Tanvi Jain. Hadamard powers of rank two, doubly nonnegative matrices. *Adv. in Oper. Theory*, 5: 839–849, 2020 (Rajendra Bhatia volume).

[133] Graham James Oscar Jameson. Counting zeros of generalised polynomials: Descartes' rule of signs and Laguerre's extensions. *The Mathematical Gazette*, 90(518): 223–234, 2006.

[134] Charles Royal Johnson and Olivia Walch. Critical exponents: Old and new. *Electr. J. Linear Alg.*, 25: 72–83, 2012.

[135] Samuel Karlin. Total positivity, absorption probabilities and applications. *Trans. Amer. Math. Soc.*, 111: 33–107, 1964.

[136] Samuel Karlin. *Total positivity.* Vol. 1. Stanford, CA: Stanford University Press, 1968.

[137] Samuel Karlin and James McGregor. Coincidence probabilities of birth and death processes. *Pacific J. Math.*, 9(4): 1109–1140, 1959.

[138] Samuel Karlin and James McGregor. Coincidence probabilities. *Pacific J. Math.*, 9(4): 1141–1164, 1959.

[139] Samuel Karlin and Yosef Rinott. Classes of orderings of measures and related correlation inequalities. I. Multivariate totally positive distributions. *J. Multivariate Anal.*, 10(4): 467–498, 1980.

[140] Samuel Karlin and Yosef Rinott. Total positivity properties of absolute value multinormal variables with applications to confidence interval estimates and related probabilistic inequalities. *Ann. Statist.*, 9(5): 1035–1049, 1981.

[141] Samuel Karlin and Zvi Ziegler. Chebyshevian spline functions. *SIAM J. Numer. Anal.*, 3(3): 514–543, 1966.

[142] Apoorva Khare. Sharp nonzero lower bounds for the Schur product theorem. *Proc. Amer. Math. Soc.* 149(12): 5049–5063, 2021.

[143] Apoorva Khare. Smooth entrywise positivity preservers, a Horn–Loewner master theorem, and symmetric function identities. *Trans. Amer. Math. Soc.*, in press, DOI: http://dx.doi.org/10.1090/tran/8563

[144] Apoorva Khare. Critical exponents for total positivity, individual kernel encoders, and the Jain–Karlin–Schoenberg kernel. *arXiv*, 2020. arXiv:math .FA/2008.05121

[145] Apoorva Khare and Terence Tao. Schur polynomials, entrywise positivity preservers, and weak majorization. *Sem. Lothar. Combin.*, 80B, Article 14, 12 pp., 2018.

[146] Apoorva Khare and Terence Tao. On the sign patterns of entrywise positivity preservers in fixed dimension. *Amer. J. Math.*, 143(6): 1863–1929, 2021.

[147] Jee Soo Kim and Frank Proschan. Total positivity. In *Encyclopedia of Statistical Sciences* (S. Kotz et al, Eds.), Vol. 14, New York: Wiley, pp. 8665–8672, 2006.

[148] Tinne Hoff Kjeldsen. The early history of the moment problem. *Historia Math.*, 20(1): 19–44, 1993.

[149] Yuji Kodama and Lauren Williams. KP solitons, total positivity and cluster algebras. *Proc. Natl. Acad. Sci. USA*, 108: 8984–8989, 2011.

[150] Yuji Kodama and Lauren Williams. KP solitons and total positivity for the Grassmannian. *Invent. Math.*, 198(3): 637–699, 2014.

[151] Andrey Nikolaevich Kolmogorov. Kurven in Hilbertschen Raum, die gegenüber einer einparametrigen Gruppe von Bewegungen invariant sind. *C.R. Doklady Acad. Sci. U.R.S.S.* 26: 6–9, 1940.

[152] Steven George Krantz and Harold Raymond Parks. *A primer of real analytic functions*. Birkhäuser Advanced Texts, Boston, MA: Birkhäuser, xiii+209 pp., 2002.

[153] Edmond Laguerre. Mémoire sur la théorie des équations numériques. *J. Math. Pures Appl.*, 9: 9–146, 1883.

[154] Thomas Lam, Alexander E. Postnikov, and Pavlo Pylyavskyy. Schur positivity and Schur log-concavity. *Amer. J. Math.*, 129(6): 1611–1622, 2007.

[155] Ai Li and Steve Horvath. Network neighborhood analysis with the multi-node topological overlap measure. *Bioinformatics*, 23(2): 222–231, 2007.

[156] Leo Liberti and Carlile Lavor. Six mathematical gems from the history of distance geometry. *Int. Trans. Oper. Res.*, 23(5): 897–920, 2016.

[157] Bernt Lindström. On the vector representations of induced matroids. *Bull. London Math. Soc.*, 5(1): 85–90, 1973.

[158] Charles Loewner. Über monotone Matrixfunktionen. *Math. Z.*, 38(1): 177–216, 1934.

[159] Charles Loewner. On totally positive matrices. *Math. Z.*, 63: 338–340, 1955 (Issai Schur memorial volume).

[160] Charles Loewner. Determination of the critical exponent of the Green's function. 1965 Symposium on Function Theory, Erevan, USSR, 1966.

[161] Charles Loewner. On schlicht-monotonic functions of higher order. *J. Math. Anal. Appl.*, 14(2): 320–325, 1966. http://dx.doi.org/10.1016/0022-247X(66)90033-3

[162] George Lusztig. Total positivity in reductive groups, Lie theory and geometry. *Progr. Math.*, Vol. 123, Boston, MA: Birkhäuser Boston, 1994, pp. 531–568.

[163] George Lusztig. Total positivity and canonical bases. In Algebraic Groups and Lie Groups (G.I. Lehrer, Ed.), *Austral. Math. Soc. Lect. Ser.*, Vol. 9, Cambridge: Cambridge University Press, 1997, pp. 281–295.

[164] Ian Grant Macdonald. *Symmetric functions and Hall polynomials.* Oxford Mathematical Monographs. New York: The Clarendon Press, Oxford University Press, (2nd ed.), 1995. With contributions by A. Zelevinsky, Oxford Science Publications.

[165] Ernest Maló. Note sur les équations algébriques dont toutes les racines sont réelles. *J. Math. Spéc.* (Ser. 4), 4: 7–10, 1895.

[166] Marvin David Marcus and Nisar A. Khan. A note on the Hadamard product. *Canad. Math. Bull.*, 2(2): 81–83, 1959.

[167] Valdir Antonio Menegatto and Claudemir Pinheiro Oliveira. Positive definiteness on products of compact two-point homogeneous spaces and locally compact abelian groups. *Canad. Math. Bull.*, 63(4): 705–715, 2020.

[168] Valdir Antonio Menegatto and Ana Paula Peron. Positive definite kernels on complex spheres. *J. Math. Anal. Appl.*, 254(1): 219–232, 2001.

[169] Karl Menger. Die Metrik des Hilbertschen Raumes. *Anz. Akad. Wissen. Wien, Math. Nat. Kl.*, 65: 159–160, 1928.

[170] Karl Menger. Untersuchungen über allgemeine Metrik. *Math. Ann.*: *Part I*: Vol. 100: 75–163, 1928; *Part II*: Vol. 103: 466–501, 1930.

[171] Karl Menger. New foundation of Euclidean geometry. *Amer. J. Math.*, 53(4): 721–745, 1931.

[172] James Mercer. Functions of positive and negative type and their connection with the theory of integral equations. *Phil. Trans. Royal Soc. A*, 209: 415–446, 1909.

[173] Charles Anthony Micchelli. Cardinal L-splines. In *Studies in spline functions and approximation theory* (S. Karlin, C.A. Micchelli, A. Pinkus, and I.J. Schoenberg, Eds.), pp. 203–250, New York: Academic Press, 1976.

[174] Theodore Samuel Motzkin. *Beiträge zur Theorie der linearen Ungleichungen.* PhD dissertation, Basel, 1933 and Jerusalem, 1936.

[175] Theodore Samuel Motzkin. Relations between hypersurface cross ratios, and a combinatorial formula for partitions of a polygon, for permanent preponderance, and for non-associative products. *Bull. Amer. Math. Soc.*, 54(4): 352–360, 1948.

[176] Theodore Samuel Motzkin. The Euclidean algorithm. *Bull. Amer. Math. Soc.*, 55(12): 1142–1146, 1949.

[177] Theodore Samuel Motzkin. A proof of Hilbert's Nullstellensatz. *Math. Z.*, 63: 341–344, 1955 (Issai Schur memorial volume).

[178] Theodore Samuel Motzkin. The arithmetic-geometric inequality. In *Inequalities* (O. Shisha, Ed., Proc. Sympos. Wright–Patterson Air Force Base, Ohio, 1965), pp. 205–224, New York: Academic Press, 1967.

[179] Oleg R. Musin. The kissing number in four dimensions. *Ann. of Math. (2)*, 168(1): 1–32, 2008.

[180] Alexander Markowich Ostrowski. Über die Funktionalgleichung der Exponentialfunktion und verwandte Funktionalgleichung. *Jber. Deut. Math. Ver.*, 38: 54–62, 1929.

[181] James Eldred Pascoe. Noncommutative Schur-type products and their Schoenberg theorem. *arXiv*, 2019. arXiv:math.FA/1907.04480

[182] Allan Pinkus. Strictly positive definite functions on a real inner product space. *Adv. Comput. Math.*, 20: 263–271, 2004.

[183] Allan Pinkus. *Totally positive matrices*. Cambridge Tracts in Mathematics, Vol. 181, Cambridge: Cambridge University Press, 2010.

[184] Jim Pitman. Probabilistic bounds on the coefficients of polynomials with only real zeros. *J. Combin. Th. Ser. A*, 77(2): 279–303, 1997.

[185] Georg Pólya and Gabor Szegő. *Aufgaben und Lehrsätze aus der Analysis. Band II: Funktionentheorie, Nullstellen, Polynome Determinanten, Zahlentheorie.* Berlin: Springer-Verlag, 1925.

[186] Tiberiu Popoviciu. Sur l'approximation des fonctions convexes d'ordre supérieur. *Mathematica (Cluj)*, 8: 1–85, 1934.

[187] Emilio Porcu, Moreno Bevilacqua, and Marc G. Genton. Spatio-temporal covariance and cross-covariance functions of the great circle distance on a sphere. *J. Amer. Statist. Assoc.*, 111(514): 888–898, 2016.

[188] Alexander E. Postnikov. Total positivity, Grassmannians, and networks. *arXiv*, 2006. arXiv:math.CO/0609764

[189] Augustine Francois Poulain. Théorèmes généraux sur les équations algébriques. *Nouv. Ann. Math.*, 6: 21–33, 1867.

[190] Robert Reams. Hadamard inverses, square roots and products of almost semidefinite matrices. *Linear Algebra Appl.*, 288: 35–43, 1999.

[191] Paul Ressel. Laplace-transformation nichtnegativer und vektorwertiger maße. *Manuscripta Math.*, 13: 143–152, 1974.

[192] Konstanze Christina Rietsch. Quantum cohomology rings of Grassmannians and total positivity. *Duke Math. J.*, 110(3): 523–553, 2001.

[193] Konstanze Christina Rietsch. Totally positive Toeplitz matrices and quantum cohomology of partial flag varieties. *J. Amer. Math. Soc.*, 16(2): 363–392, 2003.

[194] A. Wayne Roberts and Dale E. Varberg. *Convex functions*. New York, London: Academic Press [A subsidiary of Harcourt Brace Jovanovich, Publishers], Vol. 57, Pure and Applied Mathematics, 1973.

[195] R. Tyrrell Rockafellar. *Convex analysis*. Princeton Mathematical Series, Vol. 28, Princeton, NJ: Princeton University Press, 1970.

[196] Adam J. Rothman, Elizaveta Levina, and Ji Zhu. Generalized thresholding of large covariance matrices. *J. Amer. Statist. Assoc.*, 104(485): 177–186, 2009.

[197] Walter Rudin. Positive definite sequences and absolutely monotonic functions. *Duke Math. J*, 26(4): 617–622, 1959.

[198] Konrad Schmüdgen. Around Hilbert's 17th problem. *Documenta Math.*, Extra Volume, "Optimization Stories," pp. 433–438, 2012.

[199] Konrad Schmüdgen. *The moment problem.* Graduate Texts in Mathematics, Vol. 277, Cham: Springer, 2017.

[200] Isaac Jacob Schoenberg. Über variationsvermindernde lineare Transformationen. *Math. Z.*, 32: 321–328, 1930.

[201] Isaac Jacob Schoenberg. Zur Abzählung der reellen Wurzeln algebraischer Gleichungen. *Math. Z.*, 38: 546–564, 1934.

[202] Isaac Jacob Schoenberg. Remarks to Maurice Fréchet's article "Sur la définition axiomatique d'une classe d'espace distanciés vectoriellement applicable sur l'espace de Hilbert." *Ann. of Math. (2)*, 36(3): 724–732, 1935.

[203] Isaac Jacob Schoenberg. On certain metric spaces arising from Euclidean spaces by a change of metric and their imbedding in Hilbert space. *Ann. of Math. (2)*, 38(4): 787–793, 1937.

[204] Isaac Jacob Schoenberg. Metric spaces and completely monotone functions. *Ann. of Math. (2)*, 39(4): 811–841, 1938.

[205] Isaac Jacob Schoenberg. Metric spaces and positive definite functions. *Trans. Amer. Math. Soc.*, 44(3): 522–536, 1938.

[206] Isaac Jacob Schoenberg. Positive definite functions on spheres. *Duke Math. J.*, 9(1): 96–108, 1942.

[207] Isaac Jacob Schoenberg. Contributions to the problem of approximation of equidistant data by analytic functions. Part A. On the problem of smoothing or graduation. A first class of analytic approximation formulae. *Quart. Appl. Math.*, 4(1): 45–99, 1946.

[208] Isaac Jacob Schoenberg. On totally positive functions, Laplace integrals and entire functions of the Laguerre–Pólya–Schur type. *Proc. Natl. Acad. Sci. USA*, 33(1): 11–17, 1947.

[209] Isaac Jacob Schoenberg. On variation-diminishing integral operators of the convolution type. *Proc. Natl. Acad. Sci. USA*, 34(4): 164–169, 1948.

[210] Isaac Jacob Schoenberg. Some analytical aspects of the problem of smoothing. In *Studies and Essays presented to R. Courant on his 60th birthday* (K.O. Friedrichs, O.E. Neugebauer, and J.J. Stoker, Eds.), pp. 351–370, New York: Interscience Publ., 1948.

[211] Isaac Jacob Schoenberg. On Pólya frequency functions. II. Variation-diminishing integral operators of the convolution type. *Acta Sci. Math. (Szeged)*, 12: 97–106, 1950.

[212] Isaac Jacob Schoenberg. On Pólya frequency functions. I. The totally positive functions and their Laplace transforms. *J. d'Analyse Math.*, 1: 331–374, 1951.

[213] Isaac Jacob Schoenberg. On the zeros of the generating functions of multiply positive sequences and functions. *Ann. of Math. (2)*, 62(3): 447–471, 1955.

[214] Isaac Jacob Schoenberg. A note on multiply positive sequences and the Descartes rule of signs. *Rend. Circ. Mat. Palermo*, 4: 123–131 1955.

[215] Isaac Jacob Schoenberg and Anne M. Whitney. On Pólya frequency functions. III. The positivity of translation determinants with an application to the interpolation problem by spline curves. *Trans. Amer. Math. Soc.*, 74: 246–259, 1953.

[216] Issai Schur. Bemerkungen zur Theorie der beschränkten Bilinearformen mit unendlich vielen Veränderlichen. *J. reine angew. Math.*, 140: 1–28, 1911.

[217] Issai Schur. Über Potenzreihen, die im Innern des Einheitskreises beschränkt sind. *J. reine angew. Math.*, 147: 205–232, 1917.

[218] Binyamin Schwarz. Totally positive differential systems. *Pacific J. Math.*, 32(1): 203–229, 1970.

[219] James Alexander Shohat and Jacob Davidovich Tamarkin. *The problem of moments*. AMS Mathematical Surveys, New York: American Mathematical Society, 1943.

[220] Wacław Sierpiński. Sur l'équation fonctionelle $f(x+y) = f(x) + f(y)$. *Fund. Math.*, 1(1): 116–122, 1920.

[221] Barry Martin Simon. *Loewner's theorem on monotone matrix functions.* Grundlehren der mathematischen Wissenschaften, Vol. 354, Berlin: Springer-Verlag, 2019.

[222] Mark Skandera. Inequalities in products of minors of totally nonnegative matrices. *J. Algebraic Combin.*, 20(2): 195–211, 2004.

[223] Serge Skryabin. On the graded algebras associated with Hecke symmetries, II. The Hilbert series. *arXiv*, 2019. arXiv:math.RA/1903.09128

[224] Sergei Lvovich Sobolev. Sur un théorème d'analyse fonctionnelle. *Rec. Math. (Mat. Sbornik) N.S.* 4(46): 471–497, 1938.

[225] Ingo Steinwart. On the influence of the kernel on the consistency of support vector machines. *J. Mach. Learn. Res.*, 2(1): 67–93, 2002.

[226] James Drewry Stewart. Positive definite functions and generalizations, an historical survey. *Rocky Mountain J. Math.*, 6(3): 409–434, 1976.

[227] Thomas Joannes Stieltjes. Recherches sur les fractions continues. *Ann. Fac. Sci. Toulouse*, 8(4): 1–122, 1894–95.

[228] Bernd Sturmfels. Totally positive matrices and cyclic polytopes. *Linear Algebra Appl.*, 107: 275–281, 1988.

[229] Vladimir Naumovich Vapnik. The nature of statistical learning theory. *Statistics for Engineering and Information Science*, New York: Springer, 2000.

[230] Harkrishan Lal Vasudeva. Positive definite matrices and absolutely monotonic functions. *Indian J. Pure Appl. Math.*, 10(7): 854–858, 1979.

[231] Prateek Kumar Vishwakarma. Positivity preservers forbidden to operate on diagonal blocks. *Trans. Amer. Math. Soc.*, in press, DOI: 10.1090/tran/8256

[232] John von Neumann and Isaac Jacob Schoenberg. Fourier integrals and metric geometry. *Trans. Amer. Math. Soc.*, 50: 226–251, 1941.

[233] Jan Vybíral. A variant of Schur's product theorem and its applications. *Adv. Math.*, 368, Article 107140, 9 pp., 2020.

[234] David G. Wagner. Total positivity of Hadamard products. *J. Math. Anal. Appl.*, 163(2): 459–483, 1992.

[235] Hsien-Chung Wang. Two-point homogeneous spaces. *Ann. of Math. (2)*, 55(1): 177–191, 1952.

[236] Yi Wang and Yeong-Nan Yeh. Polynomials with real zeros and Pólya frequency sequences. *J. Combin. Theory Ser. A*, 109(1): 63–74, 2005.

[237] Philip White and Emilio Porcu. Towards a complete picture of stationary covariance functions on spheres cross time. *Electron. J. Statist.*, 13(2): 2566–2594, 2019.

[238] David Vernon Widder. *The Laplace transform.* Princeton Legacy Library, Princeton, NJ: Princeton University Press, 1941.

[239] Hans Sylvain Witsenhausen. Minimum dimension embedding of finite metric spaces. *J. Combin. Th. Ser. A*, 42(2): 184–199, 1986.

[240] Yuan Xu. Positive definite functions on the unit sphere and integrals of Jacobi polynomials. *Proc. Amer. Math. Soc.*, 146(5): 2039–2048, 2018.

[241] Yuan Xu and Elliott Ward Cheney. Strictly positive definite functions on spheres. *Proc. Amer. Math. Soc.*, 116(4): 977–981, 1992.

[242] Xingzhi Zhan. *Matrix theory.* Graduate Studies in Mathematics, Vol. 147, Providence, RI: American Mathematical Society, 264 pp., 2013.

[243] Bin Zhang and Steve Horvath. A general framework for weighted gene co-expression network analysis. *Stat. Appl. Genet. Mol. Biol.*, 4: Article 17, 45 pp (electronic), 2005.

[244] Fuzhen Zhang. *Matrix theory: Basic results and techniques.* Universitext, New York: Springer-Verlag, xvii+399 pp., 2011.

[245] Johanna Ziegel. Convolution roots and differentiability of isotropic positive definite functions on spheres. *Proc. Amer. Math. Soc.*, 142(6): 2063–2077, 2014.

Index

absolutely monotonic function, 127

Basic Composition formula, 38
Bernstein's theorem, 127
Boas–Widder theorem, 125, 176

Cauchy mean-value theorem for divided
    differences, 125
Cauchy–Binet formula, 34
Cayley–Menger
    matrix, 192
    matrix, alternate form, 102, 192
Chebyshev polynomials, 109, 110
chordal graph, 70
column-strict Young tableau, 220
complete homogeneous symmetric
    polynomials $h_k(\mathbf{u})$, 246, 256
completely monotone functions, 199
conditionally positive semidefinite matrix,
    103, 104, 205
cone, 12
contiguous minor, 26
correlation matrix, 9, 56, 90, 109
covariance matrix, 56
critical exponent for:
    a graph G, 64
    Loewner convexity, 79
    Loewner monotonicity, 66
    Loewner positivity, 41
    Loewner super-additivity, 68
    positive definiteness, 44

Daletskii–Krein formula, 18
Descartes' rule of signs, 33, 73
Dirac measure δ_x, 10

divided differences, 124
dominated convergence theorem (Lebesgue),
    122

elementary symmetric polynomials $e_k(\mathbf{u})$, 256
entrywise
    map/function/transform $f[A]$, 40
    map/function/transform $f_G[A]$, 59
    polynomial positivity preserver, 216
    power $A^{\circ\alpha}$, 40
    powers preserving positivity $\mathcal{H}_G$, 64

Fekete–Schoenberg lemma, 29
forward difference $(\Delta_h^k f)(x)$, 124, 176
Further questions, 71, 82, 250

Gaussian kernel T_{G_σ}, 37, 105
gcd matrix, 86
Gegenbauer polynomials, 109, 110
geometric mean $A\#B$, 87
Gram matrix, 8, 107, 109, 191

Hadamard composition of polynomials/power
    series, 19
Hamburger's theorem, 132
Hankel
    moment matrix H_μ, 10
    TN $n \times n$ matrices HTN_n, 53
Hilbert sphere S^∞, 106
Historical notes, 11, 18, 19, 26, 97, 99, 101,
    104, 121, 130, 138, 193, 200
holomorphic function, 151
homogeneous space, n-point, 191, 205
homotopy argument, 34, 74
Horn–Loewner theorem, stronger form, 114